Heinemann
Integrated Science for *CSEC*®

Byron Dawson
David Sang
Lawrie Ryan
Marcia Rainford

Advisor: Doltan Ramsubeik

Part of Pearson

Heinemann is an imprint of Pearson Education Limited, a company incorporated in England and Wales, having its registered office at Edinburgh Gate, Harlow, Essex, CM20 2JE. Registered company number: 872828

www.pearsoncaribbean.co.uk

Heinemann is a registered trademark of Pearson Education Limited

Text © Pearson Education Limited 2010

21
10 9 8 7 6 5

ISBN 978-14-0823127-2

First published 2010

Designed and typeset by Tech-Set Limited
Original illustrations © Pearson Education Limited 2009
Illustrated by Tech-Set Limited
Indexed by Indexing Specialists
Printed and bound by CPI Group (UK) Ltd, Croydon, CR0 4YY

Heinemann Integrated Science for *CSEC*

CSEC is a registered trade mark of the Caribbean Examinations Council (*CXC*). Heinemann Integrated Science for CSEC® is an independent publication and has not been authorised, sponsored, or otherwise approved by *CXC*.

Acknowledgements
The authors would like to thank the authors of Science About Us – A. Farley, W. King, N. Lambert and V. McClenan - for the use of material from this title.

The authors and publisher would like to thank the following individuals and organisations for permission to reproduce photographs:

(Key: b-bottom; c-centre; l-left; r-right; t-top)

Alamy Images: All Canada Photos 319c; Roger Allen Photography 231b; Steve Bly 358b; Bubbles Photography 371tc; Alexander Caminada 344bl; Danita Delimont 323cr; Leslie Garland Picture Library 248c; Jennie Hart 126c; imac 185t; Images & Stories 324t; Images of Africa Photobank 221tr; Imagestate Media Partners Limited / Impact Photos 318tr; Robert Harding Picture Library Ltd 157b, 158b; Medical-on-Line 51br; Phototake. Dennis Kunkel Microscopy, Inc. 8cr; Paul Ridsdale 295t; sciencephotos 221br; Sundlof - EDCO 317b; Hugh Threfall 221cr; **Corbis:** James L. Amos 242b; Nic Bothma/epa 326tl; Buddy Mays 187b; Reuters 175cr; Jeffrey L. Rotman 22b; Galen Rowell 326tr; Wendy Stone 213tr; **Dermatlas; http://www.dermatlas.org:** Chris Lehmann, MD 206b; **DK Images:** Andy Crawford 294t; **ESA:** CNES / Arianespace 361t; **Getty Images:** AFP 26b, 243t, 253br; 21t, 192c, 253bl; Domino 322cr; Glow Images 37-47 (Heading); National Geographic 351tc; NBAE 253tr; Nancy Nehring i-x (Heading); Paul Nicklen 197cr; Thinkstock 107-130 (Heading); Raphael Van Butsele 251c; **GNU Free Documentation license:** National Cancer Institute 66t; Public Health Image Library / C. Goldsmith 65b; **Heinemann:** 3-367 (Lightbulb); **Heinemann Biology for OCRA, Dawson & Honeysett:** page 4 2c; **iStockphoto:** Douglas Allen 212c; Brinkstock 300cr; creacart 259cl; Antonio D'Albore 207tc; Jim DeLillo 21b; ElementalImaging 197b; Eliandric 176c; Peter Elvidge 50b; Silvia Ganora 19 (Bran); Steve Geer 246cr; Angelo Gilardelli 259cr; Dennis Guyitt 186b; Geoffrey Holman 344cr; Eric Hood 98c; Dennis Hoyne 343c; David H. Lewis 27tr; Anna Lubovedskaya 238b; Bryan Malley 269-281 (Heading), 272tc; Dirk-Jan Mattaar 204tl; Tulay Over 239br; Ina Peters 19 (water); Photoknot 291t; Ian Poole 291b; Proxyminder 160 (Grasshopper); Christina Richards 273t; Claes Torstensson 203b; Zuki 115cr; **Jeff Moore (jeff@jmal.co.uk):** 46c, 225br; courtesy Science Department Benenden School 191b; **Jupiter Unlimited:** AbleStock.com 108 (onion); BananaStock 108 (Ginger); Brand X Pictures 1cl, 45tl, 108b (Taro corm); Comstock Images 12 (Girl), 18br, 136bl, 136br, 144b; Stockxpert 11bl, 11br, 12 (Cow), 18-36 (Heading), 19 (Butter), 19 (Fruit & veg), 19 (Liver), 48b, 64b, 92c, 108 (Potato tuber), 115cl, 160 (Maize), 185cr, 206cr, 206tr, 211b, 213br, 213cr, 232t, 239t, 246cl, 249t, 259bc, 283c, 288b, 309t, 321b, 321t, 327cr, 343cl, 343cr, 355t; Goodshoot 160 (Sun); Liquidlibrary 12 (Maize); PhotoObjects.net 18c; Photos.com 12 (Sun), 94b, 145b, 203cl, 229 (CO2); Thinkstock Images 136bc; **Mary Jo Kolb:** 109b; **NASA:** 172tr, 363t; Ames Research Center 196b; Johnson Space Center 367b; MCREL / background: William K. Hartmann, courtesy of UCLA 349b; SOHO 325tc; **NOAA:** National Weather Service (NWS) Collection 156b; **Photolibrary.com:** Animals Animals 115br; Howard Grey 244b; Philip Rosenberg 175bl; **POD - Pearson Online Database:** Brand X Pictures. Morey Milbradt 149-166 (Heading); Comstock 255-268 (Heading); Constructionphotography. com. David Potter 316-329 (Heading); Corbis 97-106 (Heading); Fancy Veer Corbis 48-53 (Heading); Digital stock 298-306 (Heading); Peter Gould 221bl; Chris Honeywell 76-85 (Heading); Imagestate. Ian Cartwright 183-199 (Heading); Imagestate. John Foxx Collection 11-17 (Heading); Photodisc 68-75 (Heading), 160 (Lizard), 160 (Snake), 211t, 282-297 (Heading), 307-315 (Heading), 358-373 (Heading); Photodisc. Don Farrall. LightWorks Studio 220-235 (Heading); Photodisc. Geostock 86-96 (Heading); Photodisc. Karl Weatherly 343-357 (Heading); Photodisc. Photolink 167-182 (Heading), 200-205 (Heading), 236-254 (Heading); Photodisc. Photolink. Annie Reynolds 330-342 (Heading); Photodisc. Photolink. E. Pollard 54-67 (Heading); Photodisc. Photolink. Jack Star 131-148 (Heading); **Press Association Images:** AP 231br; **Taran Rampersad, KnowProSE.com:** 231tl; **Rex Features:** Sipa Press 175br; **Roslin Embryology Ltd :** Bill Ritchie 111b; **David Rydevik:** 178c; **Science Photo Library Ltd:** 167b; Michael Abbey 110b; Martin Bond 322br; Dr. Jeremy Burgess 58tl; Martin F Chillmaid 263b; Martyn F. Chillmaid 300br; Patrick Dumas / Look at Sciences 307cr; John Durham 1-10 (Heading); Kenneth Eward / Biobgrafx 363cr; Cecil H. Fox 207cr; Gordon Garradd 168b; Steve Gschmeissner 8cl; Adrienne Hart-Davis 287c; Herve Conge, ISM 60c; Coneyl Jay 51t; Andrew Lambert Photography 271c; Mark Sykes 229 (Foam), 229 (Water), 235cl; Cordelia Molloy 229 (Powder); David Nunuk 282b; Paul Rapson 327tr; J. C. Revy 59b; Sinclair Stammers 218; James Steveson 51bl; Andrew Syred 206-219 (Heading); Sheila Terry 273b, 289b, 298br; TRL Ltd. 357; Emmeline Watkins 353tc; Western Ophthalmic Hospital 101c; Jerome Wexler 107b; **shutterstock:** argo74 115bl; Mikael Damkier 125b; Spe 195b; **Soreen Ltd:** The Fruity Malt Loaf 209; **SuperStock:** Angelo Cavalli 158t; **The Advertising Archives:** 72c; **U.S. Air Force :** Master Sgt. Val Gempis 203cr; **US Navy:** Mass Communication Specialist 2nd Class Jayme Pastoric 204c; **Wikimedia Commons:** From Scheme I. of his 1665 Micrographia. On permanent display in "The Evolution of the Microscope" exhibit at the National Museum of Health and Medicine, in Washington, DC 1bl; Robert Hooke, "Micrographia", 1665 1br

Cover images: *Front:* **Getty Images:** David Wasserman Nancy Nehring Tom Pfeiffer / Volcano Discovery. *Back:* **Getty Images:** Christine Balderas tr; sb162698e-1Baoba Images cl

All other images © Pearson Education

Picture Research by: Liz Moore and Sally Turner

Every effort has been made to trace the copyright holders and we apologise in advance for any unintentional omissions. We would be pleased to insert the appropriate acknowledgement in any subsequent edition of this publication.

Contents

How the book is structured

Heinemann Integrated Science for *CSEC* has been written to provide comprehensive coverage of the latest *CSEC* Integrated Science syllabus. The following features will help you achieve your best in the *CSEC* Integrated Science examinations and School-Based Assessments (SBAs):

Objectives appear at the beginning of each chapter and show the concepts you will cover in the following chapter

Activities appear throughout the teaching in each chapter, and are labelled as either **To do** or **Experiments**. The **To do** activities are quicker to perform and require fewer scientific apparatus. The **Experiments** are more involved and structured. Both types of activity will help you master the scientific principles you are being taught.

The **SBA skills** covered in each activity are indicated with icons in the margin areas. Your teacher may use this information to decided which Activities will be used to assess the different SBA skills. The icons are: (ORR) Observation, recording and reporting, (MM) Manipulation and measurement, (AI) Analysis and interpretation, (PD) Planning and design and (D), for Drawing. When the boxes are highlighted it means the skill can be tested in this particular activity.

Short exercises appear in each chapter as you proceed – your teacher can use these to check your understanding of concepts being taught.

CSEC-style questions appear at the end of the chapter. You will notice the earlier questions are multiple-choice while later ones require short answers. This is similar to the format in the *CSEC* exams.

The **SBA Guide and SBA skills-matching chart** appear in appendices at the back of the book. Your teacher may use these to decide how the SBA should be approached.

The **CD-ROM** contains interactive questions that test your knowledge in preparation for your examinations, a glossary, and answers to all questions in the book.

Further resources are available at www.pearsoncaribbean.com/heinemannintscicsec

SBA Skills Chart

Activity	Unit	Page	ORR	D	MM	PD	AI
To do – Making a model cell	A1 The Cell	3			✓	✓	
To do – Observing diffusion	A1 The Cell	5	✓				
To do – Effects of osmosis	A1 The Cell	8	✓				✓
Experiment – Investigating the concentration of potato cell contents	A1 The Cell	9	✓		✓		✓
Experiment – Testing for starch	A2a Photosynthesis and Photochemical Reactions	14	✓		✓		✓
Experiment – Is light needed for photosynthesis?	A2a Photosynthesis and Photochemical Reactions	15	✓		✓		✓
Experiment – Is carbon dioxide needed for photosynthesis?	A2a Photosynthesis and Photochemical Reactions	15	✓		✓		✓
Experiment – Is chlorophyll needed for photosynthesis?	A2a Photosynthesis and Photochemical Reactions	16	✓	✓	✓		✓
To do – Night vision	A2b Food and Nutrition	22	✓				
Experiment – Food tests	A2b Food and Nutrition	23	✓		✓		✓
To do – Energy conversion	A2b Food and Nutrition	24			✓		
To do – Information from food labels	A2b Food and Nutrition	25					✓
Experiment – Measuring the energy in a crisp	A2b Food and Nutrition	26	✓		✓		✓
To do – Identifying types of teeth	A2b Food and Nutrition	28					✓
To do – Effect of saliva on bread	A2b Food and Nutrition	30	✓				
Experiment – How do temperature and pH affect reaction time?	A2b Food and Nutrition	32	✓		✓	✓	✓
To do – Breathing and swallowing	A2b Food and Nutrition	33					✓
Experiment – Making a model gut	A2b Food and Nutrition	35	✓		✓	✓	✓
Experiment – How the lungs work	A3a Respiration and Breathing	38	✓				✓
Experiment – How much oxygen is in the air we breathe?	A3a Respiration and Breathing	39	✓				✓
To do – Rings of cartilage	A3a Respiration and Breathing	40	✓				
To do – Modelling gills	A3a Respiration and Breathing	42	✓				✓
Experiment – Carbon dioxide is released during respiration	A3a Respiration and Breathing	43	✓				✓
Experiment – Energy release during respiration	A3a Respiration and Breathing	44	✓		✓		✓
Experiment – Investigating cigarette smoke	A3b Air pollution	52	✓				✓
To do – Surface area and volume of cubes	A4a – Transport Systems in Plants and Humans	54			✓		✓
Experiment – Investigating the relationship between surface area and volume	A4a – Transport Systems in Plants and Humans	55	✓		✓		✓
To do – Transport in celery	A4a – Transport Systems in Plants and Humans	57	✓	✓			✓
To do – Observing stomata	A4a – Transport Systems in Plants and Humans	58	✓		✓		✓
To do – Cohesion and adhesion	A4a – Transport Systems in Plants and Humans	59	✓				✓
To do – Blood group survey	A4a – Transport Systems in Plants and Humans	63	✓				✓
To do – Veins in the arm	A4b – The Blood System and Diseases	69	✓				✓
Experiment – Testing your fitness	A4b – The Blood System and Diseases	73	✓		✓		✓
To do – Our breath contains water	A5 Excretion	76	✓				

Activity	Unit	Page	ORR	D	MM	PD	AI
To do – Urine changes colour	A5 Excretion	80	✓				✓
To do – Different areas of the skin sweat differently	A5 Excretion	83	✓				✓
To do – How fast can you go?	A6a The Control Systems	89	✓		✓		✓
To do – The knee jerk and pupil reflexes	A6a The Control Systems	90	✓				✓
To do – Accommodation	A6b Sight and Hearing	98	✓				
To do – Finding the blind spot	A6b Sight and Hearing	99	✓				
To do – Seeing in the dark	A6b Sight and Hearing	100	✓				✓
To do – Measuring sound	A6b Sight and Hearing	104	✓		✓		
To do – Find your frequency range	A6b Sight and Hearing	104	✓		✓		
To do – Fooling the brain	A6b Sight and Hearing	105	✓				✓
To do – Drawing organs of vegetative propagation	A7a Reproduction and Growth	108		✓			
To do – The parts of a flower	A7a Reproduction and Growth	113		✓			
Experiment – Growing pollen tubes	A7a Reproduction and Growth	114	✓		✓		
To do – Measuring growth	A7a Reproduction and Growth	116				✓	
Experiment – Investigating growth	A7a Reproduction and Growth	117	✓			✓	
To do – Extrapolating your data and predicting results	A7a Reproduction and Growth	118			✓		✓
To do – Caring for the unborn child	A7b Human Reproduction	126	✓	✓			
To do – Bigger than you think	A7b Human Reproduction	128			✓		

Activity	Unit	Page	ORR	D	MM	PD	AI
Experiment – The difference between heat and temperature	B1 Temperature Control and Ventilation	132	✓		✓		✓
To do – A model of how conduction works	B1 Temperature Control and Ventilation	133					✓
To do – Showing conduction of heat in different materials	B1 Temperature Control and Ventilation	134	✓				✓
Experiment – How to show convection currents	B1 Temperature Control and Ventilation	134	✓				
Experiment – Boiling ice	B1 Temperature Control and Ventilation	135	✓				✓
To do – What material absorbs heat the best?	B1 Temperature Control and Ventilation	136	✓			✓	✓
Experiment – Expansion and contraction of metals	B1 Temperature Control and Ventilation	137	✓				✓
Experiment – Which feels colder – water or alcohol?	B1 Temperature Control and Ventilation	142	✓		✓		✓
Experiment – A model to show how sweating works	B1 Temperature Control and Ventilation	143	✓		✓		✓
Experiment – Investigating factors that affect drying	B1 Temperature Control and Ventilation	145	✓		✓	✓	✓
Experiment – Estimating the organic content of a soil sample	B2a The Terrestrial Environment	151	✓		✓		✓
Experiment – Showing the solid components of soil	B2a The Terrestrial Environment	151	✓				✓
Experiment – Showing the presence of air in a soil sample	B2a The Terrestrial Environment	152	✓				
Experiment – Measuring the water content of a soil sample	B2a The Terrestrial Environment	153	✓		✓		✓
Experiment – Measuring water retention	B2a The Terrestrial Environment	154	✓		✓		✓
To do – Observing capillarity	B2a The Terrestrial Environment	155	✓				
To do – The Great American Dust Bowl	B2a The Terrestrial Environment	156	✓				
To do – Looking at an ecosystem	B2a The Terrestrial Environment	159	✓				

Activity	Unit	Page	ORR	D	MM	PD	AI
To do – Looking at habitats	B2a The Terrestrial Environment	161	✓				✓
To do – The local environment	B2a The Terrestrial Environment	162	✓				
To do – Drawing a food web	B2a The Terrestrial Environment	162					✓
To do – Hurricane research	B2b The Weather and Geological Events	172	✓				✓
To do – Modelling a volcano	B2b The Weather and Geological Events	174					✓
To do – A home-made seismograph	B2b The Weather and Geological Events	178			✓	✓	
To do – Using a tide table	B2b The Weather and Geological Events	180					✓
Experiment – Growing an onion by hydroponics	B3a Water and the Aquatic Environment	185	✓				
Experiment – Making a filtration plant	B3a Water and the Aquatic Environment	187	✓				
Experiment – Distilling water	B3a Water and the Aquatic Environment	188	✓				
Experiment – Sterilising river water	B3a Water and the Aquatic Environment	188	✓				✓
To do – The water cycle	B3a Water and the Aquatic Environment	189	✓	✓			
To do – Sites of water pollution	B3a Water and the Aquatic Environment	192	✓				
Experiment – Checking the density of tap water and sea water	B3a Water and the Aquatic Environment	193	✓		✓		✓
Experiment – Showing that water contains dissolved chemicals	B3a Water and the Aquatic Environment	193	✓				
Experiment – First if floats – then it stinks!	B3a Water and the Aquatic Environment	194	✓				✓
To do – Plimsoll lines	B3a Water and the Aquatic Environment	196	✓				
To do – Air resistance and friction	B3a Water and the Aquatic Environment	197					✓
Experiment – Throwing a javelin	B3a Water and the Aquatic Environment	198	✓		✓		✓
Experiment – Making a compass	B3b Activities in Water	202					✓
To do – Safety standards for fishing	B3b Activities in Water	203	✓				
To do – Multiplication of bacteria	B4 Pests, Parasites and Sanitation	208					✓
To do – Preservatives in food	B4 Pests, Parasites and Sanitation	209	✓				
Experiment – How to stop mould from growing on bread	B4 Pests, Parasites and Sanitation	210	✓				✓
To do – Waste diary	B4 Pests, Parasites and Sanitation	214	✓				
To do – Recycling plastic	B4 Pests, Parasites and Sanitation	215	✓				
To do – Researching the discovery of the cause of cholera	B4 Pests, Parasites and Sanitation	216	✓				
To do – What if?	B4 Pests, Parasites and Sanitation	218					✓
To do – Hazards in the home	B5 Safety Hazards	222	✓				✓
Experiment – Exploding tin can	B5 Safety Hazards	223					✓
To do – Electrical hazards and risks	B5 Safety Hazards	224	✓				
To do – How quickly can food become contaminated by bacteria?	B5 Safety Hazards	225					✓
To do – Chemicals in the home	B5 Safety Hazards	226	✓				
To do – Hazards in the laboratory or classroom	B5 Safety Hazards	226					✓
To do – Hazard signs	B5 Safety Hazards	227	✓				
To do – Spotting hazards around us	B5 Safety Hazards	228					✓
Experiment – Making a home-made foam fire extinguisher	B5 Safety Hazards	230	✓		✓		✓
To do – Fighting fires	B5 Safety Hazards	230					✓
To do – The right clothing for the job	B5 Safety Hazards	232					✓

Activity	Unit	Page	ORR	D	MM	PD	AI
To do – Making springs	B6 Materials	238	✓		✓	✓	✓
To do – Using solder	B6 Materials	239		✓			
Experiment – Metals plus acid	B6 Materials	240	✓		✓		✓
Experiment – What is needed for iron to rust?	B6 Materials	245	✓				✓
To do – Speeding up rusting	B6 Materials	246	✓				
To do – Rusting survey	B6 Materials	247	✓				✓
To do – Investigating the prevention of rust	B6 Materials	248				✓	
To do – Splitting wood	B6 Materials	249	✓				✓
To do – Plastics survey	B6 Materials	250	✓				
To do – Making mixtures	B7a Mixtures	258	✓		✓		✓
To do – Making mayonnaise	B7a Mixtures	260			✓		
To do – Properties of solutions, suspensions and colloids	B7a Mixtures	261	✓		✓		✓
Experiment – Stain removal	B7a Mixtures	262	✓				✓
To do – Comparing water samples for hardness	B7a Mixtures	264	✓		✓		✓
Experiment – Separation of a mixture of copper sulphate and water by simple distillation	B7a Mixtures	266	✓				✓
To do – The wonders of baking soda	B7b Acids, Bases and Salts	270	✓				
Experiment – Finding the pH of toothpaste	B7b Acids, Bases and Salts	274	✓				✓
Experiment – Using home-made indicators to identify acids and base	B7b Acids, Bases and Salts	274	✓				✓
Experiment – Neutralisation reaction	B7b Acids, Bases and Salts	276	✓		✓		✓
Experiment – Stain removal	B7b Acids, Bases and Salts	278	✓				✓
To do – Identifying dangerous substances	B7b Acids, Bases and Salts	279	✓				

Activity	Unit	Page	ORR	D	MM	PD	AI
Experiment – How well do solids conduct electricity?	C1a Electrical Circuits	285	✓		✓		✓
Experiment – Do liquids conduct electricity?	C1a Electrical Circuits	286	✓				✓
Experiment – Electric heat and light	C1a Electrical Circuits	289	✓		✓		
Experiment – Measuring resistance	C1a Electrical Circuits	291	✓		✓		
Experiment – Series and parallel circuits	C1a Electrical Circuits	292	✓		✓		
Experiment – Observing the field around a current	C1a Electrical Circuits	293	✓				✓
Experiment – Making and testing an electromagnet	C1a Electrical Circuits	294	✓		✓		✓
Experiment _ Ring that bell!	C1a Electrical Circuits	296	✓				✓
Experiment – Make a buzzer	C1a Electrical Circuits	296			✓		
To do – Power ratings	C1b Electrical Power and Energy	298	✓				
To do – Meter reader	C1b Electrical Power and Energy	304	✓				✓
To do – Billing clerk	C1b Electrical Power and Energy	304					✓
Experiment – The glowing wire	C1b Electrical Power and Energy	307	✓		✓		
To do – Comparing light bulbs	C1b Electrical Power and Energy	308	✓		✓		
Experiment – Splitting light	C1b Electrical Power and Energy	310	✓		✓		

Activity	Unit	Page	ORR	D	MM	PD	AI
Experiment – Mixing coloured lights	C1b Electrical Power and Energy	311	✓				
To do – Colours making pictures	C1b Electrical Power and Energy	312	✓		✓		
To do – Using coloured filters	C1b Electrical Power and Energy	313	✓				
Experiment – Chromatography	C1b Electrical Power and Energy	314	✓				
Experiment – Burning fuels	C2 Sources of Energy	317	✓				
To do – Finding fuels	C2 Sources of Energy	319	✓				
To do – Charting energy	C2 Sources of Energy	320	✓				
Experiment – Turning power	C2 Sources of Energy	323	✓		✓		
To do – From the Sun	C2 Sources of Energy	324	✓				
Experiment – Solar heat	C2 Sources of Energy	326	✓		✓	✓	✓
Experiment – Electricity from the Sun	C2 Sources of Energy	327	✓		✓		✓
Experiment – Testing the law of levers 1	C3 Machines and Movement	331	✓		✓		
Experiment – Testing the law of levers 2	C3 Machines and Movement	332	✓		✓	✓	
Experiments – Investigating pulleys	C3 Machines and Movement	335	✓		✓		
Experiment – Up the slope	C3 Machines and Movement	335	✓		✓		✓
Experiment – Energy on an inclined plane	C3 Machines and Movement	336	✓		✓		
To do – Efficiency of the inclined plane	C3 Machines and Movement	338			✓		✓
To do – Energetic toys	C4 Conservation of Energy	344	✓				
Experiment – Energy producing changes	C4 Conservation of Energy	345	✓				✓
Experiment – The effect of heat on copper(II) carbonate	C4 Conservation of Energy	345	✓				
Experiment – Lime power	C4 Conservation of Energy	346	✓				
Experiment – Light energy produces a chemical change	C4 Conservation of Energy	346					✓
Experiment – Bringing energy to a focus	C4 Conservation of Energy	353	✓		✓		
To do – Discovering dishes	C4 Conservation of Energy	354		✓			✓
Experiment – Observing collisions	C4 Conservation of Energy	354	✓			✓	✓
Experiment – Trolley collisions 1 – springy collisions	C4 Conservation of Energy	355	✓		✓		
Experiment – Trolley collisions 2 – sticky collisions	C4 Conservation of Energy	356	✓				
Experiment – Action and reaction	C5 Forces	359	✓		✓		✓
Experiment – Blowing gives lift	C5 Forces	361	✓				
Experiment – Experiencing gravity	C5 Forces	363				✓	
Experiment – Investigating the range of a projectile	C5 Forces	364	✓		✓		✓
To do – Spacecraft in orbit	C5 Forces	367	✓				
Experiment – Finding the centre of gravity by balancing	C5 Forces	368	✓		✓		
To do – Balancing symmetrical shapes	C5 Forces	369	✓				
Experiment – The hanging card	C5 Forces	369			✓		
Experiment – The balancing parrot	C5 Forces	371					✓
Experiment – Investigating stability	C5 Forces	372	✓				

By the end of this unit you will be able to:

- draw and label simple diagrams to show the structure of specialised plant and animal cells.
- explain the importance of the cell wall, cell membrane, nucleus, chromosomes, cytoplasm, mitochondria, vacuoles and chloroplasts.
- explain the movement of substances by diffusion and osmosis in living organisms.
- explain the processes of diffusion and osmosis using an experimental model.

Looking at cells

All living things are made of **cells**. Our eyes can see objects as small as 0.1 mm. But most cells are much smaller than this. To be able to see them we need to use a **microscope**. Light microscopes enable you to easily see objects as small as 0.002 mm.

A typical microscope has two lenses; the eyepiece lens, which magnifies ×10, and the objective lens, which magnifies ×40. This means the microscope will magnify the image ×400 (10 × 40).

The first scientist to observe cells using a light microscope was Robert Hooke. In 1665 he looked at a thin slice of cork and saw that it was made of lots of tiny boxes. The boxes reminded him of the small rooms that monks lived in called cells. So he called the small boxes 'cells'.

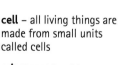

▲ **Fig A1.1** This scientist is looking at cells using a light microscope

cell – all living things are made from small units called cells

microscope – an instrument that magnifies the image of small objects

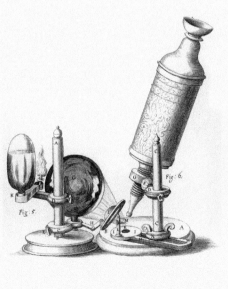
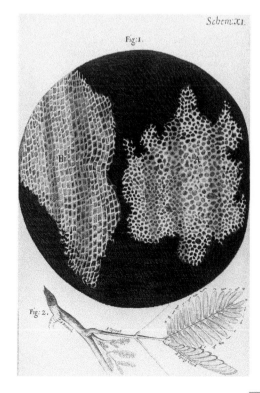

▲ **Fig A1.2** Hooke's microscope and what he saw

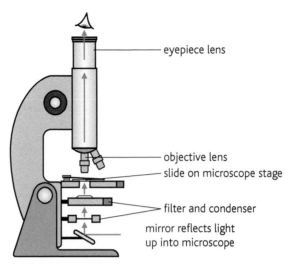

▲ **Fig A1.3** A modern light microscope

Today we have powerful electron microscopes that use a beam of electrons instead of light. They can be used to observe objects as small as 0.000 002 mm. This enables scientists to see many other much smaller objects inside the cell.

Q1 How many times would a microscope magnify if the eyepiece lens was ×15 and the objective lens ×40?

Plant cells and animal cells are different

Look at the photographs of animal and plant cells in Fig A1.4 and compare them with the drawings beneath.

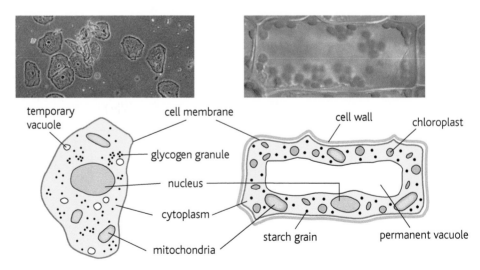

▲ **Fig A1.4** An animal cell and a plant cell

Q2 Which features are found in both animal and plant cells?

Q3 Which features are only found in a plant cell?

Parts of the cell

Different parts of the cell have different functions. All cells are enclosed by a **cell membrane**. The job of the membrane is to control what enters and leaves the cell. The contents of the cell are made from **cytoplasm** and this is where **proteins** are made and many other chemical reactions take place. Within the cytoplasm is the **nucleus**. The nucleus is where the **DNA** is stored on structures called **chromosomes**. DNA is the chemical code for all the instructions that tell the cell how to work and what to do.

Table A1.1 shows you what else is found in both animal and plant cells and what the function of each part of the cell is.

	cell organelle	function
plant cells only	cell wall	supports the cell giving it a fixed shape
	chloroplast	converts light energy into chemical energy by photosynthesis
	starch grain	a carbohydrate food store
	permanent vacuole	contains water and helps support the cell
animal cells only	temporary vacuole	only exists for a short time
	glycogen granule	a carbohydrate food store
both animal and plant cells	cell membrane	a partially permeable barrier controlling entry into and out of the cell
	cytoplasm	a substance where many chemical reactions take place
	nucleus	chromosomes that carry genetic information in the form of DNA
	mitochondria	where energy is transferred from chemical energy

Table A1.1 The parts and functions of cells

To do – Making a model cell

1 Use Plasticine or other suitable material to make a model of an animal cell. You will need to make a round nucleus first and then use a different coloured Plasticine to surround the nucleus with the rest of the cell.

2 Once you have made the cell, carefully use a sharp knife to cut the cell into two equal halves, exposing the nucleus.

3 Draw and label a diagram of the exposed section of your model cell. You can then make and draw a model of a plant cell using a third colour for the vacuole and a fourth colour for the outer cell wall. When you are making it, it may look something like Fig A1.5.

Can you think of anything else you could add to your model to make it more realistic?

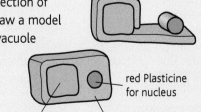

red Plasticine for nucleus

blue Plasticine for vacuole

green Plasticine for cytoplasm

▲ Fig A1.5

Movement of substances

Cells are living structures. There are many chemical reactions taking place within the cell to keep it alive. These chemical reactions need many different substances in order to work and they each produce different waste products. All of these substances and waste products need to move into and out of the cell through the cell membrane.

Some substances that...	
...enter the cell	**...leave the cell**
glucose for energy	carbon dioxide produced by respiration
oxygen for respiration	urea produced by the breakdown of proteins
amino acids to make proteins	enzymes produced by the cell
water to act as a solvent	water produced by respiration
minerals	hormones (chemical messengers)

Table A1.2

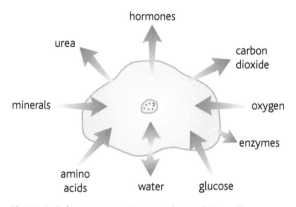

Fig A1.6 Substances moving in and out of the cell

Q4 Water can sometimes enter a cell and sometimes leave it. Use information from Table A1.2 to explain why.

Diffusion

molecule – the smallest particle of a substance made from two or more atoms

diffusion – the movement of molecules from an area of high concentration to low concentration

concentration gradient – the variation in concentration of a substance in two different areas

A **molecule** is the smallest particle of a substance made from two or more atoms. Molecules of a substance that are free to move will move from an area of where there are lots of molecules to an area where there are not many molecules. In other words the molecules spread out until they are evenly distributed. This process is called **diffusion**. This is a passive process where the molecules move down a **concentration gradient** from an area of high concentration to an area of low concentration.

Diffusion happens because the molecules have kinetic energy. They are in constant movement. Small molecules such as carbon dioxide and glucose move faster than large molecules such as amino acids and can diffuse through the cell membrane very quickly.

SBA Skills

ORR	D	MM	PD	AI
✓				

To do – Observing diffusion

1 Fill a beaker with water. Let the water settle until all movement has ceased.

2 Drop a crystal of copper sulphate or potassium permanganate into the water.

3 Leave the solution for several days, observing at regular intervals. You should see the colour diffusing through the water until the solution is a uniform colour, as shown for copper sulphate in Fig A1.7.

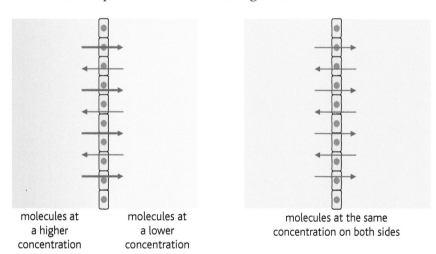

▲ **Fig A1.7** Molecules of copper sulphate diffuse throughout the water

partially permeable membrane – a membrane that only allows certain molecules to pass through

Molecules move into and out of a cell through the cell membrane by diffusion. Cell membranes are **partially permeable**. This means they will allow some substances through but not others.

net movement – the overall movement of molecules when more move in one direction than another

Diffusion into a cell is not a one way process, but because more molecules will move from where there is a high concentration to where there is a low concentration, than the other way round, we say that there is a **net movement** of molecules from high to low concentration. This continues until there is an even distribution of molecules either side of the membrane when diffusion in both directions is equal. This is shown in Fig A1.8.

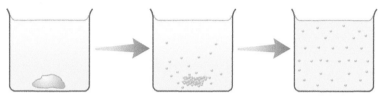

molecules at a higher concentration   molecules at a lower concentration

molecules at the same concentration on both sides

▲ **Fig A1.8** On the left of the diagram more molecules move from left to right than from right to left . On the right, there is no net movement as molecules move equally from left to right and right to left.

Q5 Explain what is meant by a 'net movement' of molecules.

Q6 Explain what is meant by a 'partially permeable membrane'.

Examples of diffusion

Diffusion is of vital importance to all living things. It is the way substances are transported in and out of cells and from one cell to another.

It takes place in the air sacs in our lungs as we absorb oxygen and get rid of carbon dioxide…

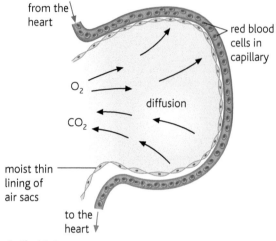

from the heart

red blood cells in capillary

O_2

diffusion

CO_2

moist thin lining of air sacs

to the heart

▲ **Fig A1.9**

… in the villi in our intestine as we absorb glucose, amino acids, fatty acid and glycerol from our food …

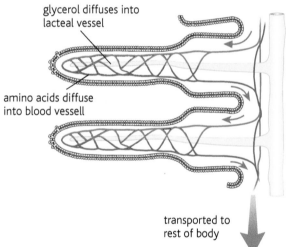

glycerol diffuses into lacteal vessel

amino acids diffuse into blood vessell

transported to rest of body

▲ **Fig A1.10**

… and through the stomata in leaves as they absorb carbon dioxide for photosynthesis and get rid of oxygen.

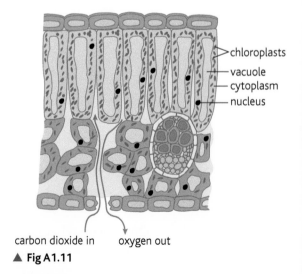

chloroplasts
vacuole
cytoplasm
nucleus

carbon dioxide in   oxygen out

▲ **Fig A1.11**

Osmosis – a special type of diffusion

osmosis – the diffusion of water molecules from a high concentration of water molecules and a low concentration of sugar molecules through a partially permeable membrane to an area of low concentration of water molecules and a high concentration of sugar molecules

Osmosis only applies to water molecules. Like diffusion, it is a passive process and relies upon the kinetic energy of the water molecules. It occurs when other substances such as sugar are dissolved in water.

Osmosis is the diffusion of water molecules from where there is a high concentration of water molecules and a low concentration of sugar molecules through a partially permeable membrane to an area of low concentration of water molecules and a high concentration of sugar molecules. This is shown in Fig A1.12.

It occurs because the sugar molecules are too large to pass easily through the membrane but water molecules, being much smaller, can pass through much more easily.

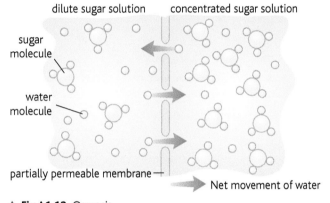

▲ **Fig A1.12** Osmosis

high water potential – a solution with a high water potential has a high concentration of water molecules, which can diffuse from the solution by osmosis

Scientists would say that the dilute sugar solution has a **high water potential**. Water moves from a solution with a high water potential, through a partially permeable membrane, to an area of low water potential.

Visking sheet – an artificial synthetic material such as cellophane that can act like partially permeable membrane

Q7 A student wrote about osmosis for homework. He stated that osmosis is the movement of glucose through a partially permeable membrane from an area of low water potential to an area of high water potential. Explain why his teacher told him that his homework was wrong.

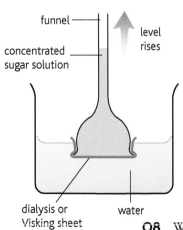

▲ **Fig A1.13**

Demonstrating osmosis

Fig A1.13 shows a simple experiment that can demonstrate osmosis in action.

1. Fill a thistle funnel with concentrated sugar solution, and enclose it with dialysis or **Visking sheet**. Visking sheet acts like a partially permeable membrane.

2. Invert the thistle funnel into a beaker of pure water. The water molecules pass through the Visking sheet by osmosis into the thistle funnel causing the level of water in the funnel to rise. It is possible to get the level of water to rise by about 10 metres using this method.

Q8 Why will the water level rise more quickly if there is a big difference in the concentrations of the water and sugar solution?

Osmosis in red blood cells

When red blood cells are placed in pure water, they burst. When red blood cells are placed in a concentrated sugar solution, they collapse and shrink.

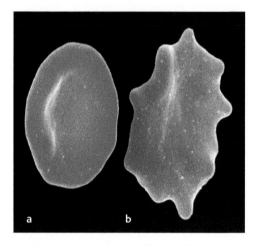

▲ **Fig A1.14** **a** Normal red blood cell
b Shrunken red blood cell

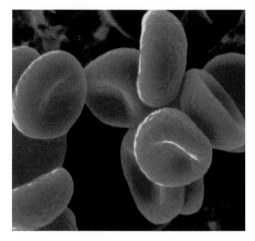

▲ **Fig A1.15** These red blood cells have swollen from too much water

Plant cells also respond like the red blood cells but, because they are surrounded by a cell wall, when they expand, they cannot burst. They just get slightly larger and much harder. This is why fresh plants are crisp when we eat them. Animal cells do not have a cell wall. So even when they absorb water by osmosis they never become crisp and crunchy.

Q9 Explain what will happen to red blood cells if they are placed in a sugar solution that is the same concentration as the contents of the red blood cells.

SBA Skills

ORR	D	MM	PD	AI
✓				✓

To do – Effects of osmosis

1 Take a bunch of flowers and place half the flowers in a jug of water. Place the other half of the flowers in a jug without any water. Leave the flowers for a few hours.

2 Use what you have learnt about osmosis to explain any differences that you can see between the two bunches.

isotonic – two different solutions with an equal concentration

hypotonic – a solution that is less concentrated than another solution

hypertonic – a solution that is more concentrated than another solution

When a plant cell is placed in a solution that is the same concentration as the cell contents we say the two solutions are **isotonic**. (Iso means equal.)

When a plant cell is placed in a less concentrated solution than the cell contents, we say the less concentrated solution is **hypotonic** and the cell contents are **hypertonic**. When this happens the cell expands slightly as water enters. Just like a bicycle tyre becomes hard when it is pumped up, this makes the cell stiff and hard. Scientists say the cell has become turgid.

plasmolysis – when the cell contents of a plant cell shrink away from the cell wall due to water leaving by osmosis

When a plant cell is placed in a more concentrated solution than the cell contents, we say the more concentrated solution is hypertonic and the cell contents hypotonic. When this happens the cell contents collapse and pull away from the cell wall as the cell loses water. Scientists call this **plasmolysis**.

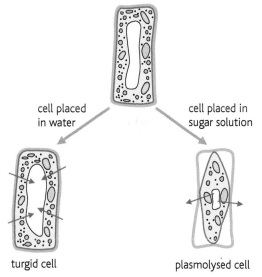

cell placed in water

cell placed in sugar solution

turgid cell

plasmolysed cell

▲ **Fig A1.16** The concentration of the solution makes cells turgid or plasmolysed

SBA Skills

ORR	D	MM	PD	AI
✓		✓		✓

Experiment – Investigating the concentration of potato cell contents

1 Cut five fresh potato chips using a cork borer. Make sure each potato chip is the same length.

2 Place each chip into a different concentration of sugar solution ranging from 0.0 mol/dm³ (pure water) to 1.0 mol/dm³ sugar solution.

3 Leave the chips for one hour, then carefully re-measure the length of each chip. Calculate the change in length.

4 Plot your results as a graph showing change in length of chip on the *y*-axis and concentration of solution on the *x*-axis. The point on your graph when the potato chips are not changing in length tells you the concentration of the potato cell contents. This is because the concentration of the sugar solution is isotonic with the potato cell contents. You should obtain a graph looking something like Fig A1.17.

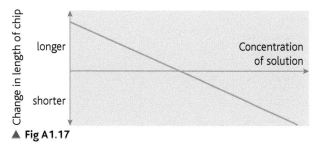

▲ **Fig A1.17**

5 Use your graph to determine the concentration of the potato cell contents.

End-of-unit questions

1 Which of the following structures are not found in an animal cell?
- **A** nucleus
- **B** cytoplasm
- **C** cell wall
- **D** cell membrane

2 Which of these cell structures is responsible for releasing energy by the process of respiration?
- **A** nucleus
- **B** cytoplasm
- **C** chloroplast
- **D** mitochondria

3 Which *two* of these statements about osmosis are correct?
- **A** It is a movement of sugar molecules through a partially permeable membrane.
- **B** It is the movement of both water and sugar molecules from a high to a low concentration.
- **C** It is the movement of water molecules from a low concentration of sugar molecules to a high concentration of sugar molecules through a partially permeable membrane.
- **D** It is the movement of water molecules from a high to a low concentration through a partially permeable membrane.

4 Maureen placed equal sized potato chips in pure water and a concentrated sugar solution. She measured the size of each chip both before and after using 2 mm squared graph paper.

Her results are shown in Fig A1.18 below.

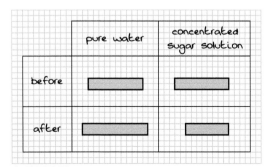

▲ **Fig A1.18**

a Complete the table to show the size of each chip.

	pure water	concentrated sugar solution
length before / mm		
length after / mm		

b Explain what is happening to the chip in the pure water.

c Explain what is happening to the chip in the concentrated sugar solution.

d Explain how Maureen could use this experiment to find the concentration of the cell sap in the potato chips.

5 Ryan was confused between diffusion and osmosis. Write down a simple explanation for Ryan describing the differences between diffusion and osmosis.

6 Doltan is training for a race. He drinks an isotonic energy drink.

a Suggest what is meant by an **isotonic energy drink.**

b During the race, Doltan's body loses water through sweat.
Explain what will happen to the concentration of Doltan's blood during the race.

c Doltan notices during the race that his breathing rate increases.
Explain what effect this will have on the diffusion of oxygen and carbon dioxide in his lungs.

d After the race Doltan feels thirsty. He has a cup of warm tea.
Doltan adds a cube of sugar to the tea but does not stir it. The tea does not taste sweet. Ten minutes later when he finishes his cup of tea, it now tastes sweet.
Use your knowledge of diffusion to explain why.

A2a Photosynthesis and Photochemical Reactions

By the end of this unit you will be able to:

- describe the process of photosynthesis.
- describe photochemical reactions, such as how light energy is transferred in photosynthesis and photography.

Photochemical reactions

Photochemical reactions use energy from light to make it work. Taking a photograph uses a photochemical reaction where energy from light reacts with silver salts, such as silver halide, and turns them black. By projecting an image onto a gelatine sheet covered in silver halide, the light parts of the image turn black and the dark parts, where there is no light, stay light. In black and white photography this process produces a negative. By shining light through the negative onto another sheet of photographic paper, a black and white photograph is produced.

▲ **Fig A2.1** How a black and white photo is formed

Photosynthesis

photosynthesis – a process by which plants convert water and carbon dioxide into oxygen and glucose using energy from sunlight

food chain – energy from the Sun transferred as food energy from a plant to a series of animals

Photosynthesis is another photochemical reaction in which energy from sunlight is used, this time to make complex organic molecules used as food.

The energy from all the food that we eat comes from the energy in sunlight that is trapped by plants by the process of photosynthesis. Even when we eat meat from animals, the energy in the meat originally came from the plants that were eaten by the animal. This is called a **food chain**. Food chains show how energy is transferred from sunlight, through plants and finally to animals and human beings.

▲ **Fig A2.2** A food chain

Q1 Write down one difference and one similarity between photosynthesis and photography.

The process of photosynthesis

carbon dioxide – a gas used in the process of photosynthesis

chlorophyll – a green chemical produced by plants that is used in the process of photosynthesis to trap light energy from the Sun

glucose – a type of sugar that is produced by photosynthesis and used in respiration as an energy source

starch – a carbohydrate food storage chemical produced by plants

oxygen – a waste product produced by the process of photosynthesis

substrate – a substance acted upon by an enzyme

Plants make food by photosynthesis. They make the food from **carbon dioxide** taken out of the air through their leaves. Light energy from the Sun is then used by a green chemical called **chlorophyll** to convert the carbon dioxide and some water into a food called **glucose**. Energy from sunlight is transferred into chemical energy in the glucose. Plants can then use the glucose to supply themselves with energy as and when they need it, store it as a chemical called **starch**, or use it as a starting point from which to make many different kinds of chemicals as the plant grows. So each time we eat some fruit or vegetables, or burn a log of wood on a fire, we are releasing the energy that the plant has transferred from sunlight.

The process of photosynthesis can be shown by using both word and chemical equations.

$$\text{carbon dioxide } + \text{ water } \xrightarrow[\text{chlorophyll}]{\text{light energy}} \text{ glucose } + \textbf{oxygen}$$

$$6CO_2 + 6H_2O \xrightarrow[\text{chlorophyll}]{\text{light energy}} C_6H_{12}O_6 + 6O_2$$

Carbon dioxide and water are the **substrates** from which the glucose and oxygen are made.

Glucose and oxygen are the products that are made by the process of photosynthesis.

Q2 Explain the importance of light and chlorophyll in photosynthesis.

A leaf – the green machine

Photosynthesis takes place inside leaves. Most cells of a leaf contain small structures called **chloroplasts**, which contain chlorophyll. This is why leaves look green in colour. Most chloroplasts are found near the upper surface of the leaf in the palisade tissue because this is where most of the sunlight shines on the leaf. Leaves are very thin and light in weight, but have a very big surface area to trap as much energy from sunlight as possible.

chloroplast – a structure found in some plant cells that contains chlorophyll used in photosynthesis

stomata – small pores on the underside of a leaf that release water vapour and oxygen and take in carbon dioxide for photosynthesis (singular stoma)

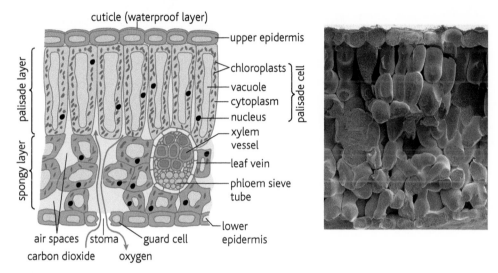

Fig A2.3 A cross section through a typical leaf

Carbon dioxide enters through the small holes on the underside of the leaf called **stomata**. Oxygen produced by photosynthesis leaves the leaf by the same route.

stomata – small pores on the underside of a leaf that release water vapour and oxygen and take in carbon dioxide for photosynthesis (singular stoma)

The carbon dioxide diffuses through the air spaces between the cells of the leaf and is absorbed into the chloroplasts. As the carbon dioxide is used by the chloroplasts, a concentration gradient is set up such that there is always more carbon dioxide in the air outside the leaf so the process of diffusion continues, ensuring a constant supply of carbon dioxide to the chloroplasts.

Water enters the roots by osmosis and is carried up the stem through xylem tissue until it reaches the leaves. It then continues to move from cell to cell by osmosis, ensuring that all the chloroplasts are supplied with water.

Q3 Explain why a leaf is thin, green and has a large surface area.

Testing for photosynthesis

The glucose that is produced by photosynthesis is converted to starch for storage until it is required. Plant leaves can be tested for starch to show that photosynthesis has taken place. The test for starch is to add **iodine solution**. Starch turns black when iodine solution is added. Our problem is that the starch is locked up inside the leaf. Simply adding iodine solution to the leaf will not show if starch is present. First we have to treat the leaf so that we can get at the starch that has been stored inside.

iodine solution – a solution of iodine in potassium iodide that turns black in the presence of starch

Experiment – Testing for starch

1 Find a green leaf that has been left in bright sunlight for a few hours. This will ensure that it has produced plenty of starch by photosynthesis.

2 Place the leaf into a beaker of boiling water until the leaf is soft. The boiling water helps to break down the cell walls so that we can get at the starch.

Next we need to remove the chlorophyll so we can easily see any colour change when we add the iodine solution.

3 Place the leaf into a test tube of hot alcohol. The alcohol is highly flammable so should not be heated directly. Instead the test tube is placed in the beaker of boiling water so that the hot water heats up the alcohol. You will need to leave the leaf in the hot alcohol until all of the chlorophyll has been removed from the leaf.

4 Next place the leaf onto a white tile and add iodine solution to the leaf. After a few seconds the leaf will turn black, showing that starch is present in the leaf and that photosynthesis has taken place.

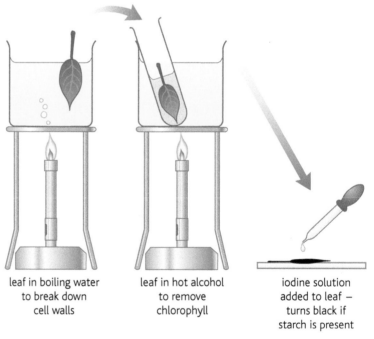

leaf in boiling water to break down cell walls

leaf in hot alcohol to remove chlorophyll

iodine solution added to leaf – turns black if starch is present

▲ **Fig A2.4** Preparing the leaf to test for starch

Q4 Explain why it is necessary to remove the chlorophyll.

_____

It is possible to use this simple experiment to investigate the conditions that are needed for photosynthesis to take place.

SBS Skills

ORR	D	MM	PD	AI
✓		✓		✓

Experiment – Is light needed for photosynthesis?

1   Make a stencil out of cardboard by cutting out the first letter of your name. It should look something like the stencil shown in Fig A2.5.

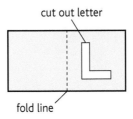

cut out letter

fold line

▲ **Fig A2.5** Making your stencil

2   Carefully fold the stencil around a leaf that is still attached to a plant. The plant should have been left in the dark for 24 hours. Fix the stencil in place using paper clips as in Fig A2.6.

▲ **Fig A2.6** Attaching the stencil

3   Leave the plant in bright sunlight for at least three hours. Then remove the leaf from the plant. Remove the stencil and carry out a starch test on the leaf.

Q5   What do you notice happens to the leaf when you have carried out the starch test?

Q6   What does this tell you about light and photosynthesis?

SBS Skills

ORR	D	MM	PD	AI
✓		✓		✓

potassium hydroxide – a chemical that absorbs carbon dioxide

sodium hydrogencarbonate – a chemical that releases carbon dioxide

Experiment – Is carbon dioxide needed for photosynthesis?

Two potted plants are placed under two bell jars on a sheet of glass. In one bell jar there is a beaker containing a solution of **potassium hydroxide**. Potassium hydroxide absorbs carbon dioxide so that the air in the bell jar will not contain any carbon dioxide. Under the second bell jar there is a beaker containing a solution of **sodium hydrogencarbonate**. Sodium hydrogencarbonate releases carbon dioxide. This means the air in this bell jar will contain lots of carbon dioxide. The bases of the bell jars are made air tight using some Vaseline.

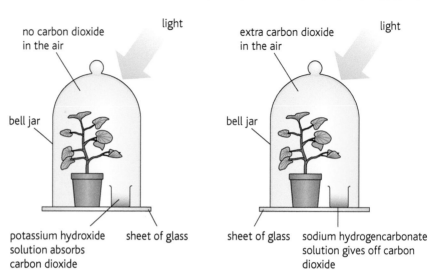

▲ **Fig A2.7** Testing the need for carbon dioxide in photosynthesis

The plants are left in bright light for 24 hours. A leaf from each plant is then tested using the same starch test as on page 14.

Q7 What do you think will happen to each leaf when the starch test is carried out?

Q8 What does this tell you about carbon dioxide and photosynthesis?

SBS Skills

ORR	D	MM	PD	AI
✓	✓	✓		✓

Experiment – Is chlorophyll needed for photosynthesis?

For this test you will need a variegated leaf that has been left attached to a plant in bright sunlight for at least three hours. A variegated leaf is a leaf that is different colours because not all of the leaf contains chlorophyll. A leaf that is both green and white works well with this test.

1 Draw a diagram of the pattern on the leaf and then carry out the starch test.

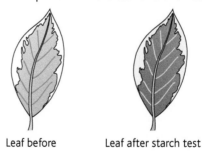

Leaf before   Leaf after starch test

▲ **Fig A2.8** The leaf before and after the starch test

2 When you have completed the test make a second drawing of the leaf.

Q9 What do you notice about the two drawings?

Q10 What does this tell you about chlorophyll and photosynthesis?

End-of-unit questions

1 Which of the following are *not* required for photosynthesis to take place?
   A carbon dioxide
   B light
   C chlorophyll
   D oxygen
   E water

2 Photosynthesis takes place in the:
   A xylem
   B phloem
   C chloroplast
   D vascular bundle

3 Which of the following chemicals is used to absorb carbon dioxide?
   A water
   B sodium hydrogencarbonate
   C iodine
   D potassium hydroxide

4 Explain the difference between respiration and photosynthesis.

5 Look at the diagram of a section through a leaf shown in Fig A2.9.

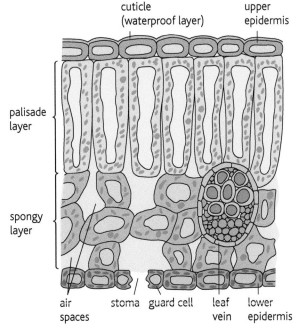

▲ Fig A2.9

a Write down the part of leaf that does the following jobs.
   A transports water
   B allows carbon dioxide into the leaf
   C where most photosynthesis takes place

b Write down the word equation and chemical equation for photosynthesis.

c Describe the steps involved in the transfer of energy during photosynthesis.

6 Michael grows tomatoes in a greenhouse. His friend grows tomatoes in a field.
   a Both lots of tomatoes grow well.
      i Describe **three** conditions needed for the tomatoes to grow well.
         1.
         2.
         3.
      ii Name the process that plants use to capture energy for growth.

   b Michael had to spend a lot of money to build his greenhouse.
      Suggest the possible benefits to Michael of growing his tomatoes in a greenhouse, rather than a field.

   c Michael and his friend sell their tomatoes at a local market.
      Michael wants to know if it is worth him building another greenhouse.
      Explain how Michael could investigate whether it is more profitable to grow his tomatoes in a field, or in a greenhouse.

A2b Food and Nutrition

The importance of food

Types of food

Food supplies all the energy that we need and all the raw materials that we need for growth. However, there are different types of food and it is important that we eat the right amounts of each type. Too little and we may suffer from food deficiencies; too much and we may become obese, and suffer from coronary heart disease. Table A2.1 below shows the different types of food, their roles in our diet and where we can find them. A healthy diet should contain all of the food types mentioned in the table.

carbohydrate – a group of chemicals made by plants for food containing carbon, hydrogen and oxygen.

protein – large food molecule made from amino acids

type of food	role of food	source of food
carbohydrates	**Carbohydrates** are converted into simple sugars such as glucose and supply us with energy during respiration.	Examples of foods rich in carbohydrates include bread, yam and sweet potato and sugar.
proteins	**Proteins** are used in the repair and growth of new cells and to make chemicals such as enzymes.	Examples of foods rich in proteins include meat, eggs, fish, beans and peas.

Table A2.1

	type of food	role of food	source of food
fat – food molecule made from fatty acids and glycerol	fats and oils	**Fats** and oils are rich in energy and so are ideal to use as an energy store in fat cells under the skin. This is why when we eat too much food we say we are getting fat.	Examples of foods rich in fats and oils include cheese, butter, milk sardines and palm oil.
roughage/fibre – indigestible plant material such as cellulose	fibre	**Fibre** or **roughage** is indigestible plant cellulose. Although our bodies cannot use it, it gives bulk to food and allows our gut to push it along the digestive tract and finally eliminate it as waste.	Examples of foods rich in fibre include bran.
vitamin – a chemical needed in small quantities by animals	vitamins	**Vitamins** are required in very small amounts for chemical reactions inside our bodies.	Examples of foods rich in vitamin C include green, red and yellow vegetables and fruits. Carrots and red fruits are rich in vitamin A and dairy produce in vitamin D. Vitamin E helps in blood clotting and is found in soya and olive oil.
mineral – inorganic substance needed in small quantities by animals and plants **iron** – a mineral needed for the haemoglobin in red blood **calcium** – a mineral needed to make strong bones and teeth	minerals	**Minerals** are inorganic substances needed in small quantities by animals and plants for good health. **Iron** is needed for the haemoglobin in red blood. **Calcium** is needed to make strong bones and teeth.	Examples of foods rich in iron include liver, spinach and wholemeal bread. Milk and dairy produce are rich in calcium.
	water	Our body is composed mostly of water. It is the solvent in which all chemical reactions take place in our bodies.	

Table A2.1 *continued*

Bread, cereals, rice or yams and potatoes provide all our carbohydrate needs along with some protein, vitamins and minerals. These carbohydrate foods are the staple diet for millions of people around the world.

A selection of fruits, green and coloured vegetables provide lots of the vitamins, minerals and roughage that we need. They are rich in vitamins A, C and E which have antioxidant properties. Scientists think that they may help to protect us from cancer.

Meat and fish provide us with protein. Oily fish in particular is rich in omega 3 which is a healthy type of oil rather than the more unhealthy saturated fats found in some animal fats, such as cheese and butter. Eating too much saturated fat may lead to conditions such as coronary heart disease.

Q1 Explain the role of protein, carbohydrate and fat in our diet.

How much food do we need?

Different people need different amounts of each type of food. Active people need more carbohydrate to supply them with energy. Children need more protein because they are growing. A woman who is breastfeeding her baby needs more calcium to make sure that there is enough for herself and her developing baby to have strong bones and teeth.

Table A2.2 shows the amount of each type of food needed by different people each day.

	energy /kJ	protein /g	iron /mg	calcium /mg
1-year-old child	3 850	14.9	7.8	525
15-year-old boy	11 500	55.2	11.3	1000
15-year-old girl	8 830	45.0	14.8	800
man	10 500	48.0	12.0	750
woman	8 100	45.0	14.8	700
breastfeeding woman	11 000	56.0	14.8	1250

Table A2.2

Q2 Suggest why a breastfeeding woman needs more calcium in her diet than a 15-year-old girl.

Q3 Suggest why a man requires more energy each day than a woman.

Food and disease

It is important that we eat a balanced diet that provides us with all the nutrients that we need. Problems can arise when we eat too little or too much of each type of food.

Kwashiorkor

kwashiorkor – a protein deficiency disease

Eating too little protein can lead to a condition called **kwashiorkor**. Children in poor developing countries often suffer from it.

We can calcuate how much protein we need to eat every day by using the following formula:

recommended daily allowance (RDA) – the recommended amount of a food type to be consumed within a day

RDA (recommended daily allowance) in grams = 0.75 × body mass.

So, a woman weighing 70 kg would require 52.5 g of protein.

PEM or protein–energy malnutrition – a condition caused by a lack of protein and energy food

A lack of protein in the diet is also usually linked to a lack of energy foods in the diet. This is sometimes called **PEM** or **protein–energy malnutrition**. Surprisingly, PEM in pregnant women can often lead to obesity in later life for the unborn child as the body stores fat very quickly when food is available, just in case times of starvation return.

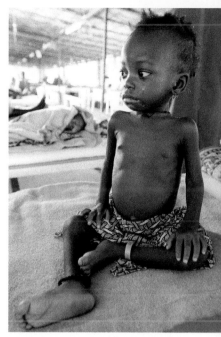

▲ **Fig A2.10** A child with kwashiorkor

Q4 Ranjit weighs 80 kg. How much protein should he eat each day?

Obesity

obesity – a condition caused by eating too much food and laying down fat deposits under the skin

Obesity can lead to fat and cholesterol deposits being laid down in arteries which can become blocked. If the artery is one that supplies the heart muscle it can cause a **heart attack** or coronary thrombosis.

heart attack – a condition caused when a blood vessel supplying heart muscle becomes blocked

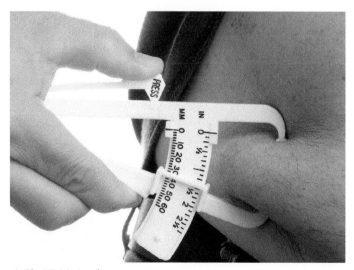

▲ **Fig A2.11** An obese person

Q5 Give one reason why eating too much food can be just as dangerous as eating too little food.

Night blindness

night blindness – a vitamin deficiency disease caused by a lack of vitamin A

Night blindness is caused by a lack of vitamin A. Vitamin A is used to produce a light-sensitive chemical at the back of our eyes that enable us to see in dim light.

To do – Night vision

When you get home tonight go from a brightly lit room out into the dark. At first you will see very little but after a few minutes your eyes will adapt and you will be able to see much better. If you do not have enough vitamin A in your diet, your eyes will not be able to do this.

SBS Skills

ORR	D	MM	PD	AI
✓				

Scurvy

scurvy – a vitamin deficiency disease caused by a lack of vitamin C

Scurvy is caused by a lack of vitamin C. Vitamin C does many jobs in our body. One of them is to make the walls of small blood vessels strong and healthy. A lack of vitamin C causes the walls of the blood vessels to break down and they begin to leak. This can have deadly effects all over the body and is noticeable in bleeding gums and teeth that fall out. Sailors on long sea voyages to the Caribbean used to suffer from scurvy. It was only when they ate fruit when they arrived or started to carry lime juice on board ship that scurvy was eliminated.

Rickets

rickets – a vitamin deficiency disease caused by a lack of vitamin D

Rickets is caused by a lack of vitamin D. Vitamin D is needed for our bodies to absorb calcium from our food. Because calcium is needed to make our bones and teeth strong, a lack of vitamin D leads to soft bones and teeth. The bones can become so soft that the bones in the legs can no longer support the weight of the body and begin to bend.

anaemia – an iron deficiency disease

Anaemia is caused by a lack of iron. Iron is needed to make haemoglobin in red blood cells, which carry oxygen around the body. This is why people suffering from anaemia often feel tired.

▲ **Fig A2.12** A child with rickets

Q6 Suggest why it is important that we eat a wide range of different types of food.

Experiment – Food tests

It is possible to test samples of food to see which of the different food types they contain.

1 Starch test

We have already seen that we can test a leaf for starch to see if photosynthesis has taken place. Iodine solution can also be used to test samples of our own food to see if any starch is present as iodine always turns black in the presence of starch.

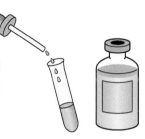

▲ **Fig A2.13** The starch test

2 Glucose and other reducing sugars test

reducing sugar – a simple sugar e.g. glucose

Add a small sample of approximately 2 cm³ of food solution to a test tube. Add an equal amount of blue Benedict's solution to the test tube. Shake the contents and then heat in a hot water bath until the contents boil, as in Fig A2.14. If a **reducing sugar** such as glucose is present the mixture will gradually change to green then yellow and finally to a brick red colour. The test is semi-quantitative. This

▲ **Fig A2.14** The reducing sugars test

means it tells roughly how much glucose is present. Green indicates very little glucose present and red indicates a greater concentration of glucose in the food.

 SAFETY: You must wear safety goggles for Tests 3 and 4 as sodium hydroxide and potassium hydroxide splashed into the eyes causes instant blindness.

3 Non-reducing sugars test such as sucrose and fructose (fruit sugar)

non-reducing sugar – a complex sugar e.g. sucrose

Add a small sample of approximately 2 cm³ of food solution, such as liquidised fruit, to a test tube. Add an equal amount of dilute hydrochloric acid to the test tube. Heat in a hot water bath and boil the contents for one minute. Neutralise the solution by adding approximately 2 cm³ of dilute sodium hydroxide solution. The **non-reducing sugar**, such as sucrose, should now be broken down into reducing sugars such as fructose and glucose, and can be tested for using Benedict's solution as described in Test 2.

4 Protein test

Add a small sample of approximately 2 cm³ of food solution to a test tube. Add an equal amount of 5% potassium hydroxide solution and stir. Then add two drops of 1% blue copper sulphate solution. If a purple colour slowly appears, then protein is present.

▲ **Fig A2.15** The protein test

5 Lipid or fat test

Add 2 cm³ of ethanol to the food solution. Shake well. Add 2 cm³ of water. Shake well. If the solution turns milky white this indicates that fat is present.

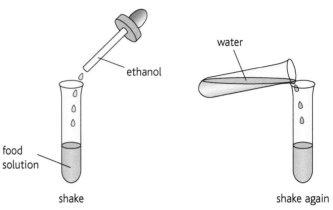

▲ **Fig A2.16** The lipid test

Q7 A student tests a food for reducing sugar. The test is negative. They boil the food with dilute hydrochloric acid and repeat the reducing sugar test. This time the result is positive. What did the original food contain?

What's in processed food?

When we cook and prepare our own food it is easy to know exactly what we are eating. However, some of the food that we eat is processed food and has been made and prepared by someone else. Processed food also contains additives. These are put into flavour the food, colour it and preserve it. They are given E numbers to help identify them.

We can see what is in processed food by looking carefully at the label on the packaging. The label tells us the amount of each different food group and how much energy is in the food. Energy in food is measured in **kilojoules (kJ)** and kilocalories (kcal). You can see from the table on page 20 that a teenage girl needs about 9000 kJ of energy in her food each day.

kilojoule (kJ) – 1000 joules (joules are the unit of energy)

The label shows the recommended daily requirements for men and woman of energy and fat. It also shows how much of each type of food is present in a 100 g serving.

Some food labels show energy content as kcal instead of kJ. You can convert kcal to kJ by multiplying the number of kcal by 4.2.

To do – Energy conversion

Convert the kcal on the label in Fig A2.17 into kJ.

SBS Skills

ORR	D	MM	PD	AI
		✓		

INGREDIENTS

Maize, Sugar, Malt Flavouring, Salt, Niacin, Iron, Vitamin B_5, Riboflavin (B_2), Thiamin (B_1), Folic Acid, Vitamin B_{12}

GUIDELINE DAILY AMOUNTS

EACH DAY	WOMEN	MEN
Calories	2000	2500
Fat	70g	95g

Official Government figures for average adults

NUTRITION INFORMATION

	Typical value per 100g	30g serving with 125ml of semi-skimmed milk
ENERGY	1550kJ 370kcal	700kJ* 170kcal
Calories	8g	7g
CARBOHYDRATES	82g	31g
of which sugars	8g	9g
starch	74g	22g
FAT	0.9g	2.5g *
of which saturates	0.2g	1.5g
FAT	3g	0.0g
PROTEIN	8g	4g
SODIUM	1g	0.35g

▲ **Fig A2.17** A food label

Q8 Look at the label in Fig A2.17. How much protein is in 250 g of the food?

To do – Information from food labels

SBS Skills

ORR	D	MM	PD	AI
				✓

1 Collect labels from different items of food packaging.

2 Work out how much of each of the food you would need each day to supply all of your nutritional needs. You may need to look at the table on page 20 to check your daily needs.

3 Identify the additives and E numbers in the foods.

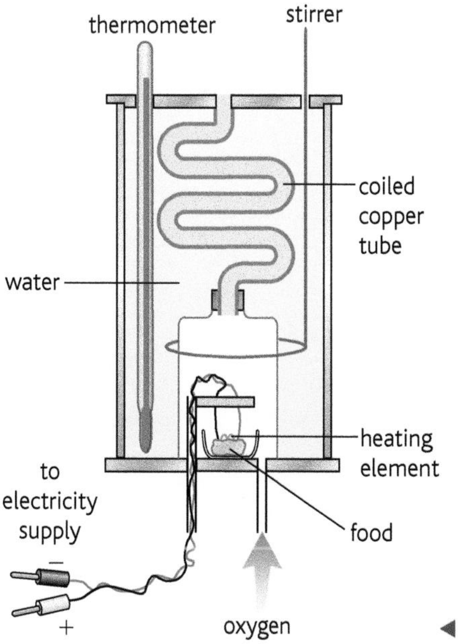

Measuring the energy in processed food

Different types of food contain different amounts of energy. We can measure how much energy is in food by burning it and recording how much the heat energy released can heat up a sample of water. The equipment we use is called a calorimeter, and is shown in Fig A2.18.

thermometer   stirrer

coiled copper tube

water

to electricity supply

heating element

food

oxygen

◀ **Fig A2.18** A calorimeter

Experiment – Measuring the energy in a crisp

You can do an experiment that works in a similar way to a calorimeter by burning a potato crisp snack under a test tube of boiling water and measuring the increase in the temperature of a known volume of water.

Make a copy of this table to record your results.

mass of potato crisp snack / g	
start temperature of water / °C	
final temperature of water / °C	
increase in temperature / °C	

Set up the apparatus as shown in Fig A2.19. The test tube should contain 25 cm^3 of water.

1 Weigh the potato crisp snack and record its mass in your table.

2 Measure the temperature of the water in the test tube and record the value in your table.

3 Make sure the potato crisp snack is firmly attached to a mounted needle. Light the potato crisp snack and immediately hold the flame under the test tube of water.

4 Wait until the potato crisp snack is burned out, then measure the temperature of the water in the test tube and record the result in your table.

5 Calculate the increase in temperature of the water and record the result in your table.

6 Use the following formula to calculate the amount of energy in joules in each gram of potato crisp snack.

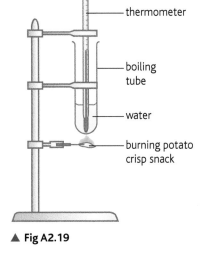

▲ **Fig A2.19**

$$\text{energy in each gram of potato crisp snack} = \frac{\text{change in water temperature} \times 25 \times 4.2}{\text{mass of potato crisp}}$$

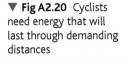

▼ **Fig A2.20** Cyclists need energy that will last through demanding distances

Dietary needs for specific activities

We have already seen that the energy needs for different people varies. The dietary needs for different activities also vary.

Endurance athletes such as cyclists and long distance runners often use a diet called 'pasta-packing'. It involves eating large amounts of energy food such as pasta, which contains lots of carbohydrates, in the days before the race. This allows the body to build up lots of energy reserves that will be used during the event.

Weight lifters, however, do not need to sustain high energy output over a long period of time so their dietary needs are different. They need to eat more protein foods to build up strong muscles so that they can lift heavier weights.

It's not just *what* we eat, but also *when* we eat that is important. Most world speed records are broken in the late afternoon and early evening when our body is performing at its best. We also process the food we eat more

▲ **Fig A2.21** Muscle growth depends on having a source of protein

efficiently during the early part of the day. Scientists recommend that we eat most of our food in the first half of the day and eat very little in the evening. This is because as the day goes on, we produce less of the hormone insulin which removes glucose from our blood.

Digestion

digestion – breaking down food by physical and chemical processes

assimilation – the process where digested food is absorbed into the blood stream

Digestion is the process whereby we break down food from large complex organic molecules into smaller and simpler inorganic molecules so that we can absorb it through the gut wall and into the blood stream. **Assimilation** is the process where digested food is absorbed into the blood stream.

Mechanical digestion

mechanical digestion – breaking down food by chewing

Mechanical digestion takes place in the mouth. It is the process where we chew our food to break it down into much smaller pieces so that we can swallow it. We do this using our teeth.

All human teeth have the same basic structure.

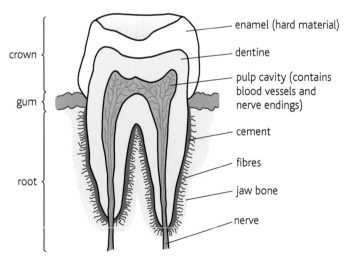

crown

gum

root

enamel (hard material)

dentine

pulp cavity (contains blood vessels and nerve endings)

cement

fibres

jaw bone

nerve

▲ **Fig A2.22** The structure of the human tooth

enamel – the very hard outer coating of the teeth

The outer coating of a tooth is made of **enamel**. This is the hardest substance found in the human body and enables us to chew and physically break down food very easily. Teeth have roots that are cemented into the jaw bone. They receive blood and nerves into the pulp cavity in the centre of the tooth.

There are four different types of tooth, each of which has a different job to perform.

1 Incisors

Incisors are chisel-shaped teeth that are used for biting and cutting into food. Rodents such as rabbits have very well developed incisors for biting grass.

2 Canines

canine – tooth used for gripping things

Canines are point teeth. They are used to pierce and grip the food so that we can tear it. Dogs have well developed canines for gripping their prey.

3 Premolars

premolar – tooth used for grinding up food

Premolars are flat ridged teeth that are used for crushing and grinding our food.

4 Molars

molar – tooth used for grinding up food

Molars are a larger version of premolars. They are also used for crushing and grinding our food. Goats have well developed premolars and molars so that they can grind the vegetation that they eat.

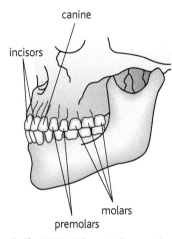

▲ **Fig A2.23** Where in the mouth the different types of teeth are located

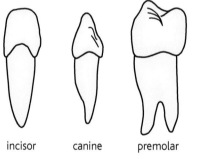

incisor   canine   premolar   molar

▲ **Fig A2.24** The different types of teeth

Q9 Explain why molars have a large flat surface but incisors have a chisel-shaped surface.

To do – Identifying types of teeth

SBS Skills

ORR	D	MM	PD	AI
				✓

1 Look at models of different types of teeth.
2 Identify each of the teeth, make a simple labelled diagram and state the job carried out by the tooth.
3 Use Fig A2.23 of the position of teeth in the jaw bone to place the teeth in the order that you would find them in a human mouth.

Dental formula

Different animals have different numbers of each type of teeth. The numbers can be expressed as a dental formula. The formula is written for one side of the mouth only. The upper row represents the teeth in the upper jaw, and the lower row, the teeth in the lower jaw.

The dental formula of humans is $\frac{2.1.2.3}{2.1.2.3}$

Keeping our teeth healthy

plaque – the build-up of bacteria and acid on the surface of the teeth

Teeth need to be cared for if they are to last. They should be carefully cleaned with a toothbrush at least twice a day. Bacteria multiply rapidly in the warm moist conditions of the mouth and form a layer of **plaque** on our teeth. This layer contains acids that can react with the enamel of the teeth. Bacteria use the food in our mouths for their own food and after a meal the pH of our mouths falls, making teeth most at risk of acid attack. This is why it is important that we should clean our teeth regularly, taking care to remove all food debris and bacterial plaque build-up from all the surfaces.

fluoride – a chemical added to water to prevent tooth decay

fluoridation – adding fluoride to water supplies to reduce tooth decay

One way of protecting teeth is to add **fluoride** to the drinking water. Fluoride is the salt calcium fluoride CaF_2. Adding this to drinking water is called **fluoridation**. Adding fluorine salts to drinking water of 1 part per million can reduce dental decay in children's teeth by over 50%. This is a controversial thing to do as some people believe that they should have a choice of whether to consume fluoride or not and adding it to drinking water gives them no choice. There is no scientific evidence that this level of fluoridation does any harm to humans whatsoever but it can reduce pain and suffering by preventing tooth decay.

Look at the graph in Fig A2.25 that shows the relationship between tooth decay and the concentration of fluorine in the water supply.

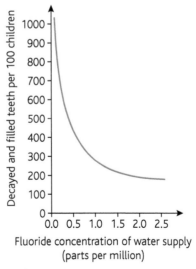

▲ **Fig A2.25**

Q10 Using the graph in Fig A2.25, suggest why it is not sensible to increase the concentration of fluorine in the water supply from 1 part per million to 2.5 parts per million.

Chemical digestion

saliva – a digestive juice containing the enzyme amylase that is released into the mouth

Chemical digestion begins as soon as we start to chew food in our mouths. This is because **saliva** containing an enzyme is added to the food.

To do – Effect of saliva on bread

1 Take a piece of bread and chew but do not swallow it. Then push the ball of chewed bread next to your cheek at the side of your mouth.

2 Leave it there for at least ten minutes. What do you notice happening to the bread after about ten minutes?

You should notice that the bread starts to taste sweet. This is because the saliva in your mouth contains an enzyme called **amylase**. Bread contains starch. Amylase breaks down or hydrolyses starch into a simple sugar.

amylase – an enzyme that breaks down carbohydrates into simple sugars

enzyme – a chemical that speeds up the break down of food without being used up in the reaction

Chemical digestion is controlled by **enzymes**. An enzyme is a chemical that speeds up a chemical reaction without becoming used up by the reaction. Think of a lock and key. The key can open or close the lock and can be used over and over again. The key does not become part of the lock or destroyed by opening the lock. It is the same with enzymes. They can break down a complex organic food molecule into smaller, simpler ones and then go and do the whole process all over again with another complex organic food molecule. Just as a key is specific to one lock, and different keys are needed for different locks, a particular enzyme can only break down one type of molecule. Different enzymes are needed for each different type of food molecule. This is known as the **lock and key model**.

lock and key model – a model to help explain how enzymes work

substrate – the molecule that is broken down by the enzyme

The molecule that is being broken down is called the **substrate**.

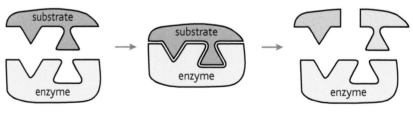

▲ **Fig A2.26** The lock and key model

The effect of temperature and pH on enzyme activity

Enzymes are not only specific in terms of what molecules they act upon, but they are also affected by temperature and pH.

Temperature

kinetic energy – energy of movement

Enzymes are chemicals. This means that like all other chemicals they gain more **kinetic energy** of movement as their temperature increases. Because they are moving faster they are much more likely to have a collision with a food molecule. This means that as temperature increases the rate of reaction also increases. The rate of reaction approximately doubles for every 10°C rise

denature – the changes in shape made to an enzyme that stops it from working

optimum – the conditions at which a reaction proceeds at its fastest rate

in temperature. However enzymes are also proteins, so once the temperature rises above 50°C the enzyme becomes **denatured** – its shape changes and it no longer 'fits the lock' to break down food molecules.

The graph in Fig A2.27 shows the rate of enzyme activity at different temperatures. The peak of the graph shows the temperature at which the enzyme works best – the **optimum** temperature. This particular enzyme stops working altogether above 60°C.

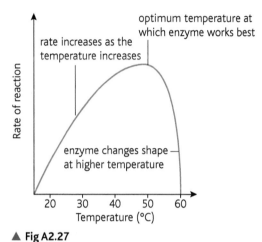

▲ **Fig A2.27**

pH

pH – a measure of how acid or alkaline a chemical is

Enzymes are also affected by **pH**, which is a measure of how acid or alkaline a substance is. As with temperature, pH causes the shape of the enzyme to change, and just as with a key, if the shape changes too much the key will not work.

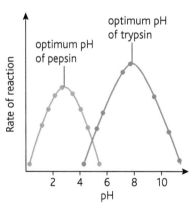

▲ **Fig A2.28**

Different enzymes have different pHs at which they work best. Pepsin prefers the acid conditions in the stomach. Trypsin prefers the alkaline conditions of the duodenum and ileum, the part of the intestine that comes after the stomach.

Q11 Use the lock and key model to explain why changing pH or temperature can affect the rate of an enzyme reaction.

SBS Skills

ORR	D	MM	PD	AI
✓		✓	✓	✓

Experiment – How do temperature and pH affect reaction time?

1 Use a pipette to add a few drops of amylase solution to a test tube. Use another clean pipette to add a few drops of dilute starch to the test tube.

2 Place 10 drops of the mixture onto a white tile and each minute add one drop of iodine solution to one of the drops of mixture.

3 The first drop will turn black showing that starch is present. However, after a few minutes the drops will no longer turn black, showing the enzyme amylase has hydrolysed the starch into a simple sugar. Make a note of how many minutes this takes.

4 Repeat the experiment, only this time either change the temperature or add a drop of dilute hydrochloric acid to the starch–amylase mixture. Make a note of how long it now takes for the amylase to hydrolyse the starch.

The journey of digestion

Our gut is over 7 metres in length and food passes along it from our mouths at one end to our anus at the other end. As the food passes along the gut, different enzymes are added to the food to break down the large molecules into smaller ones that are absorbed through the gut wall and into our blood stream.

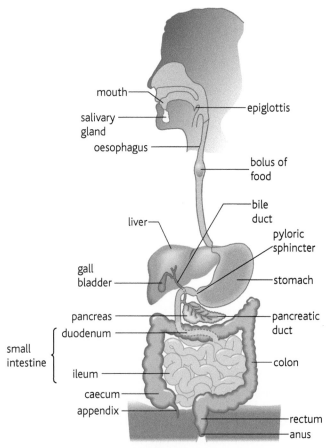

▲ **Fig A2.29** The journey of digestion

Mouth and oesophagus

Once the food has been chewed, a ball of food called a bolus is swallowed and passes down the oesophagus to the stomach. Food does not just drop down the tube. Layers of muscle squeeze the gut and push the food along. It's rather like squeezing toothpaste along the tube. A small flap of tissue called the epiglottis stops the food from going down into the lungs. This is why you cannot breathe and swallow at the same time.

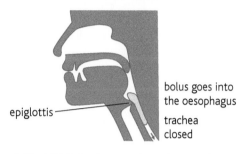

epiglottis

bolus goes into the oesophagus

trachea closed

▲ **Fig A2.30** Food passes down the oesophagus

To do – Breathing and swallowing

Try to breathe and swallow at the same time. Suggest why you cannot do it.

SBS Skills

ORR	D	MM	PD	AI
				✓

Stomach

In the stomach, food is mixed with gastric juice, which is released from pits in the stomach wall. This juice contains hydrochloric acid, and the enzymes **pepsin** and **rennin**.

pepsin – an enzyme that starts the breakdown of proteins

rennin – an enzyme that clots milk

Hydrochloric acid kills any microorganisms on the food and helps to break the food down. Pepsin starts the breakdown of proteins and rennin clots milk.

Duodenum

hydrolysis – the process of breaking down starch into simpler sugars

trypsin – an enzyme that completes the break down of protein into amino acids

amino acids – the breakdown product of proteins

lipase – an enzyme that breaks down fat into fatty acids and glycerol

fatty acids and glycerol – the breakdown products of oils and fats

The food passes from the stomach into the duodenum. Here, bile is added from the liver and gall bladder and more enzymes are added from the pancreas and the wall of the intestine. The job of bile is to emulsify fats, which means they are broken into small droplets. This provides a larger surface area for the fat-digesting enzymes to work. Bile is also alkaline and neutralises the acid from the stomach so that a different set of enzymes can get to work.

Enzymes that are added include;

* amylase to continue the **hydrolysis** of carbohydrates into simple sugars
* **trypsin** which completes the breakdown of proteins into **amino acids**
* **lipase** which breaks down fats and oils into **fatty acids and glycerol**.

Q12 Enzymes are sometimes added to washing powders. Blood contains protein. Suggest which enzymes might be added to get rid of blood stains and why the temperature of the wash should not be above 50°C.

Illeum or small intestine

Although the ileum is sometimes called the small intestine it is only small in diameter and forms the longest part of the gut, being about 6 metres in length. It is here that the enzymes continue their work and the small molecules of simple sugars, amino acids, fatty acids and glycerol are assimilated and absorbed through the gut wall and into the body.

The inner surface of the ileum is covered in folds. The folds increase the surface area for absorption of the small molecules. The folds of the ileum are also covered in small finger-like projections called **villi** and these villi are covered with even smaller microvilli. The effect of this is to produce an enormously large surface area for absorbing the small digested molecules. In fact the villi increase the surface area of the ileum by over 200 times.

villi – finger-like projections lining the wall of the gut to increase surface area for the absorption of food (singular, villus)

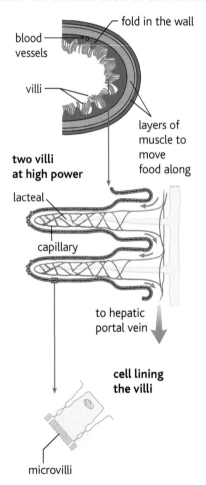

▲ **Fig A2.31** The small intestine and villi

Each villus has a blood supply and a lacteal vessel. A lacteal vessel contains lymph and is part of a system called the lymphatic system. The lymphatic system provides tissue fluid in which all cells are surrounded. When you graze yourself, the pale yellow watery liquid is tissue fluid. Glucose and amino acids are absorbed into the blood vessels and are taken straight to the liver by the hepatic portal vein. Fatty acids and glycerol are absorbed into the lacteal vessel and enter the lymphatic system.

Colon or large intestine

The indigestible remains of the food consisting of cellulose and plant fibre, along with large quantities of bacteria and water, then enter the colon. In the colon, water is absorbed into the blood.

Rectum

The final remains called faeces collect in the rectum where they are stored until sufficient pressure causes us to defecate and expel them through the anus. The correct term for this process is **elimination**. It is important not to call it excretion as this word is reserved for material that is actually produced by the body, such as carbon dioxide, and not for material that just passes through it.

elimination – getting rid of waste that is not produced by the body, e.g. faeces

SBS Skills

ORR	D	MM	PD	AI
✓		✓	✓	✓

Experiment – Making a model gut

This experiment demonstrates how small molecules such as glucose can be absorbed through the gut but large molecules such as starch cannot.

1 Set up the apparatus as shown in Fig A2.32.

2 Immediately test the water in the test tube for starch by placing a drop of the water on a white tile and adding a drop of iodine solution.

3 Now carry out a reducing sugar test on another sample of the water. What do you notice?

4 Wait for 24 hours. Repeat the two food tests on a sample of water from the test tube. What do you notice now? Explain your results.

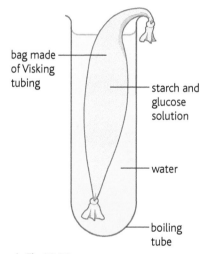

▲ Fig A2.32

5 You can modify the experiment by putting a solution of starch and amylase enzyme inside the Visking tubing and leaving out the glucose. Explain what happens.

End-of-unit questions

1 Which of these teeth are used for crushing and grinding food?
  A incisors and canines
  B premolars and molars
  C molars and incisors
  D canines and premolars

2 Which of the following pairs correctly describes an enzyme and the food that it breaks down?
  A amylase and proteins
  B pepsin and carbohydrates
  C trypsin and starch
  D lipase and fats

3 Look at the diagram of the human digestive system in Fig A2.33.

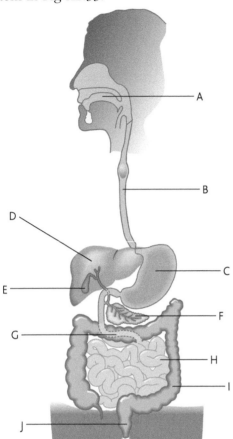

▲ **Fig A2.33**

a Choose letters from the diagram that best match the following descriptions.
   A where acid is added to food
   B where amylase is first added to food
   C where water is absorbed
   D where absorption takes place
   E where waste is eliminated

b Explain the difference between mechanical and chemical digestion.

c Digestion is about breaking down large insoluble molecules into small soluble ones. Explain why this is important.

d Annette eats a yam. Explain what happens to the starch in the yam after Annette eats it.

4 Use Fig A2.34 to explain how enzymes work.

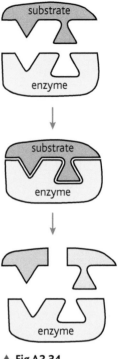

▲ **Fig A2.34**

5 Describe an experiment to show that glucose can be absorbed through the gut but starch cannot.

By the end of this unit you will be able to:

- explain the mechanism of breathing.
- distinguish between gaseous exchange and breathing.
- discuss the importance of respiration to organisms.
- compare and contrast aerobic and anaerobic respiration.
- discuss the features common to respiratory surfaces.

Respiration

respiration – a process that takes place in living cells converting glucose and oxygen to carbon dioxide and water and releasing energy

Respiration is the process by which our bodies release energy from the food that we eat. Oxygen and glucose react together to release water, carbon dioxide and energy. It is rather like photosynthesis in reverse.

Let's compare the word equation for photosynthesis with respiration.

$$\text{photosynthesis} = \text{carbon dioxide} + \text{water} \longrightarrow \text{glucose} + \text{oxygen}$$

$$\text{respiration} = \text{oxygen} + \text{glucose} \longrightarrow \text{carbon dioxide} + \text{water} + \text{energy}$$

Now let's compare the chemical equation for photosynthesis with respiration.

$$\text{photosynthesis} = 6CO_2 + 6H_2O \longrightarrow C_6H_{12}O_6 + 6O_2$$

$$\text{respiration} = C_6H_{12}O_6 + 6O_2 \longrightarrow 6CO_2 + 6H_2O + \text{energy}$$

Q1 If photosynthesis is like respiration in reverse, where does the energy for photosynthesis come from?

Breathing

breathing – the process of moving air in and out of the lungs to absorb oxygen and get rid of carbon dioxide

inhale – to breathe in

intercostal muscles – the muscles between the ribs that cause breathing by changing the volume of the lungs

diaphragm – a sheet of muscle used in breathing that separates the chest from the abdomen

trachea – the tube strengthened by cartilage through which we breathe that connects the mouth with the bronchi

exhale – to breathe out

Breathing is the process where we get oxygen from the air into our bodies and release the waste carbon dioxide that has been produced.

Breathing takes place in the lungs. When we breathe in, or **inhale**, external **intercostal muscles** between our ribs contract, pulling the rib cage upwards and outwards. The **diaphragm**, a dome-shaped muscle at the base of the lungs, also contracts and flattens. The effect of these two actions is to increase the volume of the lungs. As the volume increases, the air inside the lungs occupies a bigger volume so the air pressure inside the lungs begins to fall. As this happens, the air pressure of the air outside our body forces air through our nose and mouth, down the **trachea** and into the lungs. This is breathing in.

The reverse happens when we breathe out, or **exhale**. Internal intercostal muscles contract and pull the rib cage inwards and downwards. The diaphragm relaxes and resumes its dome shape. The volume of the lungs is now reduced, which increases the air pressure inside the lungs. Because this is now greater than the outside air pressure, air moves from the lungs, up the trachea and out of our nose and mouth. This is breathing out.

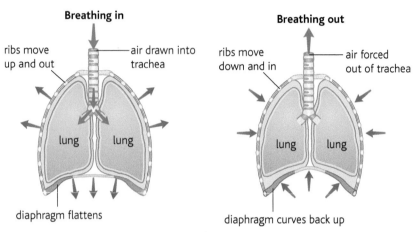

Breathing in

ribs move up and out — air drawn into trachea

lung lung

diaphragm flattens

Breathing out

ribs move down and in — air forced out of trachea

lung lung

diaphragm curves back up

▲ **Fig A3.1** The process of breathing in and out

SBA Skills

ORR	D	MM	PD	AI
✓				✓

Experiment – How the lungs work

1 Use an open-topped glass bell jar, a rubber sheet, two balloons and some glass tubing to assemble the apparatus as shown in Fig A3.2. If you do not have a bell jar you can cut the bottom off a plastic drinks bottle and use that instead.

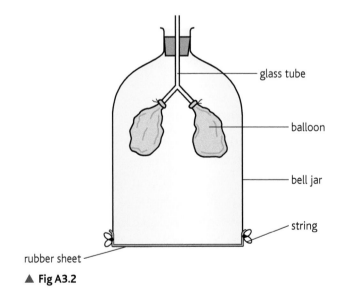

glass tube

balloon

bell jar

string

rubber sheet

▲ **Fig A3.2**

2 Pinch the rubber sheet in the middle and gently pull it down and push it up.

Q2 What do you notice is happening to the balloons? Explain what you see.

Gaseous exchange

gaseous exchange – the process of exchanging gases such as oxygen and carbon dioxide in the alveoli of the lungs

Gaseous exchange is the process by which oxygen moves from the air in our lungs and into our blood stream, and carbon dioxide moves from our blood and into the air in our lungs.

Table A3.1 shows how much nitrogen, oxygen and carbon dioxide move in and out of our body during gaseous exchange.

	percentage gases in inhaled air	percentage gases in exhaled air
nitrogen	79	79
oxygen	20	16
carbon dioxide	0.04	4

Table A3.1

Q3 The percentage composition of which gas remains unchanged during breathing in and out?

Q4 What percentage of the oxygen is absorbed into the body when we breathe in?

Q5 Suggest why the total gases in both inhaled and exhaled air do not add up to 100%.

SBA Skills

ORR	D	MM	PD	AI
✓				✓

Experiment – How much oxygen is in the air we breathe?

Set up the apparatus as shown in Fig A3.3. When the candle burns it uses up the oxygen in the air. Water enters the bell jar to replace the oxygen that has been used up by the burning candle.

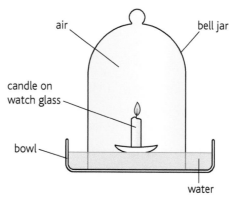

▲ **Fig A3.3**

Q6 Do the results of your experiment agree with the data in the table showing percentage gases in the air? Explain your answer.

Gas exchange in the lungs

The air enters our lungs by passing down a tube called the trachea. The trachea is kept open by stiff rings of cartilage.

To do – Rings of cartilage

Use four fingers to feel the outside of your throat. You should be able to feel the rings of cartilage that keep your trachea open.

SBA Skills

ORR	D	MM	PD	AI
✓				

bronchi – (singular bronchus) tubes that lead to the lungs

bronchioles – tiny tubes that connect the bronchi with the alveoli in the lungs

alveoli – (singular alveolus) small air sacs found in the lungs

capillaries – tiny blood vessels that carry blood to all the cells of the body

The trachea divides into two tubes call **bronchi**. One bronchus goes to each lung. The bronchi divide and divide to finer and finer tubes called **bronchioles**. Eventually these bronchioles end in millions of microscopic air sacs called **alveoli**. Many people think the lungs are like hollow balloons but in fact they look more like fine sponges and are quite solid in texture.

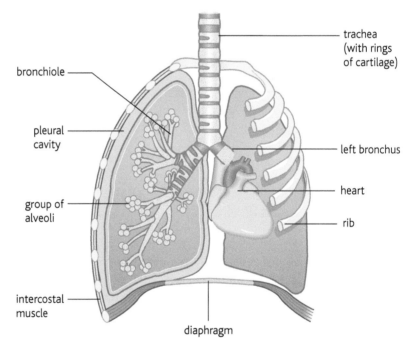

▲ **Fig A3.4** The structure of the lungs

Because there are millions of alveoli they have a very large surface area. If it were possible to lay out your lungs on a flat surface, they would cover an area bigger than a tennis court.

capillaries – tiny blood vessels that carry blood to all the cells of the body

Each alveolus is very thin. It is only one cell thick. It is also surrounded by tiny blood vessels called **capillaries** that are also only one cell thick. Because there is more oxygen in the air inside the alveoli than there is in the blood in the capillaries, oxygen passes by diffusion from the air in the lungs to the blood in the capillaries. The reverse is true for carbon dioxide; there is more carbon dioxide in the blood than there is in the air in the alveoli. This means that carbon dioxide diffuses in the opposite direction to oxygen and passes from the blood into the lungs.

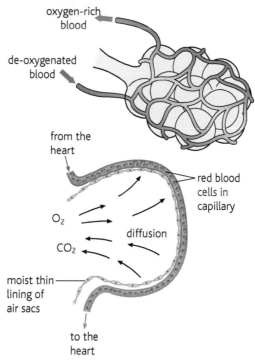

oxygen-rich blood

de-oxygenated blood

from the heart

red blood cells in capillary

O_2

diffusion

CO_2

moist thin lining of air sacs

to the heart

▲ **Fig A3.5** Gaseous exchange

Q7 Explain the difference between gaseous exchange and breathing.

Respiratory surfaces

respiratory surface – surface through which gaseous exchange takes place

Surfaces such as alveoli through which oxygen and carbon dioxide diffuse are called **respiratory surfaces**. They are not only found in alveoli. The cells inside a leaf and the gills of a fish are both examples of respiratory surfaces. All respiratory surfaces have several things in common:

- They all thin. This is to allow the oxygen and carbon dioxide to diffuse through easily.

- They have a large surface area. The larger the surface area, the more oxygen and carbon dioxide can diffuse through. The gills on fish are divided into tiny gill lamella to increase the surface area.

- They are moist. A wet surface area allows gases to diffuse through much more quickly than a dry one.

- The have a good transport system. The alveoli in our lungs and the gills of a fish have a very good blood supply to carry oxygen away from the cells of the body and return carbon dioxide for excretion.

Unlike humans that have lungs and breathe air, fish have gills and breathe under water. Why is it that humans would drown in water and fish would suffocate on land when we are both doing the same thing of getting oxygen and getting rid of carbon dioxide?

The answer lies in the structure of lungs and gills.

Alveoli in lungs are hollow and inflate when air enters. If they fill with water we cannot breathe. Gills however are like the bristles on a paint brush.

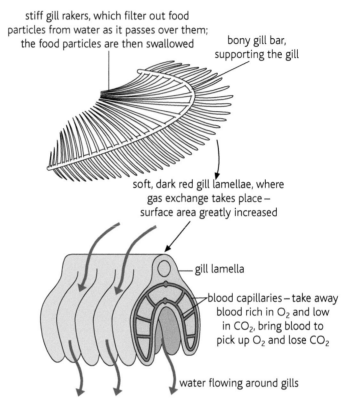

stiff gill rakers, which filter out food particles from water as it passes over them; the food particles are then swallowed

bony gill bar, supporting the gill

soft, dark red gill lamellae, where gas exchange takes place – surface area greatly increased

gill lamella

blood capillaries – take away blood rich in O_2 and low in CO_2, bring blood to pick up O_2 and lose CO_2

water flowing around gills

▲ **Fig A3.6** The structure of fish gills

To do – Modelling gills

Take a small paint brush and place it into clean water. When the paint brush enters the water, all the bristles separate and allow the oxygen-rich water to flow around and between them. Notice what happens to the bristles when you take the paint brush out of the water.

SBA Skills

ORR	D	MM	PD	AI
✓				✓

Q8 Suggest how this model helps to explain why fish suffocate when taken out of water.

The importance of respiration to living things

Once the oxygen has diffused into the blood, it diffuses into the red blood cells where it reacts with haemoglobin to form oxy-haemoglobin. It is then carried in this form by the blood to every cell in the body.

aerobic respiration – a type of respiration that uses oxygen

It is inside every cell where the chemistry of **aerobic respiration** takes place.

The word aerobic simply means 'using oxygen'.

$$oxygen + glucose \longrightarrow carbon\ dioxide + water + energy$$

$$6O_2 + C_6H_{12}O_6 \longrightarrow 6CO_2 + 6H_2O + energy$$

The blood not only supplies every cell with oxygen but also with glucose produced during digestion. These are the two substrates that are required for aerobic respiration to take place. The substances produced by respiration are called products. They are carbon dioxide and water.

The process of aerobic respiration takes place in small organelles inside the cell called **mitochondria**. As the process takes place and energy is released, the energy is stored by converting another chemical called ADP into energy-rich ATP. Many molecules of ATP are produced as each molecule of glucose reacts with oxygen. The ATP molecules then rapidly supply their energy to the cell as it is needed and get converted back to ADP ready to receive some more energy from respiration.

mitochondria – a structure found in the cytoplasm of cells in which aerobic respiration takes place

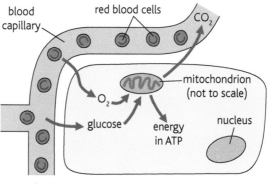

▲ **Fig A3.7** Aerobic respiration

SBA Skills

ORR	D	MM	PD	AI
✓				✓

Experiment – Carbon dioxide is released during respiration

Carbon dioxide is produced by the process of respiration in every cell of your body. Lime water turns milky when carbon dioxide is bubbled through it.

1 Set up the apparatus as shown in Fig A3.8.

2 Gently breathe in and out through the glass tube mouthpiece. As you breathe out your breath will bubble through tube B. As you breathe in, air from the atmosphere will bubble through tube A.

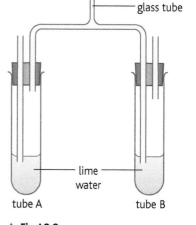

▲ **Fig A3.8**

Q9 Which contains the most carbon dioxide, air from the atmosphere or air from your breath?

SBA Skills

ORR	D	MM	PD	AI
✓		✓		✓

Experiment – Energy release during respiration

We can show that living things release energy during respiration by performing the following activity.

1 Set up two vacuum flasks as shown in Fig A3.9. One contains peas that have been soaked in water for a few hours and will begin to germinate. The other flask contains peas that have been soaked in water for a few hours and then boiled and allowed to cool to room temperature. This will kill the peas so that they cannot respire. Make sure that you have the same mass of peas in each flask.

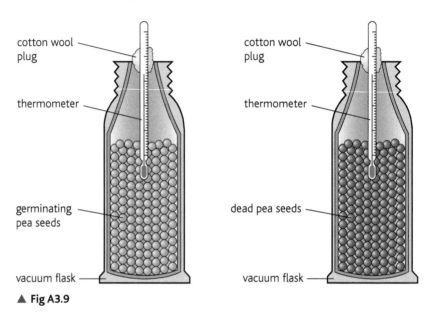

▲ **Fig A3.9**

2 Make a note of the temperature in each flask at the start of the investigation and then again after a few days.

Q10 What do you notice about the temperature in each flask?

Q11 What does this tell you about the living and non-living peas?

Q12 Explain why the investigation used vacuum flasks rather than lemonade bottles.

Anaerobic respiration in humans

Aerobic respiration always involves oxygen. However, our bodies are capable of respiring without oxygen for a short period of time. This type of respiration is called **anaerobic respiration**.

In anaerobic respiration, glucose is broken down directly into **lactic acid** which releases energy.

anaerobic respiration – a type of respiration that does not use oxygen

lactic acid – a chemical produced by anaerobic respiration that causes muscle fatigue

$$\text{glucose} \longrightarrow \text{lactic acid} + \text{energy}$$

$$C_6H_{12}O_6 \longrightarrow 2C_3H_6O_3 + \text{energy}$$

Anaerobic respiration sounds a great idea. Why bother going to all the trouble of breathing in oxygen and transporting it around the body when we can respire and release energy without it? But anaerobic respiration is not as great as it sounds. This is because it releases less energy than aerobic respiration, and it produces a waste product called lactic acid. Lactic acid causes muscle pain when we exercise too much and eventually it has to be broken down by reacting it with oxygen. This means that anaerobic respiration can only be used for short periods of time when we are exercising and our lungs and blood stream cannot provide our muscle cells with sufficient oxygen.

Once the lactic acid level has built up in our muscles we are forced to stop exercising. Even though we are no longer exercising we continue to pant and breathe deeply until we have supplied our aching muscles with enough oxygen to break down all the lactic acid. This is called the **oxygen debt**.

▲ **Fig A3.10** An athlete repaying the oxygen debt

oxygen debt – the oxygen that has to be breathed in to break down the lactic acid caused by anaerobic respiration

Table A3.2 compares the differences between aerobic and anaerobic respiration.

aerobic respiration	anaerobic respiration
uses oxygen	does not use oxygen
produces carbon dioxide and water as waste products	produces lactic acid as waste product
releases 2880 kJ of energy from each glucose molecule	releases 150 kJ of energy from each glucose molecule
occurs inside mitochondria	occurs in the cytoplasm outside of the mitochondria

Table A3.2

Q13 Describe the advantages and disadvantages of using anaerobic respiration.

Anaerobic respiration in yeast and bacteria

yeast – a single-celled fungus used to make wine, beer and bread

Yeast is a microscopic single-celled fungus. Like humans it can respire anaerobically. Unlike humans when yeast respires anaerobically it produces carbon dioxide and alcohol.

$$\text{glucose} \longrightarrow \text{carbon dioxide} + \text{ethanol} + \text{energy}$$

$$C_6H_{12}O_6 \longrightarrow 2CO_2 + 2C_2H_5OH + \text{energy}$$

If human beings respired anaerobically in the same way that yeast does we would end up drunk every time we did any exercise!

Yeast is very important economically as it is used to make bread and alcohol. The carbon dioxide gas that the yeast gives off when it is respiring is used to make bread rise before it is baked in the oven. The alcohol that it produces is used to make wine and beer by alcoholic fermentation.

lactobacillus – a bacterium used to make cheese and yoghurt

Some types of bacteria also respire anaerobically. When **lactobacillus** ferments milk it produces lactic acid, which helps to turn the milk into cheese or yoghurt.

▲ **Fig A3.11** All of these foods are produced by anaerobic respiration in yeast and bacteria

Q14 Describe the economic importance of anaerobic respiration in yeast and bacteria.

End-of-unit questions

1 Which of the following structures is *not* involved in breathing?
   A trachea   **B** aorta
   C diaphragm   **D** alveoli

2 A student used the following apparatus in an investigation into breathing. He breathes in and out through the glass tube. Which *two* of these observations and conclusions about the investigation are correct?
   A Tube A will turn milkier than tube B.
   B Inhaled air contains more carbon dioxide than exhaled air.
   C Exhaled air contains more carbon dioxid than inhaled air.
   D Tube B will turn milkier than tube A.

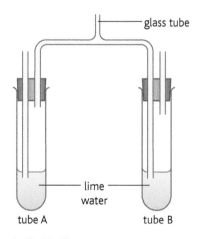

▲ **Fig A3.12**

3 Make a simple copy of this diagram of the lungs.

Breathing in

Breathing out

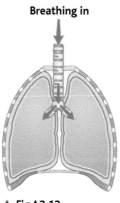
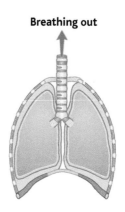

▲ Fig A3.13

a Put arrows on your diagram to show how the lungs move when breathing.
b Add labels to your diagram.
c Explain why air enters the lungs when we breathe in and leaves when we breathe out.

4 Rani is a chef. She bakes a cake using yeast.
a Write a word and chemical equation to show what the yeast will do.
b Rani mixes the yeast in the dough and immediately puts the cake into the oven. Explain why the cake does not rise.
c Describe how respiration in yeast is different from respiration in humans.

5 Describe what is meant by the oxygen debt.

6 Jane is in hospital with asthma. Her friends send her flowers.
a The nurse says she must remove the flowers from Jane's room over night. Use your knowledge of photosynthesis and respiration to suggest why the nurse removes the flowers.
b The nurse says that the asthma is caused by muscles lining the tubes in Jane's lungs contracting and making the airways smaller. Use the following diagram to explain what the nurse means.

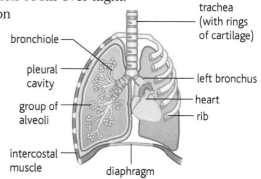

c Jane thinks her lungs are used to breathe in oxygen and breathe out carbon dioxide. Use the following table to explain why Jane is only partly correct.

	percentage gases in inhaled air	percentage gases in exhaled air
nitrogen	79	79
oxygen	20	16
carbon dioxide	0.04	4

A3b Air Pollution

Air pollution

pollutants – substances present in the environment that can harm living things

sulphur dioxide – a gaseous pollutant produced when fuels containing sulphur are burned

acid rain – rain made more acidic by the addition of gases such as carbon dioxide and sulphur dioxide

The air that we breathe into our lungs is not always clean, pure air. Sometimes it contains **pollutants**. Some of these pollutants are given below.

Types of air pollution

Sulphur dioxide

Most fossil fuels contain sulphur. When the fuel is burned, the sulphur is released into the atmosphere as **sulphur dioxide**. Once in the atmosphere the sulphur dioxide dissolves in water to form dilute sulphurous and sulphuric acid before falling as **acid rain**. The rain can turn lakes into dilute acid, killing fish. It can also kill trees and do enormous damage to buildings particularly if they are made from limestone.

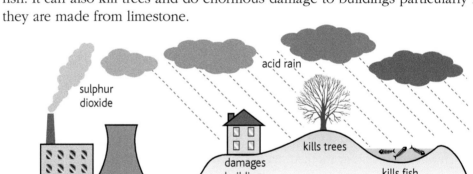

▲ **Fig A3.14** How acid rain is caused

▲ **Fig A3.15** The damage done by acid rain

Carbon dioxide

Carbon dioxide is also released into the atmosphere whenever we burn a fossil like coal, oil or natural gas. It too can dissolve in rain water to form carbonic acid and acid rain. But the real damage from carbon dioxide pollution is its effect as a greenhouse gas.

greenhouse effect – when the heat energy produced by light striking the Earth's surface is trapped by the carbon dioxide in the atmosphere

global warming – the warming of the Earth's atmosphere caused by the greenhouse effect of greenhouse gases

As carbon dioxide levels increase in the atmosphere, the carbon dioxide acts rather like glass in a greenhouse. This is called the **greenhouse effect**. It lets the short-wave light energy from the Sun through to the Earth's surface but when the energy gets transferred to the longer wavelength of heat, the carbon dioxide traps it and prevents some of it from escaping back into space. This results in a slow but gradual increase in the average temperature of the Earth's atmosphere. This is called **global warming**.

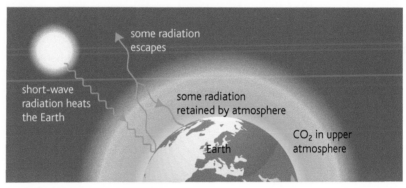

some radiation escapes

short-wave radiation heats the Earth

some radiation retained by atmosphere

Earth

CO_2 in upper atmosphere

▲ **Fig A3.16** How global warming happens

The increase in temperature may not be very noticeable to us but it is enough to have a massive impact on the Earth. It causes more violent weather such as hurricanes and also causes the polar ice caps to melt, resulting in a rise in sea levels. Scientists think that over the next 100 years sea levels could rise by as much as 1.5 metres. This is sufficient to cause world-wide flooding of low-lying land and massive migrations of human populations. The result would be an economic and humanitarian disaster.

Methane

Methane is a gas produced by the rotting of waste material and intestinal gases from cattle. Waste tips also produce large amounts of methane which, if collected, can be used as a valuable fuel. However, if left to escape into the atmosphere, it acts as an even more powerful greenhouse gas than carbon dioxide.

Q1 Explain what is meant by the greenhouse effect.

Lead

Most atmospheric lead pollution comes from burning petrol to which lead has been added. Lead is added to petrol to make it burn in the car engine more efficiently. However, it is then released into the atmosphere through exhaust gases. It has been shown that children who live close to very busy roads absorb some of this lead from the traffic fumes. Lead is a heavy metal and is very toxic. It effects nervous tissue and can cause a lowering of the IQ of affected children. Fortunately most leaded petrol is now being replaced by unleaded petrol as more and more cars are adapted to burn unleaded fuel.

Carbon monoxide

Carbon monoxide is also released into the atmosphere in traffic fumes. It is produced when a fuel is not completely burned with oxygen in the air. Faulty heaters in the home that have not been properly maintained and serviced can also produce carbon monoxide.

haemoglobin – a red pigment found in red blood cells that transports oxygen around the body

Carbon monoxide is a very dangerous gas. It is colourless, odourless and tasteless so it is very hard to know when it is present in the air we are breathing. However, if we breathe it into our lungs it quickly passes into our blood stream where it reacts with the **haemoglobin** in our red blood cells to form carboxyhaemoglobin. Unlike oxyhaemoglobin, which releases oxygen to our muscles, carboxyhaemoglobin is no longer able to carry oxygen. As more and more haemoglobin is unavailable to transport oxygen, a person quickly becomes unconscious and dies. Early warning signs of carbon monoxide poisoning are breathlessness, a bright pink skin and very bad headaches.

Lung diseases

Allergies and asthma

allergy – an inappropriate response by the body to a foreign substance

Allergy: An allergy is when the body makes an inappropriate response to a foreign substance. When some people breathe in pollen the body reacts as if the pollen was an invading microorganism and tries to get rid of it. We say that these people are allergic to pollen.

asthma – a condition of the respiratory system triggered by an allergen that causes the airways to narrow making breathing difficult

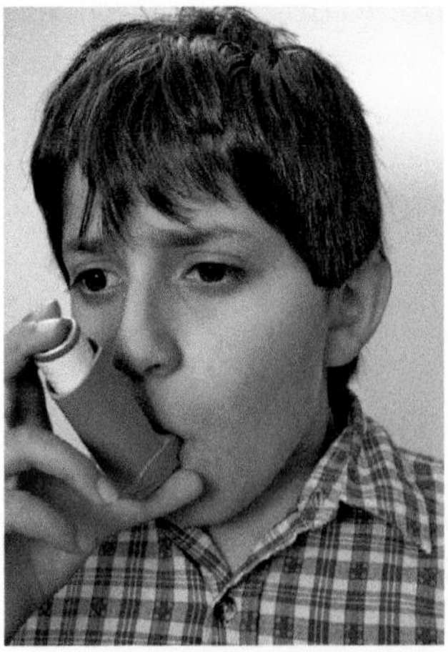

Asthma: Asthma is an allergic response to pollutants in the atmosphere. Different people are affected by different allergens but the effect on the lungs is same. Smooth muscle that surrounds our bronchiole goes into a state of spasm and contracts. This makes the airway become much smaller and consequently makes breathing much more difficult. In serious cases it can result in death.

Pollutants that cause this allergic response include sulphur dioxide, exhaust gases from traffic and natural substances such as pollen, animal fur and animal waste.

Drugs such as ventolin and seretide are used in the form of inhalers by many asthmatics to relieve and prevent the symptoms from occurring.

Doctors can measure the effect of asthma by using a peak flow meter. Patients take a deep breath and blow as hard as they can into the meter. The higher the reading, the healthier the lungs.

▲ **Fig A3.17** A child using an inhaler to treat asthma

Less serious reactions to allergens include sneezing, a runny nose and eyes and coughing. Even though these are not life-threatening they can make a person's life very miserable.

Q2 Suggest why different people are likely to suffer asthma attacks at different times of the year.

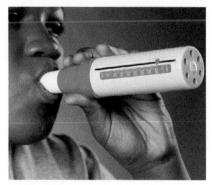

▲ **Fig A3.18** A peak flow meter

Lung cancer and asbestosis

Lung cancer has many causes but people who smoke are at much greater risk of getting it. It can also be caused by atmospheric pollutants. When materials containing asbestos are being manufactured, small fibres can be released into the air, and these fibres have been known to cause a disease of the lung called **asbestosis**. Lung cancer is very difficult to treat and can quickly spread to other parts of the body. The cancer is caused when cells within the lungs begin to divide uncontrollably. The mass of cells quickly grows and begins to invade surrounding tissue and blocks airways, making breathing difficult.

asbestosis – a condition of the lungs caused by asbestos fibres that can lead to cancer

Bronchitis and emphysema

Bronchitis is a general term for a collection of fluid in the lungs. It has many causes, such as a bacterial or viral infection, allergies or smoking. The build up of fluid can provide a breeding ground for bacteria and even if the initial cause was not an infection it can rapidly develop into one.

bronchitis – a build up of fluid in the respiratory system

Emphysema is even more serious. The walls of the alveoli break down, forming larger and larger spaces. This drastically reduces the surface area for gaseous exchange and causes extreme shortage of breath. Sufferers often have to spend much of their time breathing oxygen from a gas cylinder.

emphysema – a breakdown of alveoli cell walls reducing the surface area for gaseous exchange and allowing fluid to build up in the spaces produced

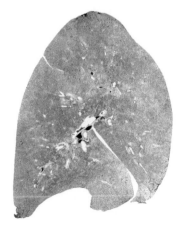

▲ **Fig A3.19** A healthy lung on the left. A lung with emphysema on the right

The role of cigarettes

Cigarette smoking is a major cause of asthma, lung cancer, bronchitis and emphysema. Cigarette smoke contains carbon monoxide, nicotine and tar plus hundreds of other harmful chemicals. Tar damages lung tissue and breaks down alveoli causing emphysema. It also causes cells to become cancerous and damages the small hair-like cilia that line the bronchioles and remove mucus and dirt from the lungs. Nicotine is a very poisonous chemical that is highly addictive. This is why smoking is hard to give up. It also raises blood pressure and heart rate and increases the risk of heart disease.

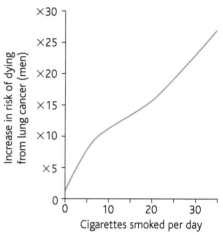

▲ Fig A3.20

The graph in Fig A3.20 shows the increased risk of dying from lung cancer for people who smoke cigarettes. About 90% of all lung cancer deaths are caused by smoking.

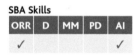

SBA Skills

ORR	D	MM	PD	AI
✓				✓

Experiment – Investigating cigarette smoke

Set up the apparatus as shown in Fig A3.21. Light the cigarette and switch on the pump.

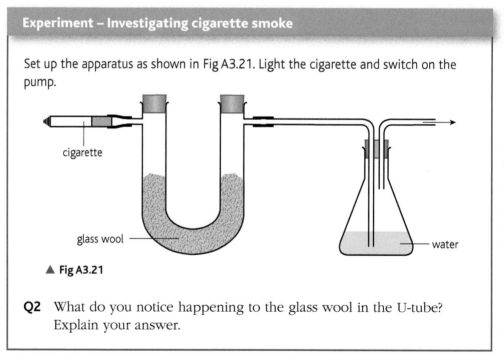

▲ Fig A3.21

Q2 What do you notice happening to the glass wool in the U-tube? Explain your answer.

passive smoking –
people who do not smoke breathing in the smoke from other people's cigarettes

Lung cancer can also be caused by **passive smoking**. This is when non-smokers breathe in the smoke from other people's cigarettes. It is for this reason that many countries in the world are now making laws to enforce smoke-free environments for workers and in public places. Many hotels in the Caribbean now advertise to attract tourists by saying that their hotels are non-smoking.

End-of-unit questions

1 Which of the following gases is an atmospheric pollutant?
   A sulphur dioxide
   B nitrogen
   C water vapour
   D oxygen

2 Which of the following carries oxygen around our bodies in the blood?
   A pepsin
   B adrenalin
   C insulin
   D haemoglobin

3 Describe the differences between acid rain and global warming.

4 The following chemicals are all air pollutants.
   • sulphur dioxide
   • carbon dioxide
   • carbon monoxide
   • lead
   • cigarette smoke

   a For each pollutant describe its effect on the environment on the human body.

   b Cigarette smoke can cause cancer and asthma. Explain the difference between these two conditions.

   c Devise a poster to encourage people not to smoke cigarettes.

5 Describe the differences between asthma, bronchitis and emphysema.

6 Scientists expect that climate change will cause global warming, raising sea levels and more extreme weather conditions.
   Climate change will affect different people in different ways.
   For each of the following groups of people, suggest how their lives could be affected by climate change:
   A Islanders living close to the sea.
   B People at risk of hurricanes.
   C Farmers who grow crops close to a desert.
   D Eskimos living on polar ice sheets.

7 Air pollution can also be produced from natural sources.
   For each of the following examples explain the environmental consequences of the atmospheric pollution:
   A Grazing animals produce large quantities of methane as they digest the vegetation that they have eaten.
   B Volcanoes produce large quantities of the gas sulphur dioxide.
   C Forest fires release large quantities of carbon dioxide into the atmosphere.

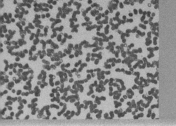

A4a Transport Systems in Plants and Humans

By the end of this unit you will be able to:

- discuss the need for transport systems in living organisms.
- relate the structures of the circulatory system in humans to their functions.
- identify the blood groups.
- explain the role of antigens in the natural and artificial control of diseases.

The need for a transport system

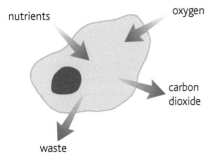

nutrients   oxygen

carbon dioxide

waste

Fig A4.1 Cells absorb all they need through their surface membrane

Single-celled organisms obtain everything they need through their surface cell membrane and get rid of their waste products the same way. Oxygen and nutrients diffuse in. Carbon dioxide and other waste products diffuse out. This is shown in Fig A4.1.

However, problems arise when organisms get bigger. The cell still has to get everything it needs through its surface cell membrane, but as the cell gets bigger its volume increases faster than its surface area.

If a cell increases in size but remains the same shape, for example a sphere, then the cell's surface area is proportional to the square of the radius and its volume is proportional to the cube of the radius.

$$\text{surface area} \propto r^2 \qquad \text{volume} \propto r^3$$

To do – Surface area and volume of cubes

SBA Skills

ORR	D	MM	PD	AI
		✓		✓

1 Look at the three cubes in Fig A4.2. One has sides of 1 cm, the second has sides of 2 cm and the third sides of 3 cm.

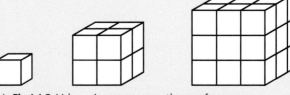

▲ **Fig A4.2** Volume increases more than surface area

2 Calculate the surface area and the volume for each of the three cubes. You can do this by counting the number of cubes, and the number of exposed surfaces for each of the three drawings.

Q1 What do you notice about the surface area and the volume of each of the cubes?

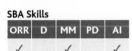

SBA Skills

ORR	D	MM	PD	AI
✓		✓		✓

Experiment – Investigating the relationship between surface area and volume

1 Get three different cubes of agar, the same size as in Fig A4.2, to which some dilute potassium permanganate solution was added when the cubes were prepared. Your teacher will have prepared them for you.

2 Place the coloured cubes into a large tray and cover with dilute hydrochloric acid, as shown in Fig A4.3.

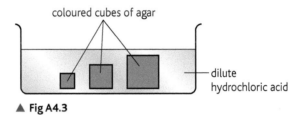

▲ Fig A4.3

3 Time how long it takes each of the cubes to turn colourless.

Q2 Explain why each cube did not turn colourless at the same time.

Q3 Explain the implications of your results for the maximum size of a cell.

The above experiment shows that there is a maximum size to the cell. If the cell gets any bigger, the surface area is just not big enough to supply all the oxygen and nutrients that the cell needs. The cell also chokes to death on its own waste products as the surface area is just not big enough to get rid of them fast enough. This limiting factor to cell size means that for normal shaped round cells, you will always need a microscope to see them.

As it is not possible to increase the size of the cell, the only way to increase the size of an organism is to increase the number of its cells. But this has both advantages and disadvantages.

The advantages are that different cells can become specialised to do different jobs. Some can contract and become muscle cells. Some can transmit information to different parts of the organism and become nerve cells. This specialisation makes the larger organism much more efficient.

The disadvantage is that the organism now needs an efficient **transport system** to carry gases, nutrients and waste around the body. In human beings, this transport system is called the **blood circulatory system**. Plants also need to be able to transport materials from one part of the plant to another.

Transport in plants

The transport system in plants starts in the roots. Water enters the roots through microscopic **root hairs** by **osmosis** (see Unit A1). This is because the cell contents of the root hair are more concentrated than the soil water. As the water enters, it dilutes the cell contents making the cell more dilute than the cell next to it. This causes water to enter the next cell by osmosis. Water

transport system – structures used by organisms to transport materials

blood circulatory system – the system made up of blood vessels and the heart that transports materials around the body

root hairs – cells found on the surface of a root that absorb water from the soil

osmosis – the movement of water molecules from a dilute solution to a concentrated solution through a partially permeable membrane

moves from cell to cell as it moves towards the centre of the root, following the **concentration gradient**. This flow of water carries along with it mineral salts that the plant needs from the soil.

concentration gradient – the variation in concentration of a substance in two different areas

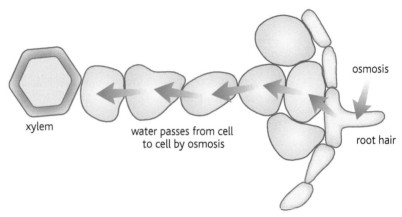

▲ **Fig A4.4** Water entering a plant's root

xylem – water-carrying tissue found in plants

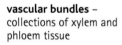

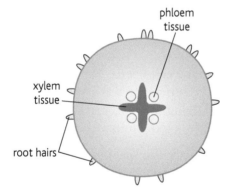

▲ **Fig A4.5** A cross section of a root showing root hairs and xylem and phloem tissue

vascular bundles – collections of xylem and phloem tissue

phloem – tissue that carries dissolved sugar down a plant stem from the leaves

The water continues to move towards the centre of the stem until it comes into contact with **xylem** tissue. Xylem cells are the dead remains of long tubular cells. They are rather like straws, stiff on the outside but hollow in the centre.

The water enters the xylem tissue. Xylem tissue is the plant's water transport system. The cells are joined together forming long tubes that extend up the roots and stem and into all the branches and leaves.

In the stem, the xylem forms collections of vascular or transport tissue called **vascular bundles**. Vascular bundles also include **phloem** tissue that transports food from the leaves, down the stem.

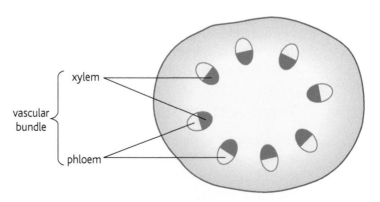

▲ **Fig A4.6** A cross section of a stem showing vascular bundles including phloem and xylem

To do – Transport in celery

1 Place plant stem such as a stick of celery with leaves attached, or a flower with a long stalk, into a beaker containing some coloured ink and leave it for a few hours.

2 Describe what you can see. Suggest why you think it has happened.

3 Now remove the stem from the ink and cut a section across it with a sharp blade.

4 Look at the section with a hand lens and draw what you can see.

5 Fig A4.7 is a section of a celery stem drawn by a student. Copy and complete the drawing to show what the student saw.

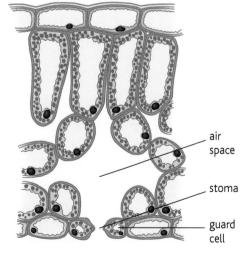

▲ **Fig A4.7**

The xylem vessels in the vascular bundles carry the water up the stem and into the leaves. At this point, the water passes out of the xylem and into to all of the cells. It then evaporates into the air spaces in the leaf before diffusing out of the leaf through the small holes called **stomata** on the underside of the leaf.

The stomata on the underside of the leaf are formed by two **guard cells**. The guard cells can open or close the stomata. During the day the stomata are usually open to allow carbon dioxide to enter the leaf for photosynthesis.

stomata – small pores on the underside of a leaf that regulate the release of water and oxygen and the absorption of carbon dioxide

guard cells – cells that control the opening and closing of stomata

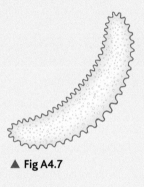

▲ **Fig A4.8** A section through a leaf

air space

stoma

guard cell

Q4 Which other gas will diffuse through the stomata and in which direction will it move?

During the night the stomata close as photosynthesis can no longer take place due to a lack of light. The closed stomata conserve water by preventing it from being lost from the leaf.

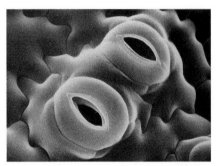

▲ **Fig A4.9** Stomata on underside of leaf

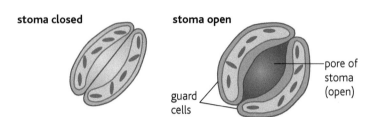

stoma closed　　　　　**stoma open**

guard cells

pore of stoma (open)

Fig A4.10 Guard cells open and close the stomata

SBA Skills

ORR	D	MM	PD	AI
✓		✓		✓

To do – Observing stomata

1 Get a fresh green leaf. You should be able to see ridges on the leaf. These are the vascular bundles that are carrying the water into the leaf.

2 Paint part of the underside of the leaf with clear nail varnish. Wait for it to dry and apply a second coat.

3 When it is completely dry, carefully peel of the layer of nail varnish and mount it on a microscope slide and look at it through a microscope. You should be able to see a cast of the underside of the leaf showing the stomata.

4 Are the stomata open or closed? It will depend upon what time of day it is and how long ago you picked the leaf. If the leaf is not freshly picked, the stomata will close to stop the loss of water from the leaf.

How water travels up a stem

How do plants get water to move up the stem and into the leaves? Is it pushed by osmosis or is it pulled? Because the xylem cells are dead and hollow it cannot be by osmosis. Osmosis stops when the water enters the xylem tissue.

adhesion – the force of attraction between different types of molecules

cohesion – the force of attraction between the same type of molecule

transpiration – the movement of water through a plant

Water is in fact pulled up the stem from the leaves. As the Sun causes water to evaporate and leave the leaves through the stomata, a continuous thread of water is pulled upwards from the leaf. This thread of water is very strong. Forces of **adhesion** keep the water attached to sides of the xylem vessels. Forces of **cohesion** stop the thread of water from snapping. So as molecules of water evaporate from the cells lining the air spaces in the leaf, the thread of water is gradually pulled upwards. This process is called **transpiration** and the energy for it comes form the heat of the Sun. Transpiration allows even giant redwoods that are up to 100 metres in height to get water to the leaves at the very top.

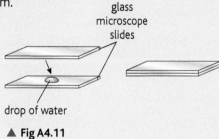

To do – Cohesion and adhesion

Water is stronger than you think. You can experience the forces of cohesion and adhesion by doing this experiment.

1 Take two glass microscope slides.

2 Add a drop of water to one of them. Place the second slide on top of the first as in Fig A4.11.

glass microscope slides

drop of water

▲ **Fig A4.11**

3 Now try to pull them apart. What you think should be easy is actually very difficult. After all, they are only being held together by water.

Q5 Explain why you cannot pull the two slides apart.

How food moves down a stem

Because food is made in the leaves, it needs to be moved down the stem for storage.

Glucose is the sugar made by photosynthesis. It is very soluble and dissolves easily in the water in the leaf cells. From the leaves the sugar solution moves into phloem. Unlike xylem, phloem consists of living cells. Like xylem it forms continuous tubes passing back down the stem in the vascular bundles. Each tubular phloem cell has an end plate called a **sieve plate**. This is because it looks like a sieve. Cytoplasm from one phloem cell connects with the cytoplasm in the next phloem cell through the holes in the sieve plate. It is through these holes that the sugar solution passes on its journey down the stem.

sieve plate – a plate with holes in found in sieve tubes

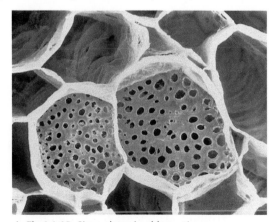

▲ **Fig A4.12** Sieve plates in phloem tissue

translocation – movement of dissolved sugar around a plant

The movement of sugar from the leaves and down the stem is called **translocation**.

Transport in humans

The transport system in humans is the blood circulatory system. Blood is not just a red coloured liquid. It is a fluid that transports substances such as glucose, oxygen, carbon dioxide, hormones and urea, all around the body.

The composition of blood

Blood is made up of plasma and three different types of cell, all of which have a different job to do.

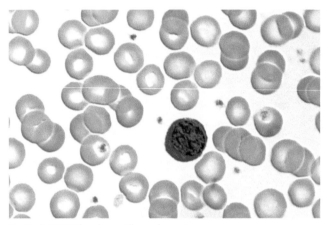

▲ **Fig A4.13** Blood seen through a microscope

plasma – liquid part of the blood

white blood cell – a blood cell that protects the body from disease and infection

platelet – a cell fragment that clots blood when we are injured

red blood cell – a blood cell that contains haemoglobin and transports oxygen around the body

Blood is mainly composed of liquid **plasma**. **White blood cells** and **platelets** are far outnumbered by **red blood cells**.

Plasma

Plasma is mainly water. It is a straw-coloured fluid that carries the different blood cells around the body. It also contains dissolved nutrients, carbon dioxide, urea and hormones. See Figs A4.13 and A4.14.

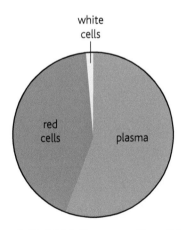

▲ **Fig A4.14** The composition of blood

Red blood cells

haemoglobin – a red chemical found in red blood cells that transports oxygen around the body

oxyhaemoglobin – haemoglobin that has combined with oxygen

The job of red blood cells is to carry oxygen from the lungs around the body to all of our cells. They can do this because they contain a red coloured chemical called **haemoglobin**. Haemoglobin is a very unusual chemical. When there is lots of oxygen in the blood, the haemoglobin combines with it to form **oxyhaemoglobin**. This is what happens at the lungs where there is lots of oxygen in the alveoli. The blood then carries the oxyhaemoglobin around the body to cells such as muscle cells, which have been using oxygen for respiration. The oxyhaemoglobin releases its oxygen for the muscle cells to use and converts back to haemoglobin. The blood is then returned to the lungs where the haemoglobin can pick up a fresh supply of oxygen.

Red blood cells are the only cells in our body that do not contain a nucleus. This is to provide extra space for haemoglobin. The shape of the red blood cell is a biconcave disc rather like a filled-in doughnut (see Fig A4.15). This is to provide a larger surface area so that oxygen can enter and leave the red blood cells more quickly.

▲ **Fig A4.15** Red blood cells

Every cubic millimetre of blood contains about 5 000 000 red blood cells. Because they do not have a nucleus and they are constantly being bumped and knocked, each cell only lasts for about six weeks. This means our bodies have to make millions of red blood cells every second to provide us with enough cells to carry all the oxygen we need.

White blood cells

Our blood contains different types of white blood cells but they all have the same job to do – protect us from invading microorganisms. White blood cells called **phagocytes** engulf and digest microbes and bacteria, as seen in Fig A4.16. White blood cells called **lymphocytes** produce chemicals to destroy microorganisms.

phagocyte – a type of white blood cell that engulfs bacteria

lymphocyte – a type of white blood cell that produces antibodies

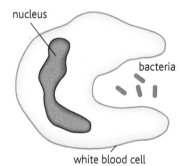

▲ **Fig A4.16** White blood cells

Platelets

Platelets are small pieces of cells. When our blood system is damaged, such as when we get a cut or a graze, platelets break open and release a chemical that causes our blood to clot. The clot prevents any further blood from being lost. It also stops any harmful microorganisms from getting into our body.

Blood groups

blood groups A, B, AB and O – four different blood groups found in humans, determined by two proteins, A and B

antigen – a protein found on the surface of some red blood cells and invading organisms

Human beings have four different **blood groups**. They are called **A**, **B**, **AB** and **O**. They are called this because red blood cells have two different proteins on their surface. These proteins are called **antigens** and there are two different types: antigen A and antigen B. People who have antigen A on their red blood cells are blood group A. Those with antigen B are blood group B. Those with both antigens are blood group AB. Those with neither are blood group O.

antibody – a chemical produced by white blood cells that destroys foreign antigens

The blood plasma also contains **antibodies**. Antibodies are also proteins. However antibodies are part of the body's defence mechanism. They react with all foreign antigens especially those on the surface of invading microorganisms. People who are blood group A contain antibody B in their plasma. People in blood group B have antibody A. Blood group AB have neither of the antibodies. People in blood group O have both antibody A and B. Table A4.1 shows the antigens and antibodies present in each blood type.

blood group	antigen on red blood cells	antibody in plasma
A	A	anti B
B	B	anti A
AB	both A and B	neither anti A nor B
O	neither A nor B	both anti A and B

Table A4.1 Antigens and antibodies in different blood groups

Q6 Suggest why a person who is blood group A does not have antibody A in their plasma.

It is important when transfusing blood into a patient that doctors and nurses use the correct blood group. If a person who is blood group A receives blood from a donor who is blood group B the antibodies in their blood will attack and clot all the type B blood that they receive in their transfusion. This would probably be fatal.

However, giving blood transfusions is not quite so simple. In certain cases it is possible to donate or receive blood from different blood groups. Because blood group AB does not have any antibodies in their plasma they can receive blood of any other type. Also because people with blood group O do not have any antigens on their red blood cells they can donate to any other blood group. Table A4.2 shows which blood type can be give to which type of recipient.

		donor			
		A	B	AB	O
recipient	A	✓	✗	✗	✓
	B	✗	✓	✗	✓
	AB	✓	✓	✓	✓
	O	✗	✗	✗	✓

Table A4.2 Compatibility of different blood groups

Q7 Explain why blood group O is called a universal donor and blood group AB is called a universal recipient.

Q8 It is possible for a person who is blood group O to give their blood to anyone, but not receive blood from everyone. Explain why.

To do – Blood group survey

1 Try to find out the blood groups of as many people in your class as you can.

2 Produce a chart to show who could give blood to whom.

SBA Skills

ORR	D	MM	PD	AI
✓				✓

Rhesus factor

rhesus – a blood protein found in some humans making them rhesus positive

Another protein antigen that is found in some people's blood is **rhesus** protein. Those people that have rhesus protein are called rhesus positive. Those that do not have it are called rhesus negative. This means that each of the blood groups (A, B, AB and O) that we have looked at so far can also be either rhesus positive or rhesus negative. This now gives us eight different blood groups: A+, A−, B+, B−, AB+, AB−, O+ and O−.

The same rule applies to rhesus as it does the A, B, AB and O blood groups. Rhesus negative blood can be given to rhesus positive recipients and rhesus positive patients can receive blood from rhesus negative donors, but rhesus positive blood cannot be given to rhesus negative recipients and rhesus negative patients cannot receive blood from rhesus positive donors.

Q9 Which of these two groups is a universal donor: O+ or O−? Explain your answer.

Q10 Which of these two groups is a universal recipient: AB+ or AB−? Explain your answer.

The rhesus factor can become a problem during pregnancy. It is possible for a rhesus negative mother to have a rhesus positive baby. During pregnancy small amounts of blood can cross the placenta so that some of the baby's blood can get into the mother's blood stream and vice versa. When some of the baby's rhesus positive blood gets into the mother's blood stream her rhesus antibodies start to destroy the baby's blood. These antibodies can cross the placenta and enter the baby's blood and start to destroy that as well. This can be very harmful for the baby. A woman's first baby may be born a blue colour due to a lack of red blood cells. Subsequent pregnancies are at even more risk as the mother is making more and more rhesus antibodies. Fortunately, when doctors are aware of the problem the danger to the baby can be avoided by giving injections to prevent the mother's antibodies being formed.

Vaccination

Antibodies A, B and rhesus are not the only antibodies in our blood. In fact there are thousands of different types of antibody, each one especially made to destroy a foreign protein that could enter our body.

The microorganisms that cause infectious diseases all carry different types of antigen to which we need a specific antibody. Whenever a new disease-causing microorganism enters our body, the race is on. It is the race between our bodies making enough of the new antibody to protect us from the organism, and the organism rapidly reproducing and destroying our body.

pathogen – a disease-causing organism

Microorganisms that cause disease are called **pathogens**. The pathogen produces poisonous chemicals called toxins that damage the cells of our body. Each pathogen has specific antigens on its surface, as seen in Fig A4.17. Our body has to produce a specific type of antibody that locks onto the antigen and kills the pathogen.

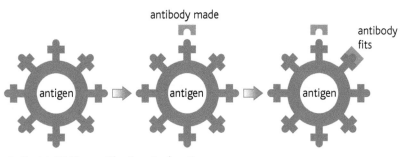

▲ **Fig A4.17** How antibodies attack antigens

Once our bodies have made the antibody we then have protection against the pathogen. This is why we catch some diseases only once, such as chicken pox. After we recover we then have the antibodies to fight off another infection. Unfortunately some pathogens, such as the one that causes the common cold, keeps changing the antigens on its surface. This means we can catch a cold more than once and each time we have to produce a new antibody.

Some diseases are so serious that we may not be able to make enough antibodies in time. This can result in death. Fortunately scientists have come up with a way of giving us protection before we even come into contact with the dangerous pathogen. They do this by **vaccination**.

vaccination – injecting a substance that encourages the body to make antibodies against a specific disease-causing organism

A vaccine is a weakened form of the disease. The vaccine may contain dead bacteria or even just little bits of bacteria. Either way it will not cause us any harm. However, our body is tricked into thinking it's the real thing and it starts to produce antibodies. Even when the level of these

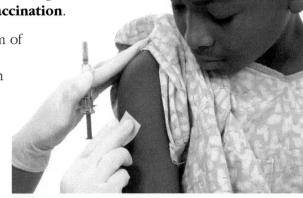

▲ **Fig A4.18** Vaccination

antibodies starts to fall in our blood, special white bloods cells called memory cells retain the ability to very rapidly make more antibodies should we urgently need them. This means that if the real pathogen arrives, we are already protected against it. This is called **active immunity**.

active immunity – an immunity that results from the body making its own antibodies

Making antibodies takes time. If we come into contact with the disease without having been vaccinated against it, it may be too late. When this happens doctors can sometimes inject us with ready-made antibodies to protect us. This is called **passive immunity**. Although this works well, the protection is short-lived because our body hasn't learned how to make it. When we recover, we will still need to have the proper vaccine.

passive immunity – temporary immunity produced by injecting antibodies collected from another person

We are said to be **immune** when our bodies can rapidly produce antibodies to protect us.

immune – having protection against a specific disease-causing organism

Q11 Look at the graph in Fig A4.19. How many days did it take to reach maximum antibody level to the first infection?

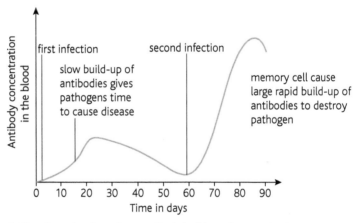

▲ **Fig A4.19** Graph to show antibody build-up after vaccination

Q12 How many days did it take to reach the same level of antibodies with the second infection? Explain why this means the person is now immune to the disease.

HIV AIDS

HIV – The Human Immunodeficiency Virus

HIV stands for the Human Immunodeficiency Virus. The virus infects white blood cells, lowering a person's immunity to illness and decreasing their ability to fight off disease. Fig A4.20 shows HIV invading a white blood cell.

The virus is spread during intimate sexual contact between two people, one of whom is carrying the virus, or HIV positive. Symptoms may not appear in an infected person for many years. This means it is possible to be infected with the virus

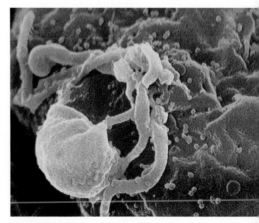

▲ **Fig A4.20** The green dots are HIV invading a white blood cell

and not be aware of it. The only way to find out is to have a blood test. However, even though the symptoms have not yet appeared, it is still possible to infect other people during intimate sexual contact.

latent phase – the time when an infected person is symptom-free

If the infected person can make new white blood cells as fast as the virus is destroying them, they remain symptom-free. This is called the **latent phase**. It may be only when a person starts to catch lots of other diseases that they become aware that they are HIV positive. When a person's immune system is lowered to the point where they catch many other infections, they are said to

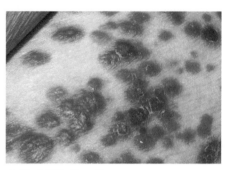

▲ **Fig A4.21** Kaposi's sarcoma

AIDS – Acquired Immune Deficiency Syndrome

have **AIDS**, Acquired Immune Deficiency Syndrome. The symptoms of AIDS vary depending upon which diseases the person catches. Tuberculosis and Kaposi's sarcoma are common diseases of AIDS.

HIV is highly effective because it hides inside and destroys the very cells that are trying to kill it, the white blood cells. Fortunately, modern drugs are far more effective in slowing and controlling the disease. But so far there is no cure. These drugs are called **anti-retroviral** drugs.

anti-retroviral drugs – these are drugs that slow down the development of retroviruses such as HIV

End-of-unit questions

1 Which statement best describes how water moves through a plant?

A Water enters the roots by osmosis, which then pushes the water up the stem.

B Water enters the roots by diffusion and then soaks up the stem.

C Water enters the roots by osmosis and is then pulled up the stem by transpiration.

D Water enters the roots by transpiration, and osmosis pulls the water up the stem.

2 Copy and complete this crossword about blood.

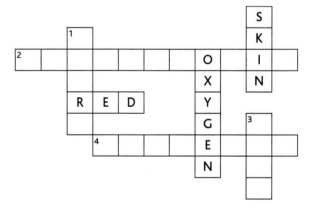

1 down: muscular pump that pumps blood
2 across: red chemical found in blood
3 down: blood vessel that carries blood back to the heart
4 across: piece of a cell that starts blood clotting

3 Blood of the correct blood group can be donated and transfused into patients who have lost blood.

  a Explain how two different blood proteins, A and B can produce four different blood groups.

  b Explain the difference between antigens and antibodies.

  c Copy and complete the table to show the compatibility of different blood groups.

		donor			
		A	B	AB	O
recipient	A				
	B				
	AB				
	O				

  d Explain why blood group AB is called the universal recipient.

4 A student used the equipment in Fig A4.22 to investigate the movement of water in the stem of a plant.

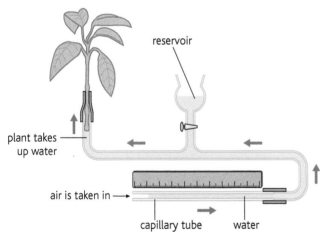

▲ **Fig A4.22**

The student carried out three different runs. Once with normal leaves, once with the top of the leaves covered in Vaseline and once with the bottom of the leaves covered in Vaseline. The results the student obtained are in Table A4.3.

	distance water moves/mm							
	0 min	1 min	2 min	3 min	4 min	5 min	6 min	7 min
leaves uncovered	0	3	7	10	14	17	20	23
top of leaves covered in Vaseline	0	2	5	8	11	15	18	21
bottom of leaves covered in Vaseline	0	1	2	4	6	7	9	10

Table A4.3

  a The apparatus measures water uptake by the shoot. Suggest why this measurement may be different from the water lost through the leaves.

  b Plot the results on a grid. Use the same grid for all three investigations.

  c Explain the results obtained by the investigations.

5 Explain the difference between HIV and AIDS.

6 Both plants and animals use a transport system to transport materials from one part of them to another.

  a Describe the similarities and differences between transport systems in animals and plants.

    Similarities:

    Differences:

  b Explain why animals require a more complex transport system than plants.

7 Hundreds of years ago, doctors who tried to transplant blood from one patient to another, often only succeeded in killing the patient.

  a Explain why these early transfusion were often unsuccessful.

  b Suggest what doctors could have done to ensure that the transfusion was more likely to be successful.

A4b The Blood System and Diseases

By the end of this unit you will be able to:

- relate the structures of the circulatory system in humans to their functions.
- explain the possible causes of hypertension and heart attacks.
- discuss the physiological effects of exercise.
- discuss the ethics and the effects of athletes using drugs to enhance their performance.

The structure of the blood transport system

Blood vessels

artery – the blood vessel that carries blood away from the heart

vein – the blood vessel that carries blood towards the heart

capillary – a microscopic blood vessel that connects arteries and veins and supplies cells with nutrients and remove waste

Our blood system is called a closed transport system. This means that the blood does not just slosh around in our bodies but is transported in a network of tubes called blood vessels. It is when one of these blood vessels is broken that we bleed. There are three different types of blood vessels: **arteries**, **veins** and **capillaries**.

Arteries

Arteries carry blood *from* the heart. Because the blood from the heart is under pressure, arteries need thick muscular walls, as shown in Fig A4.23, to withstand the high pressure each time the heart beats. The muscular wall expands to absorb the pressure then contracts between heart beats. This smoothes out the pressure between beats, ensuring a constant flow of blood to all parts of the body.

thick muscular wall

▲ **Fig A4.23** An artery

Q1 Explain why very high pressure in the blood would be a bad thing for our body.

Veins

valve – a structure found in veins that stops the backflow of blood

Veins carry blood back *to* the heart. The blood is no longer under high pressure so thick muscular walls are not needed. Instead veins have thin walls and **valves**, as shown in Fig A4.24, that prevent the blood from flowing backwards.

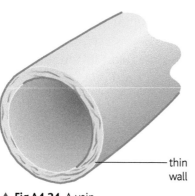

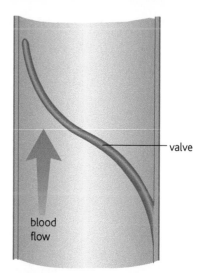

▲ **Fig A4.24** A vein

thin wall

valve

blood flow

▲**Fig A4.25** A valve in a vein

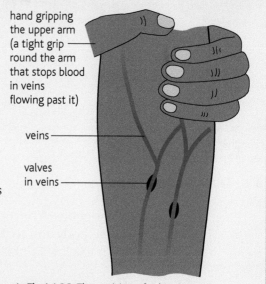

To do – Veins in the arm

Firmly grip the top of your arm as in Fig A4.26. You will shortly see the veins in your arm standing out. The small bumps that you can see are the valves. Make sure you do not maintain the grip for more than a couple of minutes.

Q2 Explain why the valves show up as small bumps on your arm.

hand gripping the upper arm (a tight grip round the arm that stops blood in veins flowing past it)

veins

valves in veins

▲ **Fig A4.26** The position of valves in veins

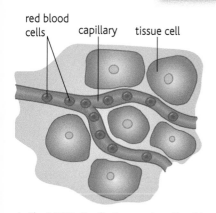

red blood cells   capillary   tissue cell

▲ **Fig A4.27** Capillaries supply cells with nutrients and remove waste

Capillaries

Capillaries are microscopic blood vessels. There are enough capillaries in one human body to go all the way around the world. Each capillary is only one cell thick and just wide enough for red blood cells to pass through easily. It is through the thin walls of the capillaries that glucose and oxygen diffuse to the cells and carbon dioxide diffuses back into the blood. Every cell of our body is never more than two cells' width away from a blood capillary.

Q3 Explain why capillary walls are only one cell thick.

A double circulatory system

Blood flows around the body in two different loops. First it goes from the heart all around the body and then back to the heart. Then it goes from the heart to the lungs to pick up fresh oxygen and get rid of carbon dioxide. It then goes back to the heart to start its journey all over again. This can be seen in Fig A4.28 and more simply in Fig A4.29.

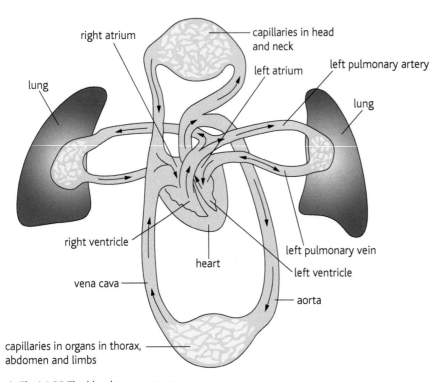

▲ **Fig A4.28** The blood transport system

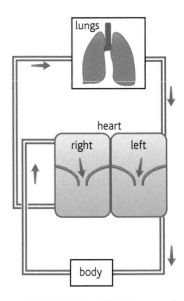

▲ **Fig A4.29** Simplified diagram of the blood transport system

The double circulation allows the heart to maintain the blood pressure both to the body and to the lungs. In a single circulation the blood pressure would drop too low by the time it reached the lungs.

Q4 Use your finger to trace the blood flowing around the body. How many times does it go through the heart before you get back to your starting point?

The heart

The heart is a four-chambered muscular pump. It is divided into two halves, left and right. Because the heart belongs to someone it is labelled as if they were labelling it on their own body. So in diagrams, the left chamber is on the right and vice versa.

left ventricle – the chamber of the heart that pumps blood around the body

Blood leaves the heart from the **left ventricle** through a blood vessel called the **aorta**. The aorta is an artery as it carries blood *from* the heart.

aorta – the main artery leaving the heart to carry blood around the body

This is the largest blood vessel in our body and is about as thick as a finger. It divides many times to eventually supply all the capillaries that go all around

vena cava – the main vein returning blood to the heart

right atrium – the chamber of the heart that receives blood from the vena cava

tricuspid valve – the valve that prevents blood flowing back from the right ventricle to the right atrium

right ventricle – the chamber of the heart that pumps blood to the lungs

pulmonary artery – the artery carrying blood from the heart to the lungs

pulmonary vein – the vein carrying blood from the lungs to the heart

left atrium – the chamber of the heart that receives blood from the lungs

bicuspid valve – the valve that prevents blood flowing back from the left ventricle to the left atrium

our body. The capillaries eventually all join up to form a vein called the **vena cava**, which enters the **right atrium** of the heart. The right atrium pumps the blood through the **tricuspid valve** into the **right ventricle**. From here, the blood is pumped along the **pulmonary artery** to the lungs. The blood is re-oxygenated and then passes along the **pulmonary vein** to enter the heart by the **left atrium**. Finally, the blood is pumped through the **bicuspid valve** into the left ventricle to start its journey all over again.

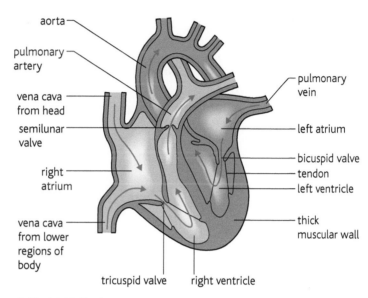

▲ **Fig A4.30** The heart

The heart beats by first contracting the muscles of the two atria. This forces blood down into the two ventricles. The muscular walls of the ventricles then contract and pump the blood out and around the body. If you listen to a heart beat you can hear the double thump as first the atria and then the ventricles contract.

An ECG machine can show the electrical impulses that cause the heart muscles to contract.

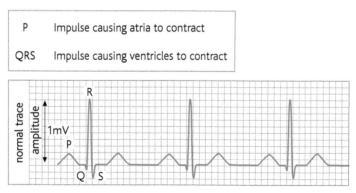

▲ **Fig A4.31** An ECG trace showing heart beat

Q5 Look at the trace in Fig A4.31. Which do you think has the more powerful contraction, atria or ventricles? Explain your answer.

Illnesses associated with the circulatory system

Sometimes things can go wrong. Hypertension and coronary heart disease are two common conditions of the circulatory system.

Hypertension

hypertension – high blood pressure

Hypertension is commonly known as 'high blood pressure'. It means that the pressure of the blood is too high, which means the heart has to work too hard, putting it at risk.

systole – when the heart is contracting

diastole – when the heart is relaxing

cholesterol – chemical found in fatty foods that can lead to a blocking of the arteries

The blood has two different pressure readings. **Systole** is when the heart is contracting, and **diastole** when the heart is relaxing between beats. Blood pressure readings are commonly given as systole over diastole. In a healthy young teenager this reading would be about 120 mmHg over 80 mmHg. However, as we get older the blood vessels become coated in **cholesterol**. This fatty deposit starts to block the blood vessels and makes it harder to pump the blood through them.

We get cholesterol in our food but our body also makes it. This is why eating a low-cholesterol diet does not always reduce the level of cholesterol in our body to a safe level.

The blocking of the arteries increases the systolic blood pressure reading. As we get older the arteries become harder and are less able to

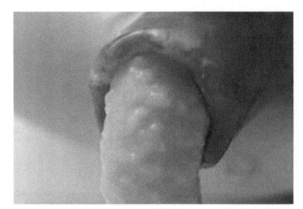

▲ **Fig A4.32** Cholesterol in an artery

expand and absorb the pressure as the heart beats. This raises the diastolic pressure between beats. A person with high blood pressure may have a reading of 170 mmHg over 100 mmHg.

Coronary heart disease

coronary artery – the artery that supplies the muscle of the heart with blood

Although the heart is full of blood, it still needs its own blood supply to feed the heart muscle with oxygen and glucose. The blood vessel that supplies the heart is called the **coronary artery**. The coronary artery divides many times so that each muscle cell of the heart can be close to a capillary blood supply. Coronary heart disease, or a heart attack, occurs when one of these small branches of the coronary artery becomes blocked.

A blockage could be caused by cholesterol or by a small blood clot that has formed elsewhere and has broken free. It is carried by the blood until it gets stuck in a small blood vessel and blocks it. All the muscle cells that are supplied by that blood vessel then die. If this is just a small area of heart muscle, the person will experience a mild heart attack and will probably survive, but if it is a large section of heart muscle the resulting heart attack can be fatal.

Q6 Suggest why a larger size clot blocking part of the coronary artery is likely to cause a more serious heart attack.

The effects of exercise

Exercise can help to prevent heart attacks and hypertension. Taking moderate exercise for 30 minutes, two or three times a week, can make a big difference to fitness and good health. Exercise has many effects on the body:

- It increases the size of the heart so that it can pump blood around the body more efficiently.
- It increases the number of blood vessels to the muscles so that they can receive oxygen and glucose more quickly.
- It increases the volume of the lungs so that they can take in oxygen and get rid of carbon dioxide more efficiently.
- It increases the strength of the intercostal muscles, making breathing easier.

SBA Skills

ORR	D	MM	PD	AI
✓		✓		✓

Experiment – Testing your fitness

1 To do the fitness test you will need a watch with a second hand or a timer, and a step or secure box 20 inches (50.8 centimetres) high.

2 Step up and down the box at the rate of 30 steps per minute – that's one step every two seconds.

3 Carry on for five minutes or until you can no longer continue. If you stop before five minutes you need to record the time you spent doing the test.

4 When you stop exercising wait one minute and then count your pulse for 30 seconds.

5 Count your pulse again at two minutes, for another 30 seconds.

6 Finally, count your pulse again at three minutes, for another 30 seconds.

7 Add the three pulse counts together. Then use this formula to calculate your fitness index:

(test duration in seconds × 100) ÷ (total number of pulse beats × 2)

8 Use Table A4.4 to see your fitness.

fitness index	rating
>90	excellent
80 – 89	good
65 – 79	high average
55 – 64	low average
<55	poor

Table A4.4 The fitness index table

To keep fit you also need to balance your energy intake and output. Our body is very good at doing this all by itself, but over a period of time it is possible to eat just a little more food than we need to supply our energy demand and we start to put on weight. In Unit A2b on Food and Nutrition we saw that different people need different amounts of energy each day. It all depends upon what sex you are, how old you are and how much exercise you take.

Enhancing athletic performance

Fitness is extremely important to athletes. They want to make sure that their performance is as good as possible. They can do this in different ways:

- *The correct diet*. Long distance runners will need more carbohydrates in their diet to provide them with all the energy they need. Weight lifters will need more protein in their diet to build up powerful muscles.
- *Training programme*. Athletes need to train to keep fit. Different types of athletes need different types of training programmes. Endurance athletes need to have good aerobic and anaerobic fitness as well as muscle strength. Weight lifters need to concentrate on muscle strength. Coaches to the athletes devise specific fitness programmes to match the needs of the athlete.

Athletic performance can also be enhanced by using less ethical measures or cheating.

anabolic steroids – drugs taken to enhance performance by developing muscle strength

- **Anabolic steroids** are a type of drug that can be used to build muscle. Muscle is very important for weight lifters and sprinters who need a sudden burst of speed. Some athletes have used steroids to help build this muscle and increase their chances of winning. This is why athletes are regularly tested for drugs. They do not know when they will be next asked for a blood sample and they risk being banned from the sport if a drugs test is positive. Taking steroids also puts your health at risk. They are a form of artificial hormone similar to the male hormone testosterone. Some of the side effects of taking steroids include feelings of aggression, nausea, increased blood pressure, liver damage, heart disease, stroke and cancer.

blood packing – enhancing performance by removing red bloods cells and them replacing them several weeks later

- **Blood packing** is another method to illegally improve performance. Blood is taken from an athlete several weeks before the race. The athlete's body replaces the lost red blood cells. Shortly before the race, the athlete's blood is returned to their body by a blood transfusion. This gives the athlete many more red blood cells than they would normally have in their blood and improves their performance.

erythropoietin – a hormone that encourages the production of red blood cells by the body

- **Erythropoietin** or EPO can also be used to increase the number of red blood cells. EPO is a hormone that makes the body make red blood cells. As it occurs naturally in the body it is difficult to detect when drugs tests are carried out.

Q7 Suggest why having more red blood cells will improve an athlete's performance.

End-of-unit questions

1 Which statement best describes the route taken by blood around the body?

   A left ventricle – aorta – right atrium – right ventricle – pulmonary artery – pulmonary vein – left atrium

   B left ventricle – right atrium – right ventricle – pulmonary vein – pulmonary artery – left atrium – aorta

   C left ventricle – right ventricle – right atrium – pulmonary artery – aorta – pulmonary vein – left atrium

   D right ventricle – right atrium – left ventricle – pulmonary artery – pulmonary vein – aorta – left atrium

2 Look at the ECG trace in Fig A4.33 below.

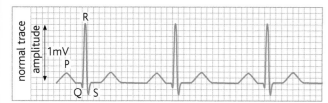

▲ **Fig A4.33**

   a Calculate the heartbeat rate from the trace.

   b At what point on the trace will the ventricles contract?

3 Copy the diagram shown in Fig A4.34.

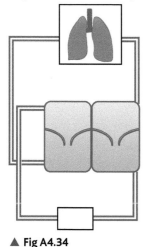

▲ **Fig A4.34**

   a Add arrows to show the direction the blood flows around the body.

   b Label the diagram as fully as possible.

4 Explain the difference between systolic and diastolic pressure in the circulation of the blood.

5 Describe three ways that athletic performance can be improved by cheating.

6 Doltan made a model of the human circulatory system. He used a pump for the heart, copper tubing for the blood vessels, and water for the blood.

   a Describe three **other** differences between Doltan's model and the human circulatory system.

   b Doltan switched the pump ion and off to represent the beating of the human heart. But no matter how hard he tried, he could not model the normal human blood pressure of 120 mm of Hg over 80 mm of Hg.
   Explain why not.

   c Doltan changed his model by including a section of copper pipe that was furred up on the inside with hard deposits.
   What aspect of the human circulatory system was Doltan modelling?
   Explain your answer

7 Blood pressure can increase with age.

   a Suggest two reasons why.
      1
      2

   b Suggest why regular exercise and a good diet can help to reduce high blood pressure.

Excretion and egestion

excretion – the process of getting rid of waste products produced by the body

egestion – the process of eliminating waste material not produced by the body

The word **excretion** means getting rid of waste that has been produced by an organism. Carbon dioxide and urea are types of waste excreted by animals. Both are produced by metabolic reactions that take place inside the body. Metabolic processes are all those chemical reactions that are taking place inside our body to keep us alive and well.

The word **egestion** means getting rid of material that has not been directly produced by the organism. Eliminating faeces is an example of egestion. The waste faeces are the cellulose left over remains of our food along with microorganisms that have reproduced in the gut. Unlike carbon dioxide and urea, neither of these has been directly produced by the body.

Excretion in humans

carbon dioxide – a gas produced by respiration and breathed out through the lungs

water – liquid produced during respiration

urea – waste material produced in the liver by the breakdown of excess amino acids

lungs – the organs of gaseous exchange and excretion of carbon dioxide

kidneys – the organs of excretion and osmoregulation

skin – the outer surface of the body

Humans excrete three different metabolic waste products, **carbon dioxide**, **water** and **urea**. The waste products are excreted through three organs of excretion, the **lungs**, the **kidneys** and the **skin**.

Carbon dioxide and water are both products of respiration and are breathed out from the lungs.

$$\text{oxygen} + \text{glucose} \longrightarrow \text{carbon dioxide} + \text{water}$$

Surprisingly, if you carefully measure the amount of water that you take in each day and compare it with the amount of water the body loses each day we lose about a cupful of water each day more than we take in as food or drink. This extra cupful of water comes from respiration. Some desert animals such as gerbils are so good at conserving water that this extra water produced by respiration is all the water they need and they never need to drink liquid water.

To do – Our breath contains water

Next time you have a cold drink, breathe onto the glass or can. The water vapour in your breath will condense on the side of the glass, causing water droplets to form.

SBA Skills

ORR	D	MM	PD	AI
✓				

Water is also lost through the skin as we sweat.

Q1 Suggest why water could be considered as both an excretory product and also an egested product.

Urea is a waste product produced in the liver. When we eat protein it is broken down by digestion into amino acids. Many of these amino acids are surplus to requirements. This is particularly true in adults who are no longer growing in size and just need enough amino acids to replace lost and damaged tissue. The liver breaks the excess amino acids down into urea, which then enters the blood. It is the job of the kidneys to get rid of the urea from the blood.

The lungs

You have already studied how the lungs get rid of carbon dioxide from our body in Unit A3a and you should read pages 40–42 again for more detail.

The lungs consist of millions of alveoli. This gives the lungs a very large surface area. Each alveolus is very thin – only one cell thick. It is also surrounded by tiny blood vessels called capillaries that are also only one cell thick. Because there is more carbon dioxide in the blood than there is in the air in the alveoli, the carbon dioxide diffuses from the blood and into the alveoli, as shown in Fig A5.1. When we breathe out, the carbon dioxide is expelled into the atmosphere.

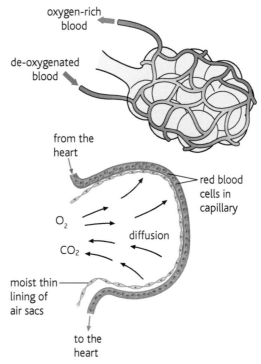

▲ **Fig A5.1** Carbon dioxide diffuses from the blood into the alveoli

Q2 Explain the importance of the blood capillaries and the alveoli being only one cell thick.

The kidneys

The kidneys have two jobs. One is to excrete urea that is produced by the breakdown of excess amino acids. The second is to control the water level of the body. This is why the kidneys are also called the organs of osmoregulation.

Our bodies have two kidneys. They are about the size of your fist and situated towards the back of your abdomen, as shown in Fig A5.2.

The outer layer of the kidney is called the **cortex** and the inner layer is called the **medulla**. At the centre of the kidney is an empty space called the **pelvis**, which collects the urine before passing it along the **ureter** to the **bladder** for storage. The urine is then periodically expelled from the bladder through a tube called the **urethra**.

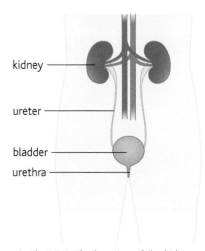

▲ **Fig A5.2** The location of the kidneys

cortex – the outer layer of the kidneys

medulla – the inner layer of the kidneys

pelvis – the space inside the kidneys into which urine drains

ureter – the tube through which urine passes from the kidneys to the bladder

bladder – the storage organ for urine

urethra – the tube through which urine passes from the bladder to the outside

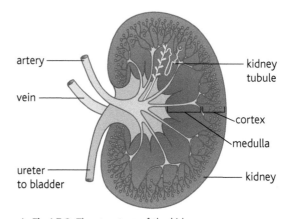

▲ **Fig A5.3** The structure of the kidneys

How the kidneys work

Each kidney consists of about 1 000 000 (one million) microscopic tubules called **nephrons** that, if laid out end-to-end, would reach over 60 kilometres.

Blood enters the kidney through the renal artery. ('Renal' is the term used for anything to do with the kidney.) The renal artery divides many times until the blood is passing through microscopic blood capillaries. Eventually each capillary forms a small knot of blood vessels called a **glomerulus**. The blood in the glomerulus is under high pressure and the small molecules present in the blood are squeezed out of the capillaries. This is called **ultra-filtration** and is shown in Fig A5.4.

nephron – a renal tubule that filters urea from the blood

glomerulus – a knot of blood vessels where ultra-filtration of the blood takes place

ultra-filtration – removal of small molecules from the blood that takes place in the kidney

These small molecules include:

- water
- salt
- glucose
- urea

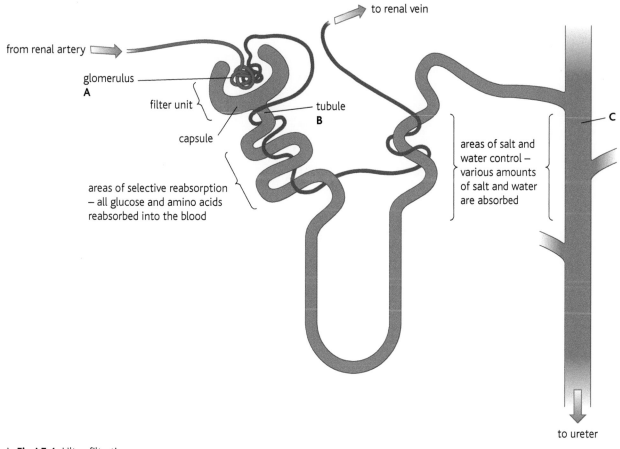

to renal vein

from renal artery

glomerulus
A

filter unit

tubule
B

capsule

C

areas of salt and
water control –
various amounts
of salt and water
are absorbed

areas of selective reabsorption
– all glucose and amino acids
reabsorbed into the blood

to ureter

▲ **Fig A5.4** Ultra-filtration

Q3 Suggest why red blood cells do not get squeezed out of the capillaries by ultra-filtration.

Larger molecules such as blood proteins are too large to pass out of the blood capillaries.

Bowman's capsule – part of a nephron that surrounds a glomerulus

Each glomerulus is surrounded by a cup-shaped capsule that is part of a nephron. The capsule is called a **Bowman's capsule**. It collects the water, containing the dissolved salt, glucose and urea that have been squeezed out of the glomerulus. The fluid or 'filtrate' then starts to pass down the tubular nephron. The first part of the nephron is folded and surrounded by more blood capillaries. The body has gone to a lot of trouble to digest carbohydrates to get glucose for food and is not about to waste all that useful glucose, so the glucose is actively transported from the tubule and back into the blood. This is called **selective reabsorption**.

selective reabsorption – the reabsorption of glucose but not urea

urine – a solution of urea in water

The filtrate then passes along the tubule, to another area where the tubules are folded and surrounded by capillaries. In this region, salt and much of the water are reabsorbed. This leaves a solution of urea in water called **urine**. The urine passes into the centre of the kidney before passing along the ureter to the bladder for storage. It finally passes out of the body through the urethra.

Table A5.1 shows the different composition of blood plasma, renal filtrate and urine shown in different parts of Fig A5.4.

	A composition of blood plasma	B composition of renal filtrate	C composition of urine
water %	90	99	97
glucose %	0.1	0.1	0
urea %	0.03	0.03	2.0

Table A5.1

Q4 Suggest why the percentage of urea is greater in the urine than in the blood and filtrate.

Water balance

As mentioned previously, the kidney is an organ of osmoregulation, which means that it balances the amount of water present in our bodies. This is quite difficult to do because not only do we take in water through food and drinks, but we lose water by urination, sweating and water vapour in our breath. If we drink too much liquid, it produces a greater volume of dilute urine. If we do not drink enough liquid, it produces a smaller volume of more concentrated urine.

To do – Urines changes colour

Notice the colour of your urine when you are thirsty and when you have had plenty to drink. What do you notice about the difference in colour? Explain your answer.

SBA Skills

ORR	D	MM	PD	AI
✓				✓

pituitary gland – found at the base of the brain, it produces its own hormones and hormones that control the other glands

ADH (Anti Diuretic Hormone) – a hormone that increases the reabsorption of water in the kidneys by making the nephron more permeable to water

negative feedback – a control mechanism where a rise in the level of one substance brings about a reduction in the level of a different substance

When our body is short of water, a hormonal gland situated at the base of our brain just above the roof of our mouths called the **pituitary gland** releases a hormone called **ADH** or the Anti Diuretic Hormone. ADH makes the kidney reabsorb more water back into the blood. This produces more concentrated urine. It does this by making the renal tubule more permeable to water so it is reabsorbed more easily. When we have drunk too much liquid the pituitary gland produces less ADH and the tubule becomes much less permeable. This means less water is reabsorbed and more passes to the bladder, making a larger volume of more dilute urine. The pituitary gland does this by monitoring the concentration of the blood. The more concentrated the blood, the more ADH is produced. The more dilute the blood, the less ADH is produced. This process is called **negative feedback**, and is shown in Fig A5.5. You will study negative feedback in more detail when you study the section on hormones in Unit A6a.

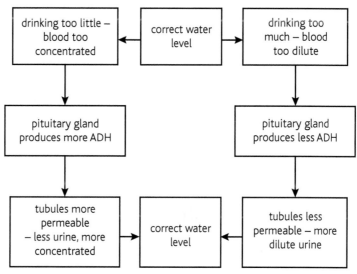

▲ **Fig A5.5** Negative feedback helps keep things in the body at a constant level

Q5 Can you think of anything else in the body that may be controlled by negative feedback?

When things go wrong

Very rarely the kidneys can fail. When this happens the person's life is at risk. The doctor can do one of two things. One is a kidney transplant, but spare kidneys are not always available. The other option is to put the patient onto a **dialysis** machine. A dialysis machine is like an artificial kidney. Patients need to use if for several hours about three days each week.

dialysis – an artificial process to filter the blood inside a kidney machine

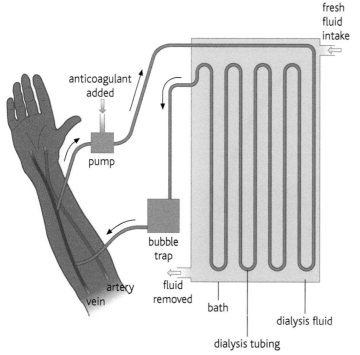

Fig A5.6 shows how a dialysis machine works. Blood passes from the patient's vein and is mixed with anticoagulants before being passed into the machine. Anticoagulants stop the blood from clotting before it is returned to the patient. The machine contains lots of dialysis tubing. It is rather like the Visking tubing you used in your osmosis experiments. The tubing is selectively permeable and surrounded by another fluid that is similar to the patient's blood plasma. Urea and excess water pass out of the blood, through the membrane and into the fluid in the machine. The cleaned blood is then returned to the patient through a filter to remove any small clots and air bubbles.

▲ **Fig A5.6** Inside a kidney dialysis machine

The skin

The skin is the largest organ in the human body and performs many functions:

- Its self-repairing surface protects us from the harsh outside environment.
- It keeps our body temperature at a constant 37°C.
- It acts as a barrier to keep out invading microorganisms.
- It excretes water and small quantities if urea from the body in the form of **sweat**.

sweat – waste liquid released onto the surface of the skin that also helps to cool the body

Sweat is produced in the sweat glands in the skin. It then passes through a duct and out of a pore onto the skin's surface. It is the evaporation of this sweat that helps to keep us cool in the hot Caribbean sun.

Table A5.2 shows how the structure of skin is related to its function

structural feature of skin	function
covers whole surface of body	keeps the organs inside protected and keeps microorganisms out
covered in hairs	helps to keep the body warm and maintain a constant internal body temperature
has sweat glands	helps to keep the body cool and maintain a constant internal temperature
has a good blood supply and can change the amount of blood that flows near its surface	helps maintain a constant body temperature
contains sensors connected to sensory neurones	provides us with the sense of touch and informs the brain of pain if the skin is being damaged

Table A5.2

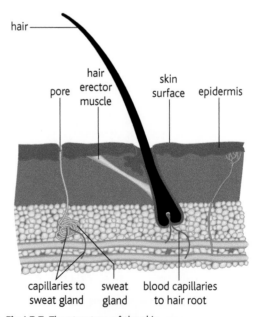

▲ **Fig A5.7** The structure of the skin

To do – Different areas of the skin sweat differently

1 Place small pieces of blue cobalt chloride paper onto different parts of your skin and fasten them in place using sticky tape.

2 Leave for several minutes until the paper begins to change colour – cobalt chloride paper is blue when dry and pink when damp.

3 Notice on which parts of the body the paper changes colour first.

Q6 What does this tell you about the distribution of sweat glands on different parts of the body?

Excretion in plants

Plants also have excretory products, but different ones from those produced by animals.

Like animals, plants respire to release energy. However, respiration in plants is far outweighed by photosynthesis and any carbon dioxide produced by respiration will be more than used up by photosynthesis. For this reason we do not normally say that carbon dioxide is an excretory product produced by plants.

oxygen – excretory gas produced by photosynthesis

An excretory product that *is* produced by photosynthesis is **oxygen**.

$$\text{water} + \text{carbon dioxide} \longrightarrow \text{glucose} + \text{oxygen}$$

We saw in Unit A2a how plants use leaves as an organ of gaseous exchange.

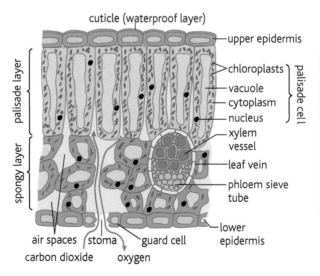

▲ **Fig A5.8** Gaseous exchange in plants

The oxygen diffuses out of the leaf through the stomata and into the atmosphere. Before green plants evolved, the level of oxygen in the atmosphere was much lower than it is today.

Unlike animals, plants cannot get rid of their other waste in urine. Instead, they store their waste in different ways, some of which may be excreted later.

One way plants do this is to store the waste in the dead materials that make up wood and bark. Unlike animals, plants do not get rid of all their waste materials, they just store them until the plant dies. This is one reason why wood is hard and strong unlike the green shoots of a young plant.

Another way that plants excrete their waste is by storing it in their leaves so that when the leaves fall from the plant, the waste is lost. The stored waste substances turn the leaves the familiar bright purple and orange colours of falling leaves.

End-of-unit questions

1 Which of the following are excretory products produced by animals?

   a oxygen, urea, water

   b water, urea, glucose

   c carbon dioxide, oxygen, glucose

   d water, urea, carbon dioxide

2 Which of the following are reabsorbed by the renal tubule?

   a glucose, water, salt

   b protein, glucose, water

   c salt, urea, glucose

   d protein, urea, water

3 Which statement best describes the effect of ADH on the kidneys?

   a ADH makes tubules less permeable so urine is more concentrated.

   b ADH makes tubules less permeable so urine is less concentrated.

   c ADH makes tubules more permeable so urine is less concentrated.

   d ADH makes tubules more permeable so urine is more concentrated.

4 Look at the diagram of the renal tubule (nephron) in Fig A5.9.

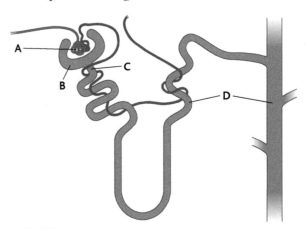

▲ **Fig A5.9**

   a Copy and label the diagram of the nephron.

   b Draw an arrow to show the direction of movement of the blood in the capillaries.

   c At what point, A, B, C or D will the concentration of urea be the highest?

   d Explain the effect of ADH on part D of the nephron.

5 Look at the graph in Fig A5.10. It shows the amount of urine and sweat produced at different temperatures.

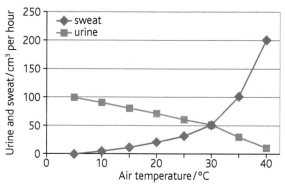

▲ **Fig A5.10**

a What is the relationship between the amounts of sweat and urine produced as the external air temperature increases?

b Suggest why urine production decreases as temperature increases.

c Suggest what effect an increasing external temperature will have on ADH production if the person does not drink.

d Suggest why urine production is higher at 0°C than at 40°C.

6 Plants excrete waste products.

Which of the following pairs of excretory product and organ is true for green plants?

A water – flower

B oxygen – root

C carbon dioxide – stem

D water – root

E oxygen – leaf

7 Animals need to drink water to survive. Water is also an excretory product. These two statements seem to be contradictory.

a Explain why water can be both needed and excreted by animals.

b Protein molecules are not normally excreted from the blod by our kidneys.
  i Suggest why not.
  ii Suggest what the implications would be if protein was found in a personís urine.

8 People with kidney failure sometimes receive a transplant that is donated by a brother or sister.

a Suggest why it is often possible for a close relative to donate a kidney to a family member suffering from kidney failure.

b Suggest why it is possible for a family member to donate a whole body organ and still be able to survive and lead a normal life afterwards.

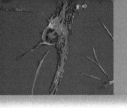

A6a The Control Systems

senses – the five senses are sight, touch, smell, taste and hearing

sensory organs – the eyes, skin, nose, mouth and ears all transmit sensory information about the outside world to the brain

Our bodies have five different **senses**. The senses and their corresponding **sensory organs** are:

- touch – skin
- smell – nose
- taste – tongue
- sight – eyes
- hearing – ears

All five of these senses carry information about the outside world into the brain. It is the brain's job to make sense of all this information and then send out instructions to the muscles in our body to make us move appropriately. This might be towards food, away from danger or just to stay and admire a wonderful sunset. All of this information, processing of data and instructions to muscles, takes place in the nervous system.

The structure and functions of the nervous system

central nervous system (CNS) – the brain and spinal cord

peripheral nervous system (PNS) – sensory and motor neurons going into and out of the brain and spinal cord

The nervous system consists of the brain, spinal cord and nerves. The brain and spinal cord form the **central nervous system** or CNS and the nerves form the **peripheral nervous system** or PNS.

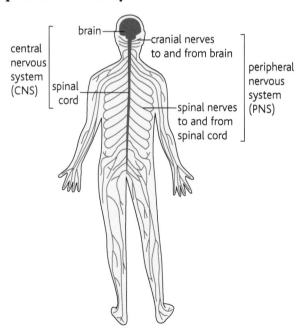

▲ **Fig A6.1** The central and peripheral nervous systems

The brain

The brain is the coordination centre for the nervous system. It gives us the ability to think, plan and be self aware. It is such an important organ that it is protected by a hard, bony structure called the skull.

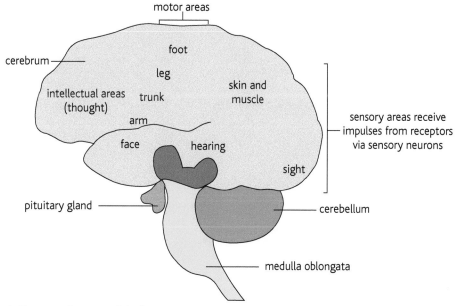

▲ **Fig A6.2** The areas of the brain

Neurons

neuron – a nerve cell

motor neuron – a nerve cell that carries out instructions from the brain

sensory neuron – a nerve cell that carries information from sense organs into the brain

receptor – a structure in a sense organ that receives a stimulus and stimulates a sensory neuron

effector – a muscle or structure that is stimulated by a motor neuron

voluntary action – an action controlled by conscious thought

Each nerve consists of many individual nerve cells or **neurons**. Neurons carry electrical impulses from one end of the neuron to the other. There are two types of neurons. **Sensory neurons** carry information from the sense organs to the brain. **Motor neurons** carry instructions out from the brain to the muscles.

Each sense organ contains **receptors**. Receptors receive a stimulus from the outside world. This may be light in the eye, sound in the ear, or touch on the surface of the skin.

The sensory neuron carries the electrical impulse into the spinal cord. Another neuron is stimulated to carry the electrical impulse up the spinal cord to the brain. Millions of neurons are then stimulated in the brain to decide what to do about the stimulus. If the stimulus is caused by a piece of food that we are holding, the decision may be to move the food towards the mouth. An electrical impulse is then sent down the spinal cord along a motor neuron. This stimulates another motor neuron that passes out of the spinal cord and goes to the muscle or **effector** in the arm. The muscle contracts and the food is moved towards the mouth. This is called a **voluntary action**.

As shown in Figs A6.3 and A6.4, a neuron transmits the impulse from dendrites, along the axon to dendrites at the other end of the neuron. Both sensory and motor neurons are surrounded by a myelin sheath. This speeds up the transmission of the impulse as it jumps from one node of Ranvier to the next.

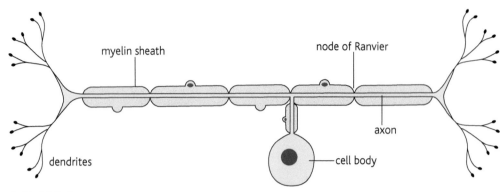

▲ **Fig A6.3** A sensory neuron

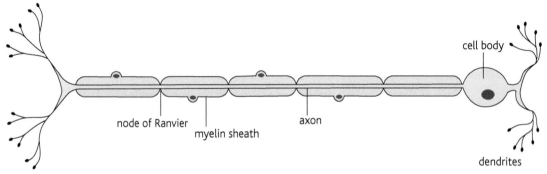

▲ **Fig A6.4** A motor neuron

Q1 Put the following words in the correct order.

brain – sensory neuron – receptor – effector – motor neuron

Synapses

Neurons do not connect one with another to pass the electrical impulse from one neuron to the next. There is a small gap between two neurons that the impulse must cross. This gap is called a **synapse**.

synapse – a gap between two neurons

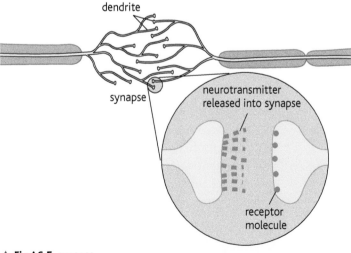

▲ **Fig A6.5** synapse

electrical impulse – a type of impulse that passes along neurons

chemical impulse – a type of impulse that crosses a synapse

Unlike neurons, which transmit an **electrical impulse**, synapses transmit a **chemical impulse**.

neurotransmitter – a chemical that transmits the impulse across a synapse

Synapses work by the stimulated neuron releasing a chemical called a **neurotransmitter** into the gap. The chemical diffuses across the gap and binds to a receptor molecule, as seen in Fig A6.5. This stimulates the next neuron to transmit another electrical impulse that travels along it. The neurotransmitter then has to be removed. Otherwise it would continue to stimulate the second neuron. Another chemical in the gap breaks down the neurotransmitter so that the synapse is ready to pass on the next impulse. A synapse acts like a gateway. Because only one neuron can release the neurotransmitter chemical the impulse can only travel in one direction.

It only takes a fraction of a second for the impulse to cross a synapse but even so, synapses slow down the speed that an impulse can be transmitted.

SBA Skills

ORR	D	MM	PD	AI
✓		✓		✓

To do – How fast can you go?

You can get a rough idea of how fast impulses pass along the nervous system by performing the following activity.

1 The class should stand in a big circle and hold the hand of each person on either side.

2 One student starts the activity by squeezing the hand of one student by their side. This student then squeezes the hand of the person next to them and the 'squeeze' is passed around the circle from one person to another.

3 A stop watch is used to time how long it takes for the 'squeeze' to pass all around the circle and back to the first student.

4 We then need to know how far the impulse has travelled. We can do this by multiplying the number of students in the circle by two. This gives us the approximate distance in metres.

5 Divide the distance in metres by the time in seconds and you have the speed of the impulse in metres per second (m/s).

Q2 What was the speed of the nerve impulse in your class circle?

When fast is not fast enough

You probably found that the nerve impulse was quite fast. But sometimes that is just not fast enough. It normally takes about 0.3 seconds for a person to react to a stimulus. However in an emergency, such as when you accidently pick up a hot object or stab yourself with a pin, it is important to remove the damaging item as quickly as possible.

reflex arc – the pathway taken by impulses in a simple reflex

intermediate or relay neuron – a neuron that passes an impulse between two other neurons

We cannot speed up the nerve impulse. But we can reduce the distance that the impulse has to travel and this can save time. We do this by using a **reflex arc**, as seen in Fig A6.6. In a reflex arc the receptor stimulates an impulse along a sensory neuron as usual. However, when the impulse enters the spinal cord, it not only connects with a neuron going up to the brain, but also with an **intermediate or relay neuron**. This neuron connects with the motor

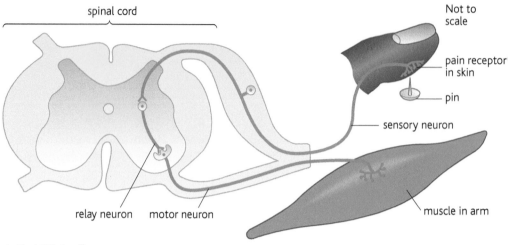

▲ **Fig A6.6** A reflex arc

neuron that sends an impulse from the spinal cord to the appropriate muscle. The muscle and the brain both receive the stimulus at about the same time. By the time the brain realises that something is painful, the hand has already been moved away from the object that is causing the pain. This is called an **involuntary action** or a reflex action. Reflex actions happen without us even thinking about them.

involuntary action – an action that is not controlled by conscious thought

To do – The knee jerk and pupil reflexes

SBA Skills

ORR	D	MM	PD	AI
✓				✓

knee jerk reflex – a reflex that stimulates the leg to straighten

pupil reflex – a reflex that opens or closes the pupil of the eye

Two other reflexes are the **knee jerk reflex** and **pupil reflex**.

The knee jerk reflex stops our leg from collapsing when we bend the knee as we walk. It acts to straighten the leg with each step.

1 Cross your knees and get a partner to tap your leg with the side of their hand, just below your knee. This may take some practice to find the right spot. When the reflex is stimulated, the lower leg jerks upwards, as shown in Fig A6.7. No matter how much you try you cannot stop the reflex from happening.

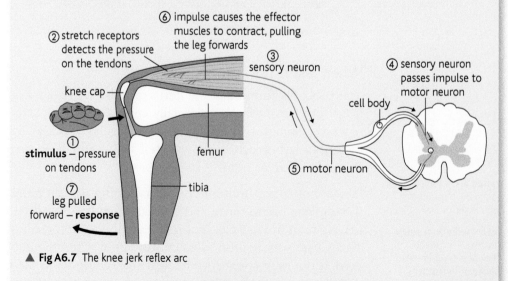

▲ **Fig A6.7** The knee jerk reflex arc

Q3 Suggest why it is impossible to stop the reflex from happening.

2 Work in pairs. One student puts their hands over their eyes for one minute. They then remove their hands and the second student observes the size of their pupil. You can also try shining a torch into the student's eye to see what effect that has on the size of the pupil.

Q4 What happens to the size of the pupil in the dark and in the light?

When things go wrong

The nervous system is a very efficient at coordinating our body. But when it is damaged the consequences can be severe. Damage to the spine in a car accident can cause complete paralysis of the body. This means that the person cannot move any part of their body below the part of the spine that is injured because the connection to the brain has been damaged. Damage to part of the brain can cause paralysis to parts of the body, speech impairment, loss of thinking skills, or even death.

Q5 Suggest why damage to the spinal cord can cause paralysis.

The functions of the endocrine system

The human body has two control systems. One is the nervous system. The other is the **endocrine or hormonal system**. **Hormones** are sometimes known as chemical messengers. They are released in one part of the body, travel via the blood stream, and have an effect in another part of the body. Hormones are released from **endocrine glands**.

Q6 Suggest why nerves bring about a faster reaction in the body than hormones.

The endocrine glands

Thyroid gland

The **thyroid gland** is butterfly shaped and situated in the neck either side of the trachea. It releases a hormone called **thyroxin**. Thyroxin controls our metabolic rate, growth and development. Metabolic rate is the rate at which respiration releases energy from the oxidation of glucose (see Unit A3a on Respiration). The more thyroxin is released, the faster our metabolic rate becomes. This is called **hyperthyroidism**. Symptoms include restlessness and a loss of body mass. An underactive thyroid gland causes **hypothyroidism**. It causes an increase in body mass and a lack of alertness. The gland can also swell, a condition known as **goitre**. This can happen when there is a lack of iodine in the diet. Thyroxin contains iodine and a lack of iodine causes the thyroid gland to swell as it compensates to try to produce sufficient thyroxin. If this happens in young children it can lead to a condition called **cretinism**.

endrocrine or hormonal system – a set of glands in the body that produce different hormones

hormones – chemical messengers produced by the body

endocrine glands – glands that produce hormones

thyroid gland – a gland that produces the hormone thyroxin

thyroxin – a hormone that controls our metabolic rate

hyperthyroidism – an overactive thyroid gland producing too much thyroxin

hypothyroidism – an underactive thyroid gland producing too little thyroxin

goitre – a swelling of the thyroid gland caused by a lack of iodine in the diet

cretinism – a condition caused in children by an underactive thyroid

Pancreas

pancreas – a gland that produces the hormone insulin

The **pancreas** is leaf shaped and located just beneath the stomach. It not only produces digestive enzymes, but also the hormone **insulin**. Insulin helps to regulate the glucose concentration in the blood. This concentration is normally about 90 mg per 100 cm³ of blood. However, after a meal this level can rise. The pancreas responds by releasing more insulin, which converts the excess glucose into glycogen which is stored in the liver and muscle cells.

insulin – a hormone that converts excess glucose in the blood into glycogen in the liver

diabetes – a condition caused by a lack of the hormone insulin

A lack of insulin is a condition called **diabetes**. In diabetics the level of glucose can rise much higher in the blood. The kidney can no longer reabsorb all the glucose and some is excreted in the urine.

Doctors can test a patient for diabetes. The first test a doctor will carry out on a patient to see if they are diabetic is to test their urine for glucose. This test can be done quite easily.

The doctor dips a clinistix into a sample of urine. The small paper tab on the end will change colour. The colour is then compared to a chart to see how much glucose is in the urine.

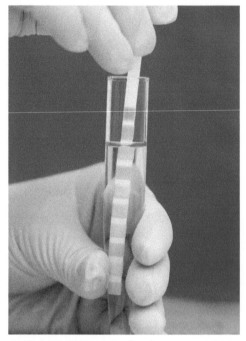

▲ **Fig A6.8** Testing urine for glucose

Testes

testes – glands that produce the hormone testosterone

The **testes** are located in sacs beneath the penis. They secrete the hormone **testosterone**. Testosterone is the hormone that causes sexual development in males. It even affects the unborn zygote by ensuring that it turns into a baby boy instead of a baby girl. At puberty it brings about all the secondary sexual characteristics of males such as a deepening of the voice, growth of a beard and pubic hair, and development of the testes to produce sperm.

testosterone – a hormone that results in male characteristics

Ovaries

ovaries – glands that produce the hormone oestrogen

The **ovaries** are located in the abdomen just above the uterus. They are small round organs that secrete the hormone **oestrogen** that brings about the secondary sexual characteristic of girls. Pubic hair begins to grow, the reproductive organs mature, and menstruation begins. You will study more about the sex organs in Unit A7.

oestrogen – a hormone that results in female characteristics

adrenal glands – glands that produce the hormone adrenalin

Adrenal glands

The **adrenal glands** are situated on the top of each kidney. They produce the hormone **adrenalin**. Adrenalin is sometimes called the 'fright or flight'

adrenalin – a hormone that gets the body ready for action

hormone as it prepares the body for action. When the adrenal glands release adrenaline into the blood system, the heart beat increases, blood flow to the muscles is increased, more glucose is released into the blood stream and the bronchioles increase in diameter to allow the lungs to get more oxygen. All of this provides the muscles with extra energy to either fight or run away. In modern life, adrenalin is often released in situations that do not require fighting or running away. For example, people who have a very stressful life can produce too much adrenalin too often, which can weaken the immune system and increase the risk of heart disease.

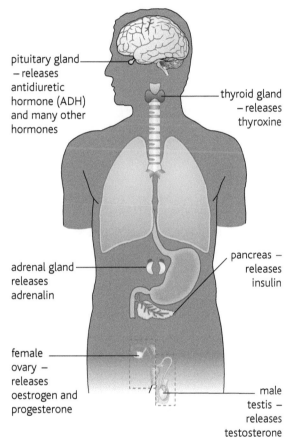

pituitary gland – releases antidiuretic hormone (ADH) and many other hormones

thyroid gland – releases thyroxine

adrenal gland releases adrenalin

pancreas – releases insulin

female ovary – releases oestrogen and progesterone

male testis – releases testosterone

▲ **Fig A6.9** The position of hormone glands

Pituitary gland

pituitary gland – a gland that produces hormones and controls the other hormone glands

The **pituitary gland** is situated just below the brain. It produces a whole range of different hormones. Some of these are hormones in their own right such as ADH. Others are hormones that regulate other glands and control how much hormone each of the glands secretes. This is why the pituitary gland is sometimes called the master or control gland.

Feedback controls

feedback – a process to control the level of a factor such as body temperature or hormone level in the blood

The pituitary gland controls other glands by a process called **feedback**.

The best way to understand feedback is to look at some examples. Fig A6.10 shows how the hormone insulin regulates the concentration of glucose in the blood.

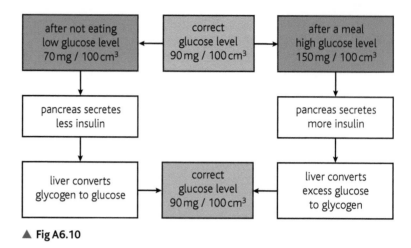

▲ **Fig A6.10**

Feedback helps keep the glucose in the blood at a constant level.

Another example of feedback can be seen in osmoregulation, which you learned about in Unit A5. You can see the flow chart for osmoregulation on page 81.

Temperature control

ectothermic – animals that absorb heat from their surroundings

poikilothermic – animals whose body temperature matches that of their surroundings

Animals such as lizards and snakes are **ectothermic**. This means they need heat from their surrounding environment to survive. They often bask in the morning sunshine to warm themselves up to increase their metabolic processes. They are also **poikilothermic**. This means their body temperature matches that of their surroundings.

Q7 Some people call poikilothermic animals cold blooded. Suggest why this is not a very good description of them.

endothermic – animals that produce their own body heat

homeothermic – animals that maintain their body temperature at a constant level independently of their surroundings

Human beings are **endothermic**. This means they generate their own heat. They are also **homoeothermic**. This means the generated heat maintains a constant body temperature that is usually warmer than the environment.

Our body temperature is kept at a constant 37°C. It is monitored by the hypothalamus in the brain. Temperature receptors around the body send messages to the hypothalamus. If the temperature rises or falls by just a few degrees we feel ill. If it changes by a few more degrees we may die. It is our skin which maintains this constant body temperature under instruction from the hypothalamus.

▲ **Fig A6.11** A snake basking in the morning sunshine

When we are too hot, tiny muscles that control the flow of blood in the capillaries of our skin relax. This allows more blood to flow near the surface of our skin. The skin becomes hotter and heat from the blood is radiated away from the body, helping it to cool down. If we are still too hot, sweat is secreted from sweat glands onto the surface of the skin. Just as you feel cold when the water evaporates from your skin when you get out of a swimming pool, when sweat evaporates it cools down the skin. Hairs on the skin lie flat to allow cool breezes to blow over the skin and speed up the evaporation of the sweat.

Q8 Suggest why we feel even colder when we get out of a swimming pool and a breeze is blowing.

When we are too cold the opposite happens. Capillary muscles contract and reduce the blood flow to the surface of the skin. The skin turns paler and feels cold as the warm blood is kept deeper from the surface to conserve the heat. Sweating comes to a stop and hairs stand erect as they interrupt the air flow and trap a layer of warmer air next to the surface of the skin. Finally, if all else fails, we begin to shiver. Shivering is rapid muscular contractions that release energy as heat to warm up the body.

Temperature control is another example of feedback control, as can be seen in the flow chart in Fig A6.13

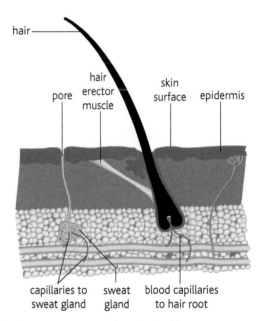

▲ **Fig A6.12** A section through the skin

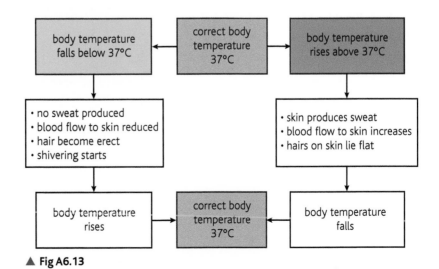

▲ **Fig A6.13**

Feedback helps keep the body at a constant temperature of 37°C.

The advantages of being endothermic are:

1 Our body is always at just the right temperature for all our metabolic processes to take place.

2 When we wake in the mornings we are active straight away; we do not have to wait for our body temperature to rise. Animals that have to warm up this way are vulnerable and at risk from attack by predators.

End-of-unit questions

1 Which of the following neural pathways is correct?

  a response – effector – motor neuron – CNS – sensory neuron – receptor – stimulus

  b CNS – sensory neuron – stimulus – motor neuron – response – effector – receptor

  c stimulus – response – CNS – effector – receptor – sensory neuron – motor neuron

  d stimulus – receptor – sensory neuron – CNS – motor neuron – effector – response

2 Which of the following carries an impulse towards the brain?

  a motor neuron

  b spinal column

  c cerebellum

  d sensory neuron

3 Explain how feedback helps to control our body temperature

4 The graph in Fig A6.14 shows the blood sugar levels of both a healthy person and a diabetic person after a meal.

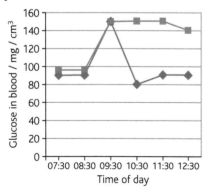

▲ **Fig A6.14**

a At what time of day did they both have a meal?

b Which graph, blue or red, represents the diabetic? Explain your answer.

c Suggest what is happening in the blood stream of the healthy person at 09:00.

d Suggest two ways in which diabetic people can control their condition.

5 Name two hormones and explain their role in the body.

6 Water balance in our bodies is an example of negative feedback.

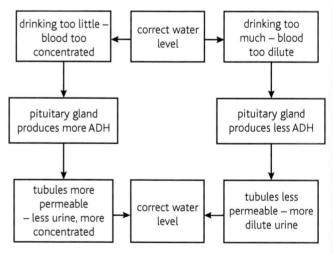

When we suffer from heatstroke it can be an example of positive feedback.
Draw a diagram like the one above to show how heatstroke can be an example of positive feedback.

A6b Sight and Hearing

By the end of this unit you will be able to:

- relate the structures of the mammalian eye to their functions.
- explain sight defects and their corrections.
- relate the structures of the mammalian ear to their functions.

The eye

The structure and functions of the eye

The eyes are one of the body's five sense organs. Their job is to convert light images to electrical impulses that are sent to the brain. They are the one sense organ that can receive information from enormous distances. When we look at a star we are seeing an object that is billions of miles away.

The eye has several similarities to a camera.

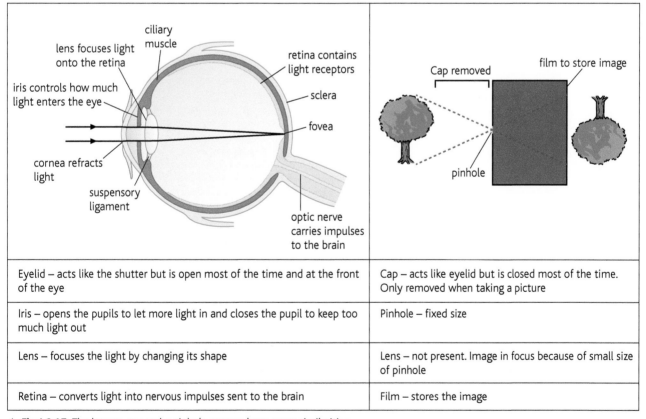

Eyelid – acts like the shutter but is open most of the time and at the front of the eye	Cap – acts like eyelid but is closed most of the time. Only removed when taking a picture
Iris – opens the pupils to let more light in and closes the pupil to keep too much light out	Pinhole – fixed size
Lens – focuses the light by changing its shape	Lens – not present. Image in focus because of small size of pinhole
Retina – converts light into nervous impulses sent to the brain	Film – stores the image

▲ **Fig A6.15** The human eye and a pinhole camera have some similarities

Q1 Make a list of similarities and a list of differences between a camera and the human eye.

How the eye works

iris – a structure that can open or close down the size of the pupil to control the amount of light entering the eye

The amount of light entering the eye is controlled by the **iris**. The iris has two sets of muscles. One set is circular. When they contract the pupil gets smaller. The other set of muscles is radial; they radiate out from the pupil. When they contract the pupil gets larger.

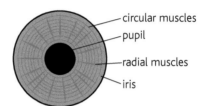

▲ **Fig A6.16** The radial and circular muscles in the iris control the size of the pupil

pupil – a hole where light can enter the eye

The **pupil** is the black centre of the eye. It is in fact a window into the eye itself. It only looks black because, just like looking through a window into a darkened room, it looks black inside the room because only a little light is getting in through the window.

As you discovered in the pupil reflex 'To do' task on page 91, when it is dark the iris opens the pupil to let in more light. When it is bright and sunny the iris closes the pupil to keep out too much sunlight.

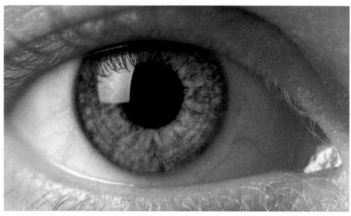

▲ **Fig A6.17** The pupil reflex

accommodation – focusing an imaging onto our retina at the back of our eyes

Light entering the eye has to be focused. This is called **accommodation**.

To do – Accommodation

SBA Skills

ORR	D	MM	PD	AI
✓				

1 Hold up a finger about 50 cm from your face and look at it. Now shift your gaze to the far side of the room. The wall will appear blurred but will rapidly come into focus.

2 Now look back at your finger. Your finger will now look blurred but will rapidly come into focus. This is because your eye is accommodating or focusing the image.

cornea – the clear curved front of the eyeball that starts the focusing of light rays

lens – the clear structure in the eye that focuses light rays onto the retina

retina – the light-sensitive layer of cells at the back of the eye

suspensory ligaments – ligaments that hold the lens in place

ciliary muscles – muscles that change the shape of the lens to focus an image

Look at Fig A6.18 to see how the eye focuses light. Light is first focused when it passes through the **cornea**. It then enters the inside of the eye and passes through the **lens**. It is the job of the lens to do the fine focusing and to produce a sharp image on the **retina**. Unlike a camera where the lens moves backwards and forwards, the lens in the eye changes its shape. The lens is held by **suspensory ligaments** that are attached to a ring of **ciliary muscles**. When the ciliary muscles contract, the ring gets smaller and the tension on the ligaments is relaxed. This allows the lens to revert to its normal, fatter shape. When the ciliary muscles relax the ring gets bigger. This puts tension on the suspensory ligaments, which stretch the lens making it thinner. The lens needs to be thin to focus light from an object that is far away. It needs to be fatter to focus light from an object that is close to the eye.

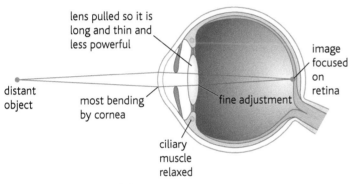
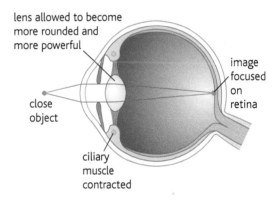

▲ **Fig A6.18** Accommodation

SBA Skills

ORR	D	MM	PD	AI
✓				

optic nerve – nerve that connects the eye to the brain

To do – Finding the blind spot

The eye is an amazing organ, but each eye is blind where the **optic nerve** leaves the retina. There are no light-sensitive cells there. We do not notice the blind spot because it is not something we look at directly and each eye has the blind spot in a different place. However, we can find the blind spot by performing a simple investigation.

1 Put your hand over your left eye.

2 Look at the X.

X   ●

3 Gradually move the book forwards and backwards. When you get the distance right you will notice that the dot disappears. This is when the light from the dot is falling on the blind spot.

rods – light-sensitive cells that work best in dim light and are sensitive to black and white

cones – light-sensitive cells that work best in bright light and are sensitive to colour

Rods and cones

The retina has two types of sensory cell. **Rods**, which only see in black and white but work well in dim light, and **cones**, which see colour but do not work very well in dim light. The cones are mainly situated at the back of the eye where the light is focused. The rods are mainly situated around the periphery of the eye and not where the light is focused. When the cones do not work properly, people experience colour blindness.

SBA Skills

ORR	D	MM	PD	AI
✓				✓

To do – Seeing in the dark

1 Go into a dimly lit room such as a bedroom at night. You will notice that it is difficult to see what colour things are.

2 When you look directly at a faint object such as a faint star in the night sky it is difficult to see it. Try looking just to one side of the star and you will see it much better.

Q2 Explain why we do not see colour in dim light and it is easier to see faint objects when we look just to one side of them.

Q3 Suggest why a lack of cones in the eye may cause colour blindness.

Sight defects and their corrections

Colour blindness is not the only eye defect. Two common conditions with the eye are **short sightedness** and **long sightedness**.

Short sightedness or myopia

short sightedness – focusing light short of the retina

long sightedness – focusing light behind the retina

concave – a lens that is thinner in the middle and thicker at the edges

diverging – a lens that is thinner in the middle and thicker at the edges and spreads out light rays

Fig A6.19 shows what causes short sightedness and how it can be corrected. Short sightedness occurs when the light rays are focused to produce an image short of the back of the eye or retina. This results in the image from distant objects being blurred. It is caused by the eyeball being too long, or the lens being too fat, or convex. It can be cured by wearing glasses with lenses that are a special shape called **concave**. Concave lenses are thinner in the middle and thicker at the edges and are called **diverging** lenses.

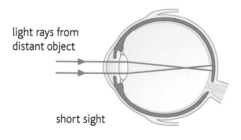

light rays from distant object

short sight

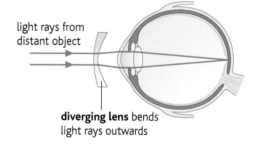

light rays from distant object

diverging lens bends light rays outwards

▲ **Fig A6.19** Correcting short sightedness

Long sightedness or hypermetropia

Long sightedness occurs when the light rays are focused to produce an image behind the back of the eye or retina. This results in the image from close objects being blurred. It is caused by the eyeball being too short, or the lens

convex – a lens that is thicker in the middle and thinner at the edges

converging – a lens that is thicker in the middle and thinner at the edges and focuses light

being too thin, or concave. It can be cured by wearing glasses with **convex** lenses. Convex lenses are thicker in the middle and thinner at the edges and are called **converging** lenses.

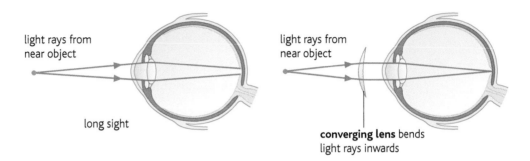

light rays from near object

long sight

light rays from near object

converging lens bends light rays inwards

▲ **Fig A6.20** Correcting long sightedness

Dazzling

Bright light can cause short-term dazzling. This causes short-term blindness. The bright light uses up chemicals such as rhodopsin on the retina and until supplies of this light-sensitive chemical have been replaced, vision does not return to normal. This normally only takes a few seconds, but can be several minutes if the light was extremely bright, such as accidently glancing at the Sun. Looking at the Sun is dangerous as the lens focuses energy from the Sun onto the retina. This can quickly burn the retina, causing permanent blindness. People who have accidently focused binoculars or telescopes on the Sun have been blinded instantly. There is no cure!

Cataracts

The Sun also gives off ultraviolet (UV) light. On bright sunny days the only way to stop UV light from entering the eye is to wear UV-protecting sunglasses. Long-term exposure to UV light causes the lens to turn milky or opaque and this stops light from passing through the lens. This condition is called a **cataract**.

cataract – a cloudiness of the lens often caused by ultraviolet light from the Sun

Surgeons can remove the defective lens and replace it with an artificial one.

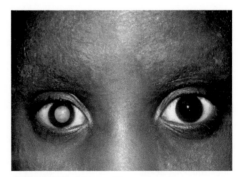

▲ **Fig A6.21** A lens with a cataract

Q4 An artificial lens will be a fixed shape. Suggest what problems this may cause to a person's vision and how the problem could be rectified.

Glaucoma

Physical injury can be caused when objects enter and damage the eye. The eye has a very good defence mechanism – it blinks.

glaucoma – pressure inside the eye that can damage the eye leading to blindness

Another type of physical injury can occur when the pressure of the fluid in the eye becomes too great. This increased pressure is called **glaucoma** and can cause damage to the optic nerve resulting in blindness. Because we have no pain receptors inside the eye, damage can be happening and we are unaware of it. Glaucoma is one of the most common causes of blindness but can be treated if caught early.

The structure and functions of the ear

The human ear has two jobs to do. The first is hearing; the second is balance and orientation. It is divided into three parts, the outer ear, the middle ear and the inner ear.

Hearing

auditory nerve – nerve that connects the ear to the brain

pinna – structure on the outside of the ear that directs sound waves into the ear

auditory canal – canal that connects the pinna to the eardrum

tympanic membrane – eardrum that passes sound waves into the middle ear

ossicles – three bones in the ear that transmit sound wave to the inner ear

malleus – one of the ear ossicle bones

incus – one of the ear ossicle bones

stapes – one of the ear ossicle bones

eustachian tube – tube that connects the middle ear to the back of the throat to equalise air pressure

Sound is caused by vibrations in the air. The vibrations or sound waves transfer sound energy to the ear. It is the job of the ear to transfer this energy into nerve impulses that are carried from the ear, along the **auditory nerve**, to the brain.

The complicated shape of the **pinna** directs the sound waves down the **auditory canal** of the outer ear. The sound waves then reach a membrane called the **tympanic membrane** or eardrum and cause the eardrum to vibrate. The eardrum passes the vibrations into the middle ear along three small bones called the ear **ossicles**, the smallest bones in our body. These bones are called the **malleus**, the **incus** and the **stapes**.

Both the outer ear and the middle ear are filled with air. Air pressure is constantly changing; weather systems are often called high and low pressure systems, and when a new weather system arrives the pressure will change. It is important that the pressure on either side of the eardrum is equalised as it changes. This is done by the middle ear being connected to the back of the throat by a tube called the **eustachian tube**. When we swallow, the tube opens and allows air to enter or leave the middle ear to equalise the pressure. When we have a cold and a blocked

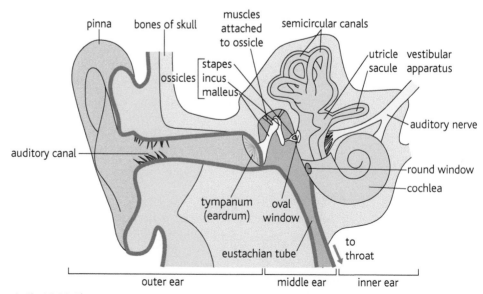

▲ **Fig A6.22** The ear

eustachian tube, or when air pressure changes quickly as when we go up in an aeroplane, the air pressure is not immediately equalised and we feel the pressure on our eardrum. Our ears pop when we swallow and the pressure is equalised.

The stapes rest upon another membrane called the oval window. This membrane separates the middle ear from the inner ear, which is filled with fluid, not air. The vibrations of the oval window cause pressure waves to enter a spiral- or snail-shaped organ called the **cochlea**. Fig A6.23 shows how the pressure waves pass all along the cochlea to its end and then return to the round window and re-enter the middle ear where the energy is lost.

cochlea – part of the inner ear where sound waves are converted to electrical impulses to be sent to the brain

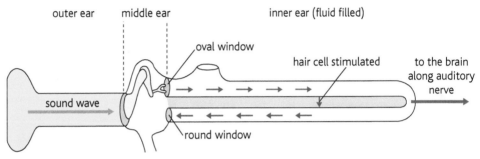

▲ **Fig A6.23** How we hear

As the pressure waves pass along the cochlea, they cause small hairs to vibrate, which in turn stimulate nerves cells. The nerve cells then pass impulses along the auditory nerve to the brain. The brain then interprets these impulses as sound.

Q5 Suggest why our ears pop when we come down in an aeroplane just before we land.

Loudness and pitch

Loud sounds transfer more energy than quiet sounds. If the energy is too great the extremely sensitive hair cells in the cochlea can be damaged permanently. Loudness is measured in decibels (dB). Table A6.1 shows the different noise levels of some sounds.

quietest sound that can be heard	0 dB
quiet whisper	30 dB
normal conversation	65 dB
traffic noise inside car	85 dB
sustained level that can cause hearing loss	95 dB
rock concert	115 dB
pain begins	125 dB
gun blast next to ear	140 dB
short term exposure causing permanent hearing loss	140 dB

Table A6.1

To do – Measuring sound

Use a decibel meter to measure the loudness of different sources of noise around the school. Use the information you collect to make a table similar to the one in Table A6.1.

Q6 Explain why loud sounds can damage our hearing.

Pitch is how we perceive the frequency of a sound. Frequency is the number of sound waves per second and is measured in hertz (Hz). A sound with a high frequency has a high pitch, and a sound with a low frequency has a low pitch. Our ears can hear between about 20 Hz and 20 000 Hz. As we get older we lose the ability to hear sounds of a very high frequency and the range can drop to about 16 000 Hz.

To do – Find your frequency range

If your school has a signal generator (a device that produces sound of a set frequency) you can use it to determine what the frequency range of your hearing is. Most students will be the same at hearing their lowest frequency of about 20 Hz. At the higher range, some students will detect higher frequencies than other students.

Balance and orientation

semicircular canals – part of the inner ear that gives us our sense of balance

vestibular apparatus – part of the inner ear that gives our sense of orientation

utricle – part of the vestibular apparatus concerned with orientation

sacculus – part of the vestibular apparatus concerned with orientation

The ear also controls our balance and orientation. Balance takes place in the three **semicircular canals** and orientation in the **vestibular apparatus**, the **utricle** and the **sacculus**.

The semicircular canals are responsible for balance. They are each arranged at right angles to one another in three different planes. They are filled with fluid and the fluid will move in the canals depending upon which direction we are moving. For example, if we move forward the fluid will move backwards in the vertical front-to-back and the horizontal canals but it cannot move in the vertical side-to-side canal. Different directions of movement cause the fluid to move in each canal differently. The movement of fluid is detected in each canal by an ampulla. The fluid moves a group of hairs that stimulates a nerve to send impulses to the brain. Depending upon which ampulla is stimulated, the brain can detect which way we are moving. Fig A6.24 opposite shows how this works.

The utricle and sacculus are responsible for orientation. They also contain sensory hair cells. These hair cells are attached to a small object called an otolith. As we tilt our head the otolith falls under the influence of gravity and bends the sensory hairs. These stimulate nerves to transmit an impulse to the brain to tell the brain in which direction we are leaning.

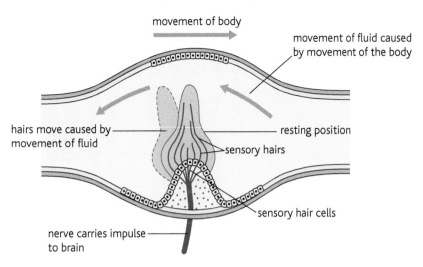

movement of body

movement of fluid caused by movement of the body

hairs move caused by movement of fluid

resting position

sensory hairs

sensory hair cells

nerve carries impulse to brain

▲ **Fig A6.24** How we sense movement

To do – Fooling the brain

Make sure you do this with a partner to protect you and have somewhere soft to fall if you should fall over.

1 Stand upright and spin round quickly several times. This will get the fluid in your semicircular canals moving.

2 Stop spinning and stand still. Just like stirring a cup of tea with a spoon, the fluid in your semicircular canals will carry on moving when you stop spinning. This movement of fluid will stimulate ampullae and tell the brain that you are still moving. Your eyes are telling your brain that you are not moving. This conflicting information will confuse the brain and you feel dizzy.

Q7 Suggest why damage to our ears may affect our sense of balance as well as our sense of hearing.

End-of-unit questions

1 Which of the following structures is found in both the eye and the ear?

  a nerve

  b lens

  c retina

  d ossicle

2 Which statement correctly matches eye structure with function?

  a retina – focuses light on to the back of the eye

  b lens – carries nerve impulses to the brain

  c cornea – adjusts the shape of the lens

  d iris – adjusts the size of the pupil

3 Which pathway correctly describes the route sound takes when entering the ear?

  a eardrum – cochlea – pinna – ossicles – oval window

  b pinna – eardrum – ossicles – oval window – cochlea

  c ossicles – pinna – oval window – cochlea – eardrum

  d oval window – ossicles – pinna – cochlea – eardrum

4 Explain the accommodation changes that take place in the eye as an object approaches it.

5 The eye and the ear are both sense organs.

  a State the three other sense organs found in the human body.

  b Touching and dropping a hot object is a reflex reaction. Describe the pathway taken by the nerve impulses in the reflex.

  c The graph in Fig A6.25 shows the relative numbers of rods and cones on the retina of an astronomer who is looking at stars.

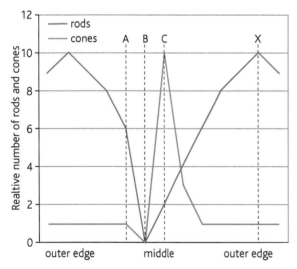

▲ **Fig A6.25**

i Which part of the graph, A, B or C is the position of the blind spot?

ii At which part of the graph, A, B or C is the person looking directly at a star?

iii At which part of the graph, A, B or C would the person see a very faint star better?

iv Explain why the person would find it difficult to tell the colour of the star if the light is falling on area X.

6 Bright light can cause short-term dazzling. This causes short-term blindness. The bright light uses up chemicals such as rhodopsin on the retina and until supplies of this light-sensitive chemical have been replaced, vision does not return to normal.

  a Use this information to help explain why our eyes are able to see more detail after being in the dark for about 15 minutes.

  b Carrots contain vitamin A. Carrots also help us to see in the dark.
Suggest a hypothesis to explain how carrots help us see in the dark.

7 As we get older we lose the ability to hear sounds of a very high frequency and the range can drop to about 16 000 Hz. High frequency sounds help us distinguish between different sources of sound.

  a Suggest why as we get older it becomes harder to hear what someone is saying to us in a noisy room.

  b When we go to a party we can often hear our name mentioned across the room even when the conversations in the room are very loud. Suggest how this can happen and why it is important to us to be able to do it.

A7a Reproduction and Growth

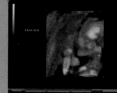

asexual reproduction – reproduction without using sex cells (gametes)

sexual reproduction – reproduction using sex cell (gametes)

Living things need to reproduce in order for the species to survive. There are two types of reproduction: **asexual reproduction** and **sexual reproduction**. Sexual reproduction is the joining of two gametes, one male and one female. This combines characteristics from both parents and brings about variation in the offspring. In asexual reproduction the organism makes a copy of itself and there is no variation.

The advantages of using asexual reproduction are that is quick and easy as it only involves one individual. New habitats can be colonised very quickly and organisms can rapidly take advantage of new food supplies. The disadvantage is that all the offspring are genetically identical to their parent. This is not a good thing for survival as evolution selects the differences between individuals that are better adapted to survive in the environment. If there are no differences between the individuals and the environment changes, none of them will have an advantage to be able to survive.

Asexual reproduction

clones – genetically identical organisms

Asexual reproduction happens when a cell divides to make two identical copies of itself. This happens during growth, but sometimes the two cells go on to produce two new identical individual organisms. Individuals that have identical cells to their parents are called **clones**.

Asexual reproduction in plants
Vegetative propagation

Vegetative propagation is a common form of asexual reproduction in plants. It occurs when a plant grows a new part that can develop into a separate new plant. Some methods of vegetative propagation are described below.

runner – a stem that grows new plants from its lateral buds

Runners are stems that grow from a parent plant. The stem produces new plants that take root in the soil. Eventually the stem or runner rots away, leaving the new plants growing independently. The strawberry plant and African violets are examples of plants that have runners.

▲ **Fig A7.1a** A strawberry runner

bulb – a condensed plant that stores food and grows new plants by vegetative propagation

Bulbs are a condensed underground stem surrounded by lots of fleshy leaves. They store food for the next season's growth. Buds at the bases of the leaves can each grow into a new plant. The onion is a good example of plant that has bulbs.

▲ **Fig A7.1b** An onion bulb

corm – a condensed stem that stores food and grows new plants by vegetative propagation

terminal bud – the bud at tip of stem

Corms are like bulbs with a short underground stem but there are no fleshy leaves and the stem is swollen to store food for next season's growth. The **terminal bud** gives rise to next season's growth and lateral buds can grow to produce new corms. Crocus and Taro or Dasheen are examples of plants that grow corms.

▲ **Fig A7.1c** A taro corm

tuber – a swollen stem that stores food and produces new plants by vegetative propagation

lateral bud – a bud along the side of a stem

Tubers are swollen tips of underground stems. The tip swells as a food storage organ and individual **lateral buds** can grow into new plants. The potato is an example of a tuber.

▲ **Fig A7.1d** A potato tuber

rhizome – an underground stem that stores food and grows new plants by vegetative propagation

Rhizomes are underground stems where the whole stem swells to store food. Each lateral bud can grow to form a new plant. Ginger is an example of a plant that grows rhizomes.

▲ **Fig A7.1e** A ginger rhizome

Q1 Explain why runners, bulbs, corms, tubers and rhizomes are all clones.

 To do – Drawing organs of vegetative propagation

Examine and draw an example of a runner, bulb, corm, tuber and rhizome. Label each diagram and explain how it reproduces by vegetative propagation.

SBA Skills

ORR	D	MM	PD	AI
	✓			

Artificial propagation

cutting – an artificial method to produce new plants by removing the side shoots or stems and growing them

Cuttings are one of the most common ways that people can reproduce plants asexually. A small stem is cut from the parent plant. Any leaves from the base of the cutting are removed. The stem is then planted into soil. Cells at the cut surface of the stem begin to divide and within a few days new roots start to grow near the base of the cutting. The cutting then grows into a new plant. Not all cuttings survive. The process is more likely to succeed if the cutting is dipped into a hormone rooting compound before being planted in the soil.

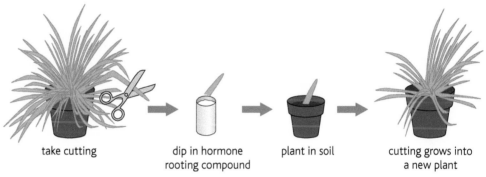

take cutting   dip in hormone rooting compound   plant in soil   cutting grows into a new plant

▲ **Fig A7.2** Taking cuttings

graft – joining the stem of one plant to another growing plant

scion – the stem that is grafted onto a growing plant

stock – the growing plant onto which a stem is grafted

A **graft** is a cutting, called a **scion**, which instead of being planted, is placed in a slit in the stem of a completely different plant called the **stock**. The joint is bound up to hold it place. Cells begin to divide and within a few days the two pieces of stem have joined and started to grow and produce new shoots and branches.

It is even possible to grow oranges and lemons on the same tree using this method – see Fig A7.4.

▲ **Fig A7.3** Making a graft

▲ **Fig A7.4**

Q2 Explain why the scion and the stock are *not* clones of each other.

tissue culture – artificial asexual reproduction of plants using single plant cells

Tissue culture is a more technical process. It can be used to make huge numbers of a plant. When plant breeders used to develop new varieties, such as the blue rose, it would take many years before they had taken sufficient cuttings for them to start to sell them and make a profit. It is now possible to have thousands of new plants within just a few months.

A tip is taken from a growing plant shoot. It is sterilised and placed in a sterile growing medium such as agar gel. The cells divide to produce lots of tiny plantlets. These grow, develop roots and when separated and re-potted, each one grows into a new complete plant.

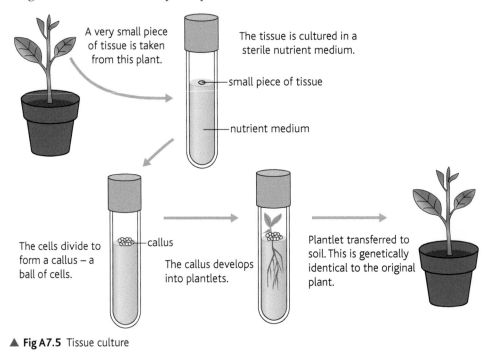

A very small piece of tissue is taken from this plant.

The tissue is cultured in a sterile nutrient medium.

small piece of tissue

nutrient medium

The cells divide to form a callus – a ball of cells.

callus

The callus develops into plantlets.

Plantlet transferred to soil. This is genetically identical to the original plant.

▲ **Fig A7.5** Tissue culture

Q3 Suggest why the tissue taken from a plant is grown in sterile conditions.

Asexual reproduction in animals

Some animals can reproduce using asexual reproduction.

amoeba – a single-celled microscopic animal that lives in ponds and reproduces by binary fission

The single-celled organism called the **amoeba** uses this type of reproduction. The amoeba is a microscopic animal that lives in slow moving water such as pond or ditches.

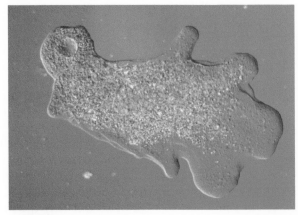

▲ **Fig A7.6** An amoeba

binary fission – asexual reproduction by a cell dividing and splitting into two identical cells

It divides by a process called **binary fission**. First the chromosomes in the nucleus are copied. Then the nucleus divides. Finally the whole organism splits into two, each with a new nucleus.

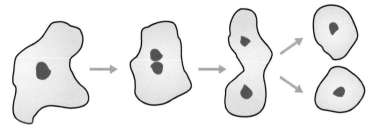

▲ **Fig A7.7** Binary fission in amoeba

Cloning

Animals can also be copied by using the artificial method called **cloning**. Dolly the sheep was the fist animal to be produced by cloning.

cloning – the process of producing clones

- An ovum was taken from a female donor sheep. Its nucleus was removed.
- The DNA was removed from a body cell of another sheep.
- The DNA was then inserted into the empty ovum of the donor sheep.
- The ovum was then returned to the donor sheep and allowed to grow and develop in its womb.
- The lamb that was born, called Dolly, was identical to the sheep that had donated the DNA.

This is shown in Fig A7.8.

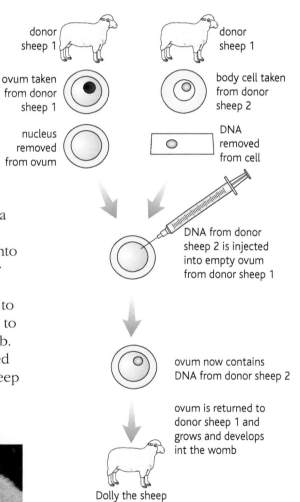

donor sheep 1

ovum taken from donor sheep 1

nucleus removed from ovum

donor sheep 1

body cell taken from donor sheep 2

DNA removed from cell

DNA from donor sheep 2 is injected into empty ovum from donor sheep 1

ovum now contains DNA from donor sheep 2

ovum is returned to donor sheep 1 and grows and develops int the womb

Dolly the sheep

▲ **Fig A7.8** The process of genetic cloning used to create Dolly the sheep

◀ **Fig A7.9** Dolly the sheep and her creator, Bill Ritchie

Q4 Many people think that humans should not be cloned. Suggest why. What do you think?

Sexual reproduction

Sexual reproduction always involves the **fusion** (joining together) of a male and a female **gamete** (sex cell). The disadvantages of sexual reproduction are that two individuals are needed and the process is not always guaranteed to succeed. The advantages are that this type of reproduction produces variation in the offspring. Because the offspring are all different, when the environment changes, one of the offspring are likely to have a characteristic that will help them to survive. Darwin called this 'Natural Selection' and evolution only works because sexual reproduction provides variation that makes some individuals better adapted for survival than others.

Sexual reproduction in plants

Sexual reproduction occurs in flowering plants. The flower is the structure where sexual reproduction takes place.

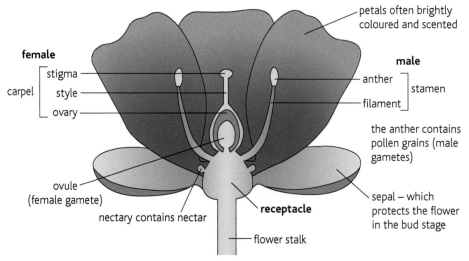

▲ **Fig A7.10** The structure of a flower

Unlike humans, many flowers contain both male and female reproductive organs and may fertilise one another. However, they usually mature at different times so the male sex cell of one flower fertilises the female sex cell in a different flower, often on a different plant.

Flowers that contain both male and female reproductive organs are called **hermaphrodites**.

Pollination

The male sex cell is transferred to the female organ inside a pollen grain. This protects the sex cell and is called **pollination**. Once the pollen has landed on the female **stigma**, pollination is complete.

In flowers where the male and female pollinate each other, it is called **self-pollination**. When the pollen is transferred to a different flower of the same species, it is called **cross pollination**.

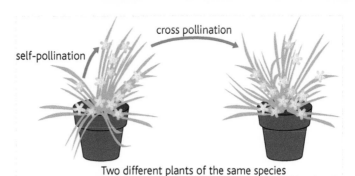

▲ **Fig A7.11** Self-pollination and cross pollination

Self-pollination results in a form of inbreeding as there is no exchange of genetic material between different plants. Cross pollination results in greater variation in the offspring as genetic information is being exchanged between two different plants.

wind pollination – transferring pollen using the wind

Sometimes the pollen is transferred by being blown by the wind. This is called **wind pollination**. In this case, pollination is a haphazard process and most of the pollen is lost, never reaching another flower.

insect pollination – transferring pollen using insects

Some plants have produced brightly coloured flowers with a scent and sugary food called nectar to attract insects. As the insects fly from flower to flower they transfer the pollen with them. This is called **insect pollination**.

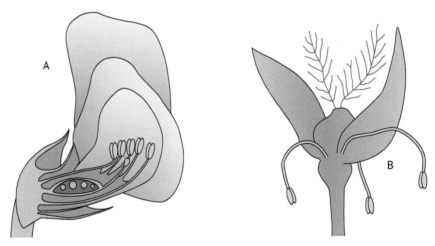

▲ **Fig A7.12** Insect and wind pollinated flowers

Q5 Which of these flowers, A or B are insect pollinated? Explain your answer.

To do – The parts of a flower

1 Make labelled drawings of three different flowers. It may help if you remove some of the petals and cut the centre of the flower in half.

2 For each flower, label the main parts and state how the flower is pollinated.

SBA Skills

ORR	D	MM	PD	AI
	✓			

Fertilisation

fertilisation – the process of male and female gametes fusing together

ovary – the female part of the flower that contains the ovule

ovule – found in the ovary, the ovule contains the female sex nucleus

style – part of the female organ in a flower that connects the stigma with the ovary

pollen tube – the tube along which the male nucleus passes from the pollen grain to the ovule

Fertilisation is the process whereby the male gamete fertilises the female gamete. This is not as easy as it sounds because the male gamete is inside the pollen grain and the female gamete is inside the **ovule** in the **ovary**. In reality it is the male sex nucleus that fertilises the female sex nucleus. For them to join, the pollen grows a tube that passes the male nucleus through the stigma and down the **style**, as seen in Fig A7.13. The **pollen tube** then enters the ovary and reaches the ovule. The male and female nuclei then fuse and fertilisation is complete.

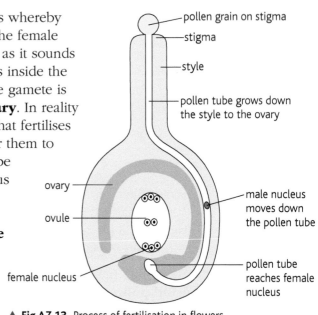

▲ **Fig A7.13** Process of fertilisation in flowers

Experiment – Growing pollen tubes

Your teacher will provide you with 2% and 5% solutions of sucrose. Different plants require different concentrations so it is best to try both 2% and 5%.

1 Place one drop of 2% or 5% sucrose solution onto a microscope cover slip.

2 Place a ring of Plasticine onto a microscope slide.

3 Dust pollen from a flower onto the drop of sucrose solution.

4 Invert the cover-slip and place it on the Plasticine ring so that the drop of sucrose solution hangs beneath the cover slip.

5 View the pollen grains using a microscope.

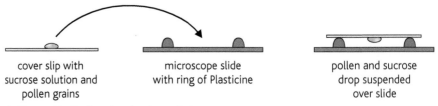

cover slip with sucrose solution and pollen grains

microscope slide with ring of Plasticine

pollen and sucrose drop suspended over slide

▲ **Fig A7.14** The hanging droplet technique

Some pollen tubes, like the primrose family, should grow during the course of the lesson. Other pollen may take a little longer. Adding 0.01% boric acid to the sucrose solution may also help the growth of the pollen tubes.

Using high power, you should be able to see the male nucleus that is used to fertilise the female nucleus.

Seeds and fruits

After fertilisation the ovule develops into the seed and the ovary wall into the fruit, as shown in Fig A7.15.

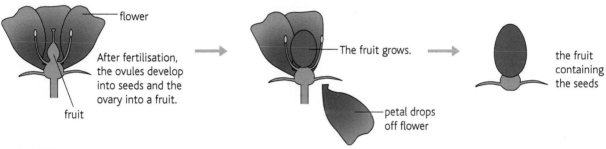

flower

After fertilisation, the ovules develop into seeds and the ovary into a fruit.

fruit

The fruit grows.

petal drops off flower

the fruit containing the seeds

▲ Fig A7.15

Different plants have different structures for their fruit. Some fruits contain just one seed, others contain many seeds. Some are fleshy and edible and brightly coloured. This is to attract animals and birds to eat them. The seeds then pass out with the animal's faeces and have their own pile of fertiliser with which to grow a new plant.

Others have seeds that are dispersed by the wind and have feathery structures like wings to help them be blown away so that when they germinate they will not compete with their parents for light and space.

▲ **Fig A7.16** Fruit formation

▲ **Fig A7.17** Nispero seeds are dispersed in animal faeces

▲ **Fig A7.18** Seeds blown by the wind

▲ **Fig A7.19** How is this bur seed dispersed?

Q6 The bur is a fruit produced by a plant. It has lots of sharp hooks. Look at the picture of the bur. Suggest how this fruit is dispersed from the parent plant.

germination – growth of a seed

Germination in plants

A seed is a food store so that the new plant can grow and develop until it has produced its own green leaves and begun to photosynthesise. The seed needs warmth, moisture and owygen to germinate. A radical (root) begins to grow downwards into the soil and a plumule (shoot) begins to grow upwards towards the light. It quickly produces green leaves and begins to photosynthesise.

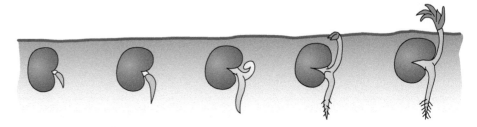

▲ **Fig A7.20** Roots grow down and shoots grow up

Growth patterns in different organisms

Measuring growth sounds straightforward, but in practice it can be quite tricky.

SBA Skills

ORR	D	MM	PD	AI
			✓	

To do – Measuring growth

Write down three different ways in which you could measure growth in a human being.

Q7 Which of these ways do you think is the most accurate way of measuring growth? Explain your answer.

The correct answer to **Q7** is that none of them will give a true picture of how much someone is growing. The best we can do is get an approximate answer to how much growth has taken place.

Let us look at some ways we can measure growth in plants.

- Measuring height – This should be done on a regular basis, say once each week and the data should be written down.

- Measuring diameter of the stem – This should also be done on a regular basis. Thin stems are tricky to measure accurately. Use a piece of fine string to wrap round the stem and then measure the length of the string.

fresh mass – the mass of a growing living plant

- Measuring **fresh mass** – This involves carefully removing the plant from the soil and washing traces of soil from the roots. The whole plant is then weighed on an accurate scale.

dry mass – the mass of a plant when all the water has been removed

- Measuring **dry mass** – This is the most accurate way of determining how much growth has taken place. The plant is removed from the soil, the roots washed and then the plant is heated in an oven at approximately 100°C to remove all of the water. The dried material that is left behind is then weighed.

- Counting the number of leaves – This can be used to see how much growth has taken place over a period of time. The number of leaves is counted each week and the data recorded. You need to be consistent week by week in deciding when a small bud opening is large enough to be called a leaf.

- Measuring the surface area of a leaf – This can be done using graph paper. The leaf is placed on a sheet of graph paper and the outline drawn with a pencil. The number of squares enclosed in the outline is then counted. Part squares are counted as a complete square if more than half of the square is enclosed by the leaf outline.

Q8 Suggest why measuring the height of a plant may not be an accurate measure of the plant's growth.

Q9 Suggest why measuring the thickness of a stem may not be an accurate measure of a plant's growth.

Q10 Suggest why measuring fresh mass is not always a useful way of measuring growth.

Q11 Suggest why measuring dry mass is not always a useful way of measuring growth.

SBA Skills

ORR	D	MM	PD	AI
✓			✓	

vermiculite – an inert material in which plants can be grown

Experiment – Investigating growth

Let us try putting some of these techniques into practice.

1 Get some seeds of such as balsam, beans or maize.

2 Soak the seeds in water for one hour.

3 Fill a jam jar full with **vermiculite** or some other inert material.

4 Place 2 cm depth of water in the jam jar.

5 Push the seeds down the side of the jam jar so that they are above the water level and clearly visible through the glass of the jam jar, as in Fig A7.21.

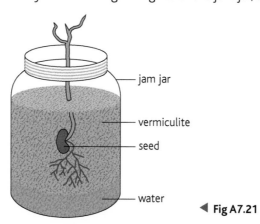

jam jar

vermiculite

seed

water

◀ **Fig A7.21**

6 Use the different ways of measuring growth described to measure the growth of your seedlings over a period of about two weeks.

Extrapolating data

extrapolate – to use a graph to make predictions

Scientists use a technique called **extrapolating**, which means using the data you have to predict further results. For example you can extrapolate a graph by extending the graph line and predicting where the line should be continued to.

SBA Skills

ORR	D	MM	PD	AI
		✓		✓

To do – Extrapolating your data and predicting results

1 When you have collected several weeks worth of data, plot the data as a graph showing growth on the *y*-axis and time on the *x*-axis. Then extrapolate the data by continuing the graph line to predict where you think the growth will have reached after a further two or three weeks. Continue to collect data for the next two or three weeks to check just how accurate your prediction was.

2 Look at the graph showing the growth of the human population on page 129. Copy the graph and extrapolate the curve to show how many people will be living on planet Earth in the next 100 years.

3 Measure the height of students at your school. Collect the data based on the age of the students and calculate an average height for each age group. Plot the average height for each age group on a graph. Use the graph to extrapolate the predicted heights of the students in two years' time.

End-of-unit questions

1 Which of the following is *not* an organ used in asexual reproduction in plants?

  a runner   **b** corm

  c pollen   **d** rhizome

2 Cloning is an artificial method of reproducing an organism asexually. Which of these statements about cloning Dolly the sheep is correct?

  a A male and a female cell are fused together.

  b The nucleus from an ovum of a donor sheep is placed in an ovum of another sheep.

  c The clone contains DNA that is different from both of the donor sheep.

  d The nucleus of the ovum from one sheep is replaced by the nucleus from an ovum of a second sheep.

3 Use Fig A7.22 to explain how fertilisation takes place in plants.

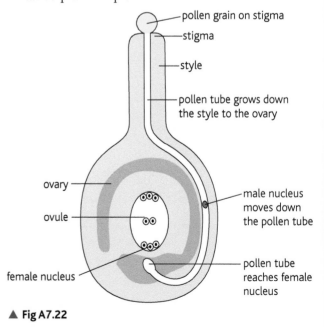

▲ **Fig A7.22**

pollen grain on stigma
stigma
style
pollen tube grows down the style to the ovary
ovary
ovule
male nucleus moves down the pollen tube
female nucleus
pollen tube reaches female nucleus

4 Explain the difference between pollination and fertilisation.

5 Copy and label the diagram of a flower in Fig A7.23. Annotate what each part does.

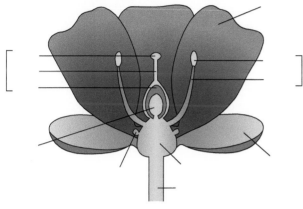

▲ **Fig A7.23**

6 There are different ways of measuring the growth of plants.

a Use the data on the height of a plant to plot a graph.

time in weeks	height / cm
1	3.4
2	5.3
3	8.5
4	11.2
5	14.9
6	18.0

i Use the graph to determine the height of the plant at $3\frac{1}{2}$ weeks.

ii Predict what the height of the plant will be after 7 weeks.

iii Explain why your answers to the previous two questions cannot be completely accurate.

b Fig A7.37 shows the outline of a leaf on 2 mm square graph paper.

i Calculate the surface area of one side of the leaf.

ii Suggest why your calculation cannot be completely accurate.

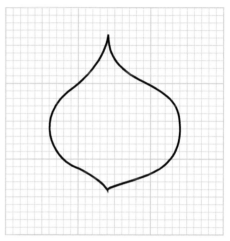

▲ **Fig A7.37**

7 Animals can also be copied by using the artificial method called cloning.
Dolly the sheep was the first animal to be produced by cloning. Cloning can also produce stem cells.
Explain how stem cells may be used in the future for each of the following.

a Growing new tissue.

b Treating diseases such as diabetes.

8 After pollination and fertilisation, fruits develop to help disperse the seeds.
Name three examples of different types of fruit.
In each case say how the fruit develops and how the seeds are dispersed.
Fruit 1
Fruit 2
Fruit 3

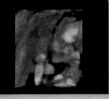

By the end of this unit you will be able to:

- describe the process of sexual reproduction in humans.
- describe the menstrual cycle.
- discuss fertilisation, development of the fetus and birth.
- discuss the advantages and disadvantages of various methods of birth control.
- discuss the importance of pre- and post-natal care of mothers and babies.
- compare growth patterns in selected organisms.
- discuss the need for human population control.

spermatozoa/sperm – the male sex cell in animals

ovum – the female sex cell in animals

testis – the male organ that produces sperm

seminiferous tubule – where sperm are produced

epididymis – where sperm are stored

prostate gland – the gland that makes fluid that mixes with sperm to make semen

seminal vesicles – the gland that stores the fluid that mixes with sperm to make semen

Sexual reproduction in humans

Unlike some flowers, the sexes in humans are always separate, male and female. In humans, the male gamete is called a **spermatozoa** or sperm for short and the female gamete is called an **ovum**. The sperm is transferred from the man to the woman during sexual intercourse and fertilisation takes place inside the woman.

The male reproductive system

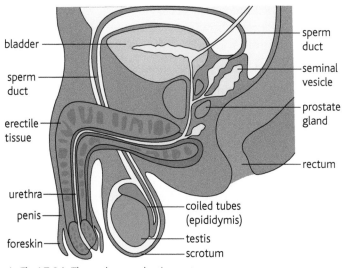

▲ **Fig A7.24** The male reproductive system

Sperm are produced in the **testis**. The testis consists of coiled tubules called **seminiferous tubules** in which the sperm are produced. Mature sperm pass into coiled tubules in the **epididymis** where they are stored until used during intercourse. During intercourse the sperm pass along the sperm duct and mix with fluid from the **prostate gland** and **seminal vesicles**. The fluid provides food for the sperm, starts them swimming and contains an alkali to neutralise

vagina – part of the female sex organs where sperm are deposited during sex

penis – the male sex organ

ejaculate – semen containing sperm that is released from the penis during sex

the acid conditions found in the female **vagina**. The sperm are then ejaculated from the erect **penis** and into the vagina of the female. This fluid is called the **ejaculate**.

Q1 Some animals such as frogs, simply release eggs and sperm into the water. Suggest why releasing the sperm inside the woman's body means that fertilisation is more likely to happen.

Female reproductive system

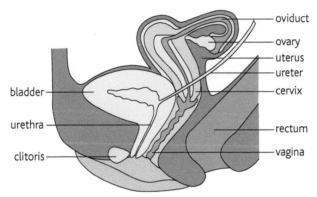

▲ **Fig A7.25** The female reproductive system

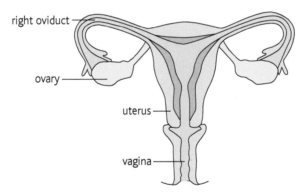

▲ **Fig A7.26** The female reproductive system

oviduct – the tube that carries the ovum from the ovary to the uterus

uterus – the female sex organ where a baby develops during pregnancy

menstruation – the process of losing blood and tissue from the vagina every 28 days

menstrual cycle – a 28-day cycle during which the lining of the uterus breaks down and then re-grows ready to receive a fertilised ovum

Ovum or eggs are produced in the female ovary. An ovum is released approximately once every 28 days, alternating between the two ovaries.

The ovum passes into the **oviduct** and microscopic hairs called cilia push the ovum along the oviduct and into the **uterus**.

If the ovum has not been fertilised, the wall of the uterus then breaks down and a process called **menstruation** starts.

The menstrual cycle

The **menstrual cycle** starts in human females at about 12 years of age. Each cycle lasts for about 28 days. It begins on Day 1 with the breakdown of the uterus wall. During menstruation, tissue and blood from the uterine wall is shed and passes out of the vagina. This lasts for about four or five days. The uterus wall then starts to rebuild the lost tissue and grow more blood vessels ready for the next ovum in case it has been fertilised.

The ovum is released on about Day 14 of the cycle. If the ovum is not fertilised, the cycle begins all over again on about Day 28.

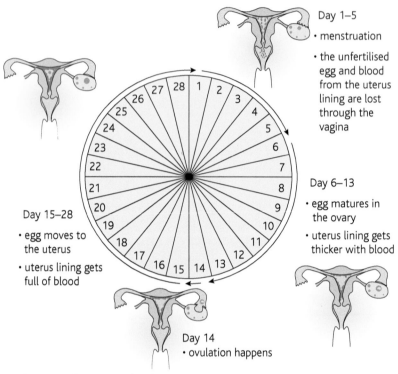

▲ **Fig A7.27** The menstrual cycle

Q2 Menstruation is controlled by hormones. Suggest how the hormones get from the pituitary gland at the base of the brain to the uterus.

The control of the menstrual cycle

oestrogen – repairs and thickens the lining of the uterus

progesterone – prevents menstruation from starting

The sex hormones **oestrogen** and **progesterone** control the menstrual cycle. Oestrogen repairs and thickens the lining of the uterus. Progesterone thickens it even more and prevents menstruation from starting. It is the changing levels of these two hormones in the blood that control menstruation.

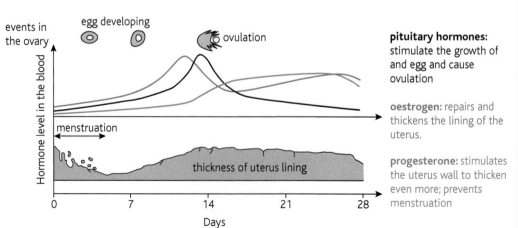

▲ **Fig A7.28** How hormone levels affect the uterus lining

Fertilisation

cervix – the opening into the uterus from the vagina

zygote – a fertilised ovum

embryo – a zygote that has started to divide into a ball of cells

If intercourse has taken place, the sperm deposited in the vagina will have swum up through the **cervix** and into the uterus. Some of the stronger sperm will continue to swim up the oviduct ready to meet and fertilise the ovum. The fertilised ovum or **zygote** then starts to divide. By the time it reaches the uterus it consists of a ball of cells called an **embryo**, as shown in Fig A7.29. The embryo then implants itself inside the wall of the uterus and is surrounded by blood vessels that can supply it with nutrients and oxygen. The menstrual cycle now stops and for the next nine months the embryo develops until the baby is ready to be born.

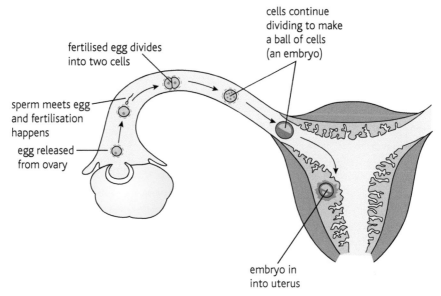

cells continue dividing to make a ball of cells (an embryo)

fertilised egg divides into two cells

sperm meets egg and fertilisation happens

egg released from ovary

embryo in into uterus

▲ **Fig A7.29** Fertilisation

Menopause

The menopause is when menstruation no longer happens and the woman is not capable of becoming pregnant. It happens because the ovaries have far fewer eggs and the level of oestrogen production falls. It can occur most commonly between the ages of 48 to 55. The fall in the level of oestrogen can also cause hot flushes and insomnia in some women.

Development

fetus – the name giving to the growing baby in the uterus

placenta – the organ that is attached to the uterus and through which the fetus gets its nutrients and oxygen

umbilical cord – the cord connecting the baby to the placenta

amniotic membrane – the membrane that surrounds the growing baby and contains the amniotic fluid

amniotic fluid – a fluid that protects the growing baby

By eight weeks the embryo looks like a tiny human being. It is now called a **fetus**. It quickly grows a **placenta**, which is attached to the wall of the uterus. Oxygen and nutrients pass from the mother's blood into the blood of the fetus by crossing the placenta. The blood then passes along the **umbilical cord** and into the fetus. Waste materials such as carbon dioxide and urea pass back to the placenta and cross into the mother's blood. The blood supply of the mother and fetus are kept separate but are in close proximity to each other, allowing exchange of materials between the two.

The fetus is protected by being inside a fluid-filled bag called the **amniotic membrane**. The fluid is called **amniotic fluid**.

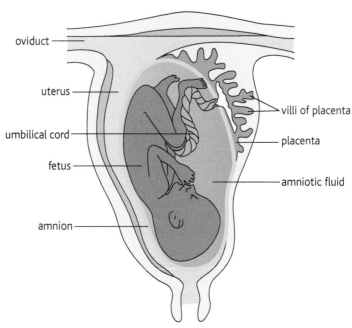

▲ **Fig A7.30** A developing baby

Q3 Suggest why it is not a good idea for the mother of the unborn baby to drink alcohol and smoke.

Birth

Shortly before birth the baby turns so that the head is facing in a downwards position. Just before birth the amniotic membrane breaks. The fluid is released through the vagina and the baby is ready to be born. Muscles in the uterus begin to contract and push the baby downwards. This is called **labour**. Shortly afterwards the baby is born. The placenta remains attached to the uterus wall and the doctor or nurse ties and cuts the umbilical cord, which still attaches the baby to the placenta. After about half an hour the placenta also passes out with the rest of the umbilical cord. This is called the **afterbirth**.

labour – the process of giving birth

afterbirth – the placenta and umbilical cord that is released from the mother's body after the baby has been born

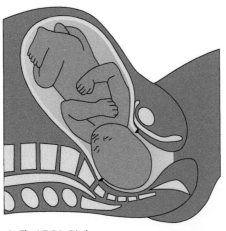
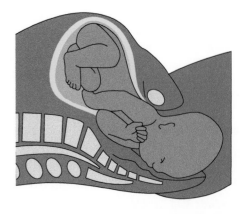

▲ **Fig A7.31** Birth

Pre- and post-natal care

To ensure the safe delivery of the baby, it is important that the mother is carefully monitored both during the pregnancy and after the baby is born.

Pregnant mothers usually attend pre-natal clinics where doctors and nurses monitor the mother's health and the development of the baby. The mother will get advice about what foods to eat and the importance of a healthy diet.

Some of the checks carried out on the mother include:

- A pregnancy test to confirm that the mother is pregnant.

- Blood tests to check to see if the mother is rhesus negative (see Unit A4a) and what blood type she is or if she is anaemic (lacking red blood cells). The blood is also checked for various diseases such as HIV and hepatitis.

- A urine test to check for glucose in case the mother is diabetic.

- An ultrasound scan to check on the development of the baby (see Fig A7.32). The ultrasound scan will provide all of the following information.

  1 The due date of the baby.

  2 Fetal heartbeat and movement.

  3 Whether it is a multiple pregnancy.

  4 Whether there are any abnormalities such as heart defects.

  5 Whether the amniotic fluid surrounding the child is adequate.

  6 The position of the placenta and the embryo.

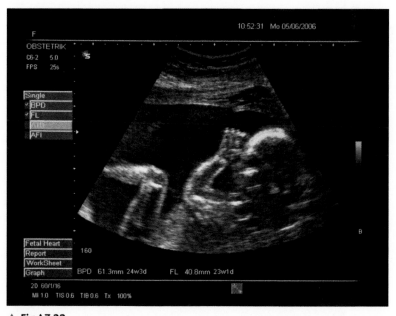

▲ **Fig A7.32**

- Blood pressure test to check that the mother does not have high blood pressure or pre-eclampsia. If left untreated pre-eclampsia can result in poor growth of the baby and, in serious cases, the death of the mother.

- Tests to check that the baby does not have any serious genetic defects.

Pregnant mothers also need to be aware of any risks they face when they are pregnant. Mothers who come into contact with rubella (German measles) run the risk of their baby being born deformed, deaf or suffering brain damage. For this reason, young girls may be vaccinated against rubella (even though it is usually a mild disease) so that later in life when they are pregnant, their unborn child is not at risk.

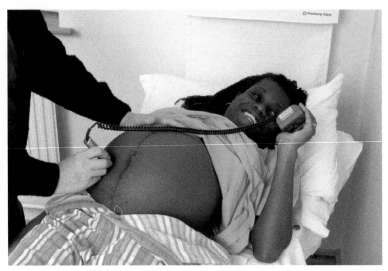

▲ **Fig A7.33** A mother-to-be having a check-up

SBA Skills

ORR	D	MM	PD	AI
✓	✓			

To do – Caring for the unborn child

Collect leaflets about pre- and post-natal care from your local health centre. Use the leaflets to produce a poster explaining to mothers-to-be what they should do to care for their unborn child.

Once the baby has been born the mother and baby still attend the clinic for post-natal checks. These are to see if the baby is thriving and growing in a healthy way. The baby will be weighed and the weight compared against a chart that shows the ideal weight range for a baby of that age and gender.

Mothers will also get advice on the advantages of breastfeeding and having the baby immunised against different diseases. Breast milk not only contains all the minerals and nutrients that the growing baby needs, it also contains antibodies from the mother that will give the baby protection against diseases in the first few months of its life.

To do – Snappy phrases

You may have heard the expression 'breast is best'. Try to think up some other different snappy phrases that could be used to encourage mothers to breastfeed their babies.

Pregnant mothers and babies are very rarely given drugs or X-rays. This is because the baby's body is growing very rapidly, and rapidly dividing cells are much more susceptible to drugs and X-rays. This is why ultrasound is used to examine unborn babies. The sound waves do not harm the baby. Ultrasound is often called a non-invasive procedure.

Q4 Suggest what 'non-invasive' means.

Birth control

Not every child is wanted and teenage parents may not always be able to give the baby the love and care that it needs. Unwanted pregnancies can be prevented by using the correct method of birth control. Some methods are more reliable than others and some methods are not very safe at all. Table A7.1 show just how reliable some of these methods are.

Green is very reliable, orange quite reliable and red unsafe.

method	advantages	disadvantages
not having sex	100% effective	can be very frustrating
condom – acts as barrier preventing sperm entering the vagina	reliable if used properly and give some protection against some sexually transmitted diseases	may reduce enjoyment and sensitivity
diaphragm – acts as barrier by preventing the sperm entering the cervix	reliable if used with a spermicide to kill sperm	awkward to use and can impede spontaneity
contraceptive pill – stops the ovum being released from the ovary	very reliable	need to remember to take it and it may have side effects not effective when taken with some antibiotics
IUD (intrauterine device) – stops the embryo from implanting and developing in the uterus	quite reliable	may interfere with menstrual cycle
rhythm method – only have sex during the times of the month when the sperm cannot fertilise the ovum	natural and no devices or drugs used	very unreliable as monthly cycle may be irregular
sterilisation – usually done to the male to cut the sperm duct so that sperm cannot reach the penis. It is called a vasectomy. In women it is called a tubal ligation	once done no more worries!	is not often reversible so you cannot change your mind later

Table A 7.1

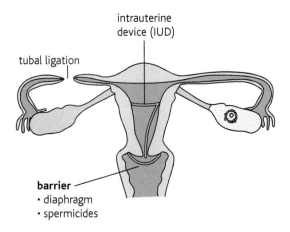

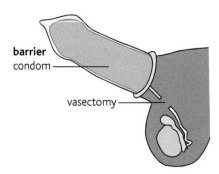

▲ **Fig A7.34** Some methods of birth control

Human population

The human population is doubling every forty to fifty years. This may not sound a lot but an increase on this scale is not sustainable on planet Earth.

> **To do – Bigger than you think**
>
> **1** Starting with the number 1, guess what number you would have if you doubled it twelve times.
>
> **2** Now do the sums and double number 1 twelve times, e.g. 1 – 2 – 4 – 8 etc.
>
> Surprised by the result? Just imagine what will happen to the Earth's population if it doubles every forty years. This type of growth where a population continues to double in the same period of time is called *exponential growth*.

SBA Skills

ORR	D	MM	PD	AI
		✓		

In reality the Earth's population cannot continue to grow at this rate. Either humans must begin to limit their own numbers or the human race will run out of resources such as food, fuel (oil and coal), and raw materials such as rare metals. When this happens famine and disease or the breakdown of society will reduce the numbers of the population to a more sustainable level.

Q5 Look at Fig A7.35. During which century has most of the growth of the human population taken place?

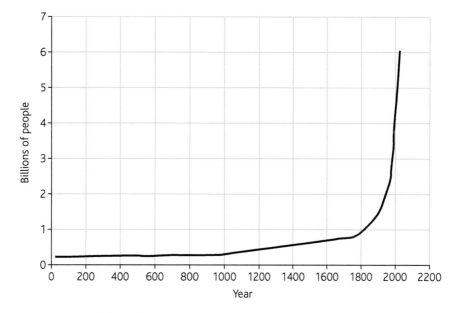

▲ **Fig A7.35** World population

If you have access to a computer, log onto the website http://math.berkeley.edu/~galen/popclk.html. It gives an estimate in real time of how the world's human population is increasing.

End-of-unit questions

1 Which of the following is not found in females?

a ovary

b fallopian tube

c prostate

d uterus

2 Which of the following is the most effective method of birth control?

a condom

b coil

c spermicide jelly

d the pill

3 Which of the following is the correct path taken by a sperm in the human male reproductive system?

a epididymis – seminiferous tubules – urethra – sperm duct

b seminiferous tubules – epididymis – sperm duct – urethra

c sperm duct – seminiferous tubules – urethra – epididymis

d urethra – sperm duct – epididymis – semiferous tubules

e epididymis – sperm duct – seminiferous tubules – urethra

4 Fig A7.36 shows the 28 days of the human female menstrual cycle.

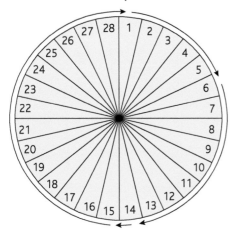

▲ **Fig A7.36**

a Make a simple copy of the chart and add the following data.
  i start of menstruation
  ii ovulation

b An ovum can only survive for about four days after it is released from the ovary if it is not fertilised. Human male sperm can also live for up to four days after being ejaculated into the vagina. Use this information to shade in your chart in red to show the time of the cycle when sexual intercourse is most likely to result in pregnancy.

c Describe three different ways in which pregnancy can be prevented. Explain the reliability of each method and how it works.

5 The contraceptive pill may contain the hormones oestrogen and progesterone.

a Suggest a mechanism to explain how the contraceptive pill may work.

b Testosterone is the human male sex hormone.
   Suggest why it is not possible to make a similar contraceptive pill for males using the hormone testosterone.

6 Some aquatic animals simply release their sperm and eggs into the water and fertilisation occurs by chance. Some amphibians release their eggs and then lay the sperm on top of the eggs. Birds have internal fertilisation and retain the developing embryo inside a protective shell. Humans have internal fertilisation and retain the embryo for nine months after which the parents nurture the child for many more years.
Suggest why humans expend so much effort in the care of their offspring.

By the end of this unit you will be able to:

- describe the methods of heat transfer and their applications.
- explain how thermostatically controlled household appliances operate.
- describe the features of thermometers and how they work.
- explain the cooling effect of evaporation.
- explain the effects of temperature and relative humidity on body functions.
- explain the need for proper ventilation.

Heat and temperature

heat – a form of energy

Heat is a form of energy. The faster molecules move, the more heat energy they have. The slower they move, the less heat energy they have. Heat is a very useful form of energy when there is a lot of it concentrated in one place such as when fuel burns. However, as it spreads out it becomes much less useful. **Fuels** are a useful concentrated form of chemical energy that can be converted into heat when they burn.

fuel – a useful concentrated form of chemical energy that can be converted into heat when burned

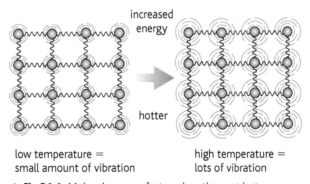

increased energy

hotter

low temperature = small amount of vibration

high temperature = lots of vibration

▲ **Fig B1.1** Molecules move faster when they get hot

temperature – a measure of how hot something is

It is easy to get temperature and heat confused, but they are two different things. **Temperature** is a measure of how hot something is. We measure temperature by using a scale such as the Celsius scale. 0°C is the temperature at which water freezes. 100°C is the temperature at which water boils.

A man called Kelvin invented the Kelvin scale for measuring temperature. It had the same divisions as the Celsius scale but started at −273°C with 0 K. In theory, molecules stop moving and have no heat energy at −273°C. Scientists call this temperature **absolute zero**.

absolute zero – the lowest possible temperature, −273°C

Q1 What temperature would the Kelvin scale read at 0°C?

SBA Skills

ORR	D	MM	PD	AI
✓		✓		✓

Experiment – The difference between heat and temperature

1 Get a beaker of water that is at room temperature. Place a thermometer in the beaker of water and measure its temperature.

⚠ **SAFETY:** Thermometers are expensive and break easily, leaving sharp slivers of dangerous glass.

2 Carefully heat a pin in a Bunsen flame until it is glowing red hot and then drop the pin into the beaker of water. When the pin has cooled down measure the temperature of water in the beaker again. Calculate by how much the water has increased in temperature.

3 Get a new beaker of water at room temperature and measure its temperature with your thermometer.

4 Carefully heat a large ball bearing in an oven set to 60°C. Carefully place the hot ball bearing into the beaker of water. Leave for about one minute then record the temperature of the water. Calculate by how much the water has increased in temperature.

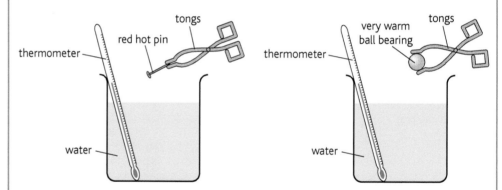

▲ **Fig B1.2** The difference between heat and temperature

You should find that even though the pin was at a much higher temperature than the ball bearing, it was the water heated by the ball bearing that showed a greater increase in temperature. This is because the ball bearing was not as hot as the pin, but being larger, it contained more heat energy.

Q2 Which do you think contains more heat, and which has the higher temperature, a swimming pool full of warm water, or a cup of hot coffee? Explain your answer.

Heat transfer

heat transfer – heat being transferred from place to place through solids, liquids and gases

Heat can be transferred from place to place. This can happen in all three different states of matter – solids, liquids and gases. Scientists call this **heat transfer**. Heat tends to transfer from where there is a lot of it to where there is not much of it.

conduction – a form of heat transfer in solids

convection – a form of heat transfer in liquids and gases

radiation – a form of heat transfer by electromagnetic radiation through empty space

Heat can be transferred in three different ways: by **conduction**, by **convection** and by **radiation**.

Conduction

Conduction is a form of heat transfer that takes place in solids. It happens because molecules that are hot vibrate more than cold molecules. The hot molecules vibrate and bump into the colder molecules next to them causing the colder molecules to vibrate more. This is how the heat is passed from molecules to molecule, moving from a hot area to a colder area.

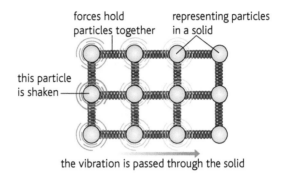

forces hold particles together

representing particles in a solid

this particle is shaken

the vibration is passed through the solid

▲ **Fig B1.3** Conduction of heat

To do – A model of how conduction works

SBA Skills

ORR	D	MM	PD	AI
				✓

1 Get ten students to stand in a line very close to each other.

2 The first student then starts 'shaking about' or vibrating. This represents heat energy. The vibrations soon get passed along the line of students until the last student in the line is also 'vibrating'.

▲ **Fig B1.4** How conduction of heat works

You should notice that the last student in the line is 'vibrating' less than the first student in the line. This is because some of the energy has been lost as it was passed from student to student.

conductors – a solid, such as a metal, that is good at transferring heat

insulator – a substance that is not good at transferring heat

Some solids are very good at transferring heat by conduction. They are called **conductors**. Metals are good examples of conductors. Some solids are not very good at transferring heat by conduction. They are called **insulators**. Wood, wool, animal fur, polystyrene and fibre glass are all good insulators.

Q3 You are building a house. You want to stay cool in the summer and warm in the winter. What type of material would you use to build it, a metal conductor or an insulator such as wood and fibre glass? Explain your answer.

To do – Showing conduction of heat in different materials

SBA Skills

ORR	D	MM	PD	AI
✓				✓

1 Place a stick of wood and a metal rod so that one end of each is in a beaker of hot water.

2 Carefully touch the other end of the metal rod and stick to see which one warms up first.

Q4 What did you discover? Explain your answer.

Convection

convection current – when hot liquids and gases rises while cold liquids and gases fall

Convection is a type of heat movement that takes place in liquids and gases but not in solids. It happens because hot liquids and gases rise and cold liquids and gases fall. This movement in liquids and gases is called **convection currents**. Have you noticed when you open a fridge door that the cold air falls out of the fridge and you can feel it on your feet but not on your head? And when you open an oven door you feel the heat on your face but not on your feet? This is convection in action.

SBA Skills

ORR	D	MM	PD	AI
✓				

Experiment – How to show convection currents

1 Fill a beaker of water two-thirds full. Leave the beaker to settle until all of the water in the beaker has stopped moving.

2 Carefully drop a crystal of coloured dye such as potassium manganate(VII) or copper sulphate into the water.

3 Carefully place a small flame from a Bunsen burner or candle under the beaker at the point where the crystal is.

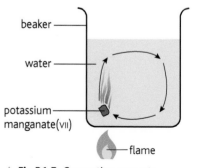

▲ **Fig B1.5** Convection currents

You should be able to see, by following the path of the coloured dye, that at the point where the water is heated it starts to rise and cold water starts to fall to take its place.

Convection currents happen because as the molecules get hotter they move about more. This causes them to spread out. As they spread out the liquid or gas becomes less dense. This means it will float on denser liquids or gases, so it starts to rise. The denser liquid or gas starts to fall.

Convection currents are found all around us. Our weather is produced because of convection currents. You may have noticed that on Caribbean islands the breeze during the day is often a cooling breeze from the sea. But at night the breeze is a warm breeze blowing off the land towards the sea. This happens because of convection currents in the atmosphere.

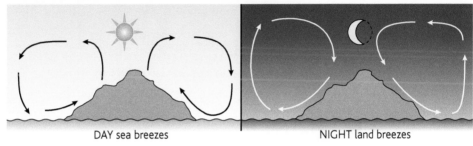

DAY sea breezes   NIGHT land breezes

▲ **Fig B1.6** Sea and land breezes

During the day, the Sun warms the land more than it warms the sea. As the land heats up, convection currents develop, and warm air above the land rises into the atmosphere. This air is replaced by colder air above the sea, blowing onto the land to replace the warm air that has risen.

At night, the land cools down faster than the sea. As soon as the sea is warmer than the land, the air above the sea starts to rise by convection. The breeze then moves from the land and towards the sea.

Q5 If you were a glider pilot looking for lift, would it best to fly over land or sea during the day? Explain your answer.

SBA Skills

ORR	D	MM	PD	AI
✓				✓

Experiment – Boiling ice

1 Get a large test tube or boiling tube and fill it two-thirds full of water.

2 Wrap an ice cube in some metal gauze and place it in the test tube.

3 Use a Bunsen burner to boil the water in the top of the test tube. If you do this carefully you should be able to have ice at the bottom of the test tube and boiling water at the top of the test tube.

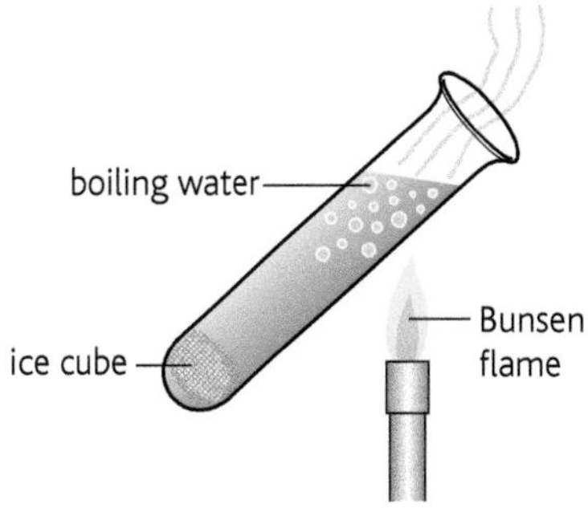

boiling water

ice cube

Bunsen flame

▲ **Fig B1.7** Ice and boiling water

Q6 Suggest why the ice did not melt in the boiling water.

Radiation

Radiation is very different from conduction and convection in that it does not need molecules to transfer the heat. The heat is transferred by an electromagnetic wave and can travel through empty space. When you stand in front of a bonfire on the beach you can feel the heat from the fire on your face. This is radiated heat.

▲ **Fig B1.8** The radiated heat can be felt from the bonfire

Radiated heat is best absorbed by dark surfaces and reflected by shiny surfaces.

SBA Skills

ORR	D	MM	PD	AI
✓			✓	✓

To do – What material absorbs heat the best?

1 Place two sheets of paper in the sunshine, one black and one white. Place a sheet of aluminium foil next to them.

2 Leave all three sheets for about half an hour.

3 Feel each of the three sheets to see which is the hottest and which is the coldest.

Q7 Suggest why the black sheet is the hottest.

Three students were asked beforehand which sheet they thought would be the coldest.

I think foil will be the hottest because it is a metal and metals are good conductors

I think the white will be the coldest because it reflects the radiated heat

I think foil will be the coldest because it is shiny and will therefore absorb the heat

Mary   **Susan**   **Deepak**

Q8 Which of the three students gave the correct answer? Explain why they were correct.

Q9 If you were making a solar panel to trap heat energy what colour would you paint it? Explain your answer.

Q10 Suggest which form of heat transfer is used for the Earth to receive heat energy from the Sun. Explain your answer.

Thermostats and thermometers

expand – to get bigger

contract – to get smaller

When objects get hot they **expand**. When they cool down they **contract**. But some substances, such as metals, expand and contract more than others.

Materials expand when heated because as the molecules move more rapidly they take up more space.

Look at the picture below. It shows four people dancing. In this picture they are dancing slowly. They take up little space.

In the next picture people are dancing quickly. They take up much more space. People do not get bigger when they dance more quickly. They just take up more space.

▲ **Fig B1.10** Hot molecules take up more space

SBA Skills

ORR	D	MM	PD	AI
✓				✓

Experiment – Expansion and contraction of metals

1 Set up the apparatus as shown in Fig B1.11.

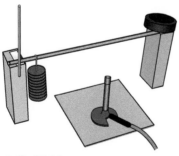

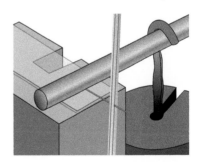

▲ **Fig B1.11**

2 Rest the metal rod on the two wooden supports and fix one end by resting a heavy weight on it. The other end of the rod is resting on a needle on a glass microscope slide. This end of the rod is able to move. The needle has a straw attached to it to show any movement in the metal rod. As the rod is heated and expands, the needle rotates and the straw moves counter clockwise. When the metal rod is allowed to cool and contract the straw moves in a clockwise direction.

Q11 How could you use this method to show that different types of metal expand and contract by different amounts when heated and cooled?

The principle of materials expanding when heated and contracting when cooled is responsible for making some **thermostats** and **thermometers** work.

Thermostats

thermostat – a device that can maintain a specific temperature by switching on or off an electrical circuit connected to a heater or cooling system

thermometer – a piece of apparatus for measuring temperature

bi-metal strip – a strip consisting of two different metals that expand by different amounts when heated, causing it to bend. It can be used to switch electrical circuits on or off.

Thermostats are devices that switch circuits on or off depending on the temperature of the surroundings. A simple type of thermostat uses a **bi-metal strip**. The strip consists of two different metals such as steel and copper, which are joined together along their whole length. Copper expands more than steel when heated. This causes the bi-metal strip to bend as one side of the strip expands more than the other. When the strip cools, it bends back to normal.

▲ **Fig B1.12** The bi-metal strip bends as one metal expands more than the other metal when heated

Q12 Which side of the bi-metal strip in the pictures in Fig B1.12 is made of copper, and which made of steel, top or bottom? Explain your answer.

The bi-metal strip can be used to make or break electrical circuits. Look at the picture of the electrical iron in Fig B1.13.

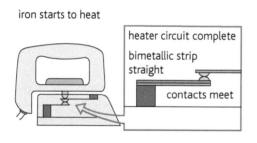

iron starts to heat

heater circuit complete

bimetallic strip straight

contacts meet

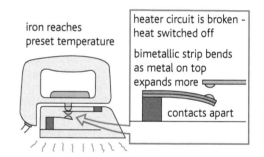

iron reaches preset temperature

heater circuit is broken – heat switched off

bimetallic strip bends as metal on top expands more

contacts apart

▲ **Fig B1.13**

When cold, the bi-metal strip is straight and the iron can be switched on. As the iron heats up, the strip bends and breaks the electrical circuit. This switches the iron off. As it cools the strip straightens and once more completes the circuit, switching the iron on again. This switching on and off enables the iron to stay at a constant temperature, without getting too hot or too cold.

The same process is used in many household items where it is important that the temperature is maintained at a constant level. Examples include electric cookers, fridges and room heating thermostats. Even gas ovens contain a bi-metal strip in their thermostat. In a gas oven the thermostat operates a device that opens or closes a gas valve to regulate how much gas is being burned. If the oven gets too hot the valve is closed down and less gas is burned so the oven cools. The reverse happens to heat the oven up.

Q13 Think about the size of the gap between the bi-metal strip and its contact. Suggest how this could be used to set the thermostat to switch the device on or off at a fixed temperature.

Thermometers

Thermometers also work by expansion and contraction. But in thermometers it is expansion and contraction of liquids instead of solids. Thermometers are used to measure the temperature of their surroundings. The liquid in the thermometer can be coloured water, coloured alcohol, or even the metal mercury that is a liquid at room temperature.

Alcohol is cheaper than mercury and remains a liquid at temperatures as low as −114°C. Mercury, however, has a higher boiling point and remains a liquid at temperatures over 300°C. So the choice of whether to use water, alcohol or mercury depends upon the range of temperatures for which the thermometer is going to be used.

Q14 Suggest why the water or alcohol in a thermometer is coloured.

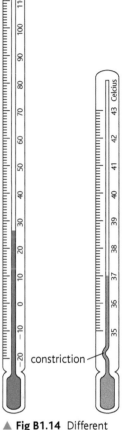

▲ **Fig B1.14** Different types of thermometers

laboratory thermometer – a thermometer used in laboratories that usually measures temperatures between -10°C and 110°C

clinical thermometer – a thermometer used to measure the temperature of the human body

digital thermometer – a thermometer that produces a digital readout

Three common types of thermometer are the **laboratory thermometer** the **clinical thermometer** and the **digital thermometer**.

Laboratory thermometer

The laboratory thermometer is a sealed glass tube containing a reservoir or bulb full of liquid.

When the liquid is heated it expands. This causes the liquid to move up from the bulb and into the sealed tube. The hotter its surroundings, the more it expands and the more it moves up the tube. The glass tube has a scale printed along it showing the temperature of the thermometer. This is set at 0°C (freezing point of water) and 100°C (boiling point of water). The distance between these two points is then divided into 100 equally sized spaces. Because the amount the liquid expands at a given temperature is always the same, the scale on the thermometer always reads the correct temperature. Most laboratory thermometers work over a temperature range from −10°C to 110°C.

Q15 Suggest what would happen if you put the bulb of a laboratory thermometer into a Bunsen flame.

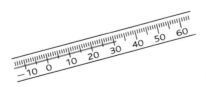

◄ **Fig B1.15** A laboratory thermometer

Q16 Look at the picture of the laboratory thermometer in Fig B1.15. What temperature is the thermometer reading?

Clinical thermometer

The clinical thermometer is designed to measure the temperature of the human body. Body temperature should be about 37°C but when we are ill the temperature can change. The thermometer is normally placed into the mouth or under the armpit to measure the body's temperature. The clinical thermometer is like the laboratory thermometer but with two important differences.

1 It is made to work over a much narrower temperature range around 37°C. This means it can be more accurate in measuring the human body temperature.

2 The thermometer has a constriction in the tube through which the liquid passes. This is to stop the liquid contracting back down into the bulb when it cools down after being taken out of the mouth. Otherwise we could only read the temperature scale while the thermometer was still in the mouth. The thermometer has to be reset each time it is used by shaking it hard to force the liquid back down into the bulb.

Look at the picture of the clinical thermometer in Fig B1.16.

▲ **Fig B1.16** A clinical thermometer

Q17 What temperature is the thermometer reading? Is the person ill or well?

The digital thermometer

The digital thermometer produces a digital readout on a screen. It is much easier to use than alcohol and mercury thermometers.

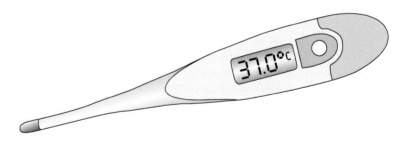

▲ **Fig B1.17** A digital thermometer

maximum and minimum thermometer – a thermometer used to measure the maximum and minimum temperatures over a period of time

Maximum and minimum thermometers

We are not always there to read the temperature shown by a thermometer. It is also useful to know what the highest and lowest temperatures were recorded over a period of time, for example over 24 hours. To find this out we would use a **maximum and minimum thermometer**.

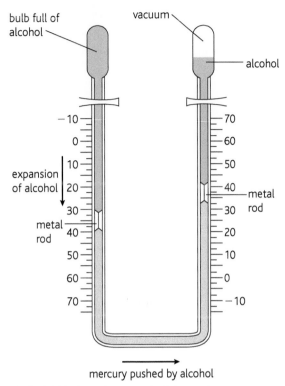

▲ **Fig B1.18** A maximum and minimum thermometer

The thermometer consists of a U-shaped tube with a glass bulb at both ends, as shown in Fig B1.18. One side of the U measures the minimum temperature and the other side measures the maximum. The bulb above the minimum side is full of alcohol and the bulb above the maximum side has a vacuum.

The alcohol is really the liquid that measures the temperature. It expands and contracts as the temperature changes and pushes the mercury along the tube. It is the position the mercury is pushed to that indicates the temperature on both of the scales.

As the mercury moves it pushes two metal rods. They record the furthest point that the mercury is pushed to in each of the maximum and minimum tubes. Once the temperatures have been read, the metal rods can be reset using a small magnet.

Q18 Look at the picture of the maximum minimum thermometer in Fig B1.18. What were the maximum and minimum temperatures since the thermometer was last reset?

Cooling by evaporation

Our body temperature is about 37°C. Whatever the temperature outside our bodies, our body temperature remains constant. Even in hot parts of the world like the Caribbean, when the air temperature goes above 37°C, our body temperature does not change.

We can keep our body temperature constant because our body cools itself by sweating. When it is very hot we can sweat several litres of fluids each day. It is important that we replace this lost fluid by drinking lots of liquid when it is hot.

Latent heat of vaporisation

evaporate – to turn from a liquid into a gas

latent heat – heat used by molecules to turn from a liquid into a gas

When water **evaporates**, it changes state from a liquid to a gas. Even though the liquid water and the water vapour are at the same temperature, the water vapour contains more energy. This energy allows the water molecules to break free of the water and become a gas. This hidden heat energy is called **latent heat**. As the water evaporates, the molecules absorb the heat energy by conduction from their surroundings, making the surroundings feel cold.

This is why even on a hot day, you feel cold when you get out of the swimming pool. The wet water on your body evaporates and uses heat from your body to give the water molecules enough energy to turn into a gas.

Liquids that evaporate faster than water feel even colder.

SBA Skills

ORR	D	MM	PD	AI
✓		✓		✓

▲ **Fig B1.19** The energy for water molecules to leave the liquid and evaporate into a gas is called latent heat and is absorbed from the surroundings

Experiment – Which feels colder – water or alcohol

1 Place a drop of water and a drop of alcohol on the back of your hand. Even though they were both at room temperature, which feels colder?

The alcohol should feel colder as it evaporates faster than water so absorbs heat energy more quickly.

Let's see just how cold it can get.

2 Place about 1 cm depth of alcohol in a small beaker. Place the beaker in a puddle of water on your desk. Now blow air through a straw so that it bubbles through the alcohol. This will make the alcohol evaporate even more quickly. Take care not to breathe in the alcohol vapour.

Within a couple of minutes you should find that your beaker has become frozen to your desk.

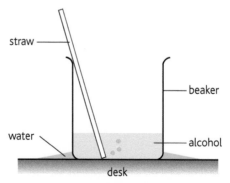

▲ **Fig B1.20** Evaporating alcohol

Q19 Explain how the beaker became frozen to the desk.

Scientists often use models to simplify how things work and make them easier to understand. We can make a model to show how sweating cools the body down.

SBA Skills

ORR	D	MM	PD	AI
✓		✓		✓

Experiment – A model to show how sweating works

1 Set up two flasks as shown in Fig B1.21. Both should contain the *same amount* of hot water at the *same temperature*. One should be covered in dry cotton wool and one in wet cotton wool.

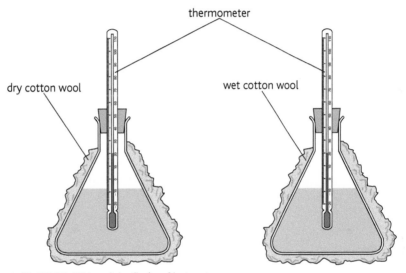

▲ **Fig B1.21** Wet and dry flasks of hot water

2 Record the temperature of each flask every minute. Use a table like this to record your results.

time / minutes	temperature of water in flasks / °C	
	dry cotton wool	wet cotton wool
1		
2		
3		

3 Plot your results as a graph with 'time in minutes' on the *x*-axis.

Q20 What can you conclude from your investigation?

Q21 Explain why both flasks contained the same volume of water at the same temperature.

Staying cool is not always easy

You may have noticed that it is not always so easy to keep cool. It all depends on how much water vapour is in the air. This called the **humidity** of the air. Humidity is often expressed as a percentage called **relative humidity**. Relative humidity is the amount of water vapour in the air as a percentage of how much water vapour the air can hold. The higher the relative humidity, the slower the evaporation rate of our sweat. When the air is dry the sweat evaporates very quickly and we feel cool. When the air is humid the sweat evaporates slowly and we feel 'hot and sticky'.

It is not just humidity that affects how quickly sweat evaporates. Wind speed and temperature also affect evaporation.

Wind speed – the windier it is, the faster sweat evaporates as the water molecules get blown away. When the air is still the water molecules can stay close to the skin, preventing more water molecules from evaporating.

Temperature – the higher the temperature, the more kinetic energy the molecules have, which speeds up the rate of evaporation.

Q22 Using what you know about evaporation, suggest what weather conditions would make for a good drying day after you have washed some clothes.

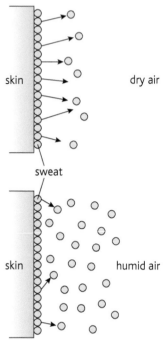

▲ **Fig B1.22** Sweat evaporates faster on dry days

▲ **Fig B1.23** Sweating helps us maintain body temperature efficiently

When we exercise we generate heat. During respiration, glucose and oxygen react together to provide us with energy. Some of this energy is released as heat. This heat has to be lost from the body to keep our body temperature at 37°C. This is why we sweat when we exercise. It is to get rid of the extra heat generated as our metabolic rate increases.

Q23 Suggest one way we could keep warm when we are cold without wearing more clothes or turning up the heating.

humidity – how much water vapour is in the air

relative humidity – the amount of water vapour in the air expressed as a percentage of how much water vapour the air can hold

SBA Skills

ORR	D	MM	PD	AI
✓		✓	✓	✓

Experiment – Investigating factors that affect drying

1 Get three pieces of the same type and size of cloth. Add the same amount of water to each piece.

2 Place one in the warm, one in windy conditions and the last one in cold, damp conditions.

3 See which one takes the longest to dry. You will need to find a way of comparing the dryness of each piece of material.

Explain how you managed to compare the dryness of each piece of material.

Ventilation

Ventilation is important. It provides us with breathable air at a comfortable temperature and humidity. Modern buildings often use humidifiers and air conditioners to maintain a comfortable environment. Buildings that are not ventilated properly can cause illness in the people that work in them. This is sometimes called 'sick building syndrome'. Good ventilation ensures that buildings do not smell stale and that dust and pollutants are removed.

humidifier – a device that adds water vapour to the air in a room or building to make it more humid

Humidifiers put water vapour into the air when the air is too dry. Even though it is easier to stay cool in dry air, some people find dry air uncomfortable. Humidifiers evaporate water and return the humidity to a more comfortable level.

air conditioner – a device that removes water vapour from the air and also controls the temperature of the air in a room or building

Air conditioners take water vapour out of the air when it is too humid and also regulate the temperature of the air by producing warmed or cooled air.

▲ **Fig B1.24** An air conditioner

The design of the building is also important to produce adequate ventilation. In our homes we use windows and chimneys to ventilate the rooms.

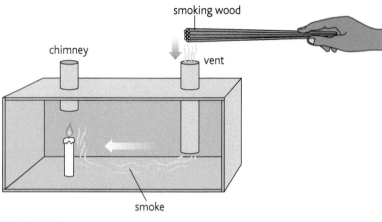

▲ **Fig B1.25**

In Fig B1.25 the candle is causing hot air to rise up the chimney. The smoke from the wood shows how cooler air is drawn into the room through open vents and window to replace the hot air that has gone up the chimney.

Fans can also be used to extract stale air and cause fresh air to be drawn into a room, as shown in Fig B1.26.

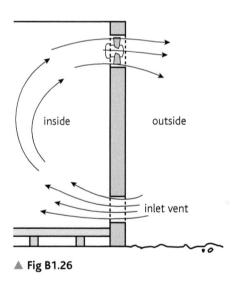

▲ **Fig B1.26**

Q24 Write down three features of a room that promote ventilation.

End-of-unit questions

1 Which of the statements about heat and temperature are correct?

  a Temperature is a type of energy; heat is a measure of how hot something is.

  b Temperature and heat are a measure of how hot something is.

  c Neither temperature nor heat are a type of energy.

  d Heat is a type of energy; temperature is a measure of how hot something is.

2 Which statement about heat transfer is correct?

  a Convection is cold air rising and hot air falling.

  b Conduction is hot air rising and cold air falling.

  c Radiation is heat transfer through space by electromagnetic radiation.

  d Radiation is heat being passed from molecule to molecule as they bump into each other.

3 Which *two* of the statements about expansion are correct?

  a All substances expand by the same amount when heated.

  b Substances expand because the molecules get bigger when heated.

  c Substances expand because the molecules move about more and take up more space.

  d Not all substances expand by the same amount when heated.

4 Temperature can be measured using a thermometer.

  a Name three different thermometers. Explain how each of the thermometers is used.

  b Look at the diagram of a thermometer in Fig B1.27. Explain how it works.

  c Explain the difference between a thermometer and a thermostat.

  d Many thermostats contain a device called a bi-metal strip. Explain how a bi-metal strip works.

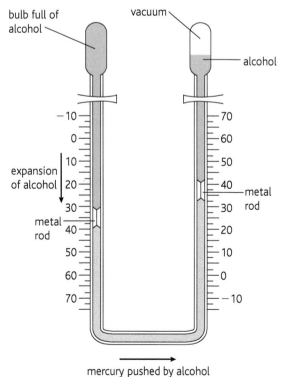

▲ **Fig B1.27**

5 The apparatus in Fig B1.28 can be used to show that as water evaporates it takes heat from its surroundings.

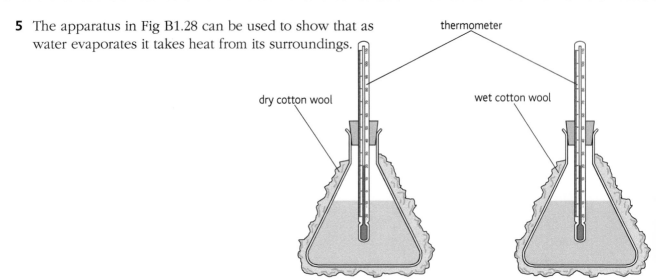

▲ **Fig B1.28**

a Explain how you would carry out an experiment to show the cooling effect of evaporation.

b Explain why evaporation of water has a cooling effect.

6 Ventilation in buildings is important.

a Explain why ventilation in buildings is important.

b Look at Fig B1.29. Explain how it can be used to show ventilation in buildings.

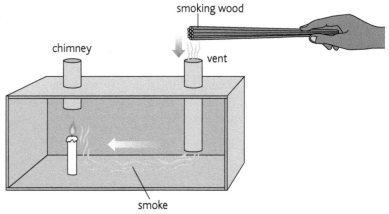

▲ **Fig B1.29**

7 The Atlantic Ocean is a large mass of cold water.
A coal fired power station burns a thousand tonnes of coal to produce electricity.

a Explain why the Atlantic Ocean contains more heat energy than burning a thousand tonnes of coal even though the water is quite cold.

b Refrigerators produce a low temperature by evaporating a solvent. Suggest how this produces the low temperatures inside the refrigerator.

8 Gliders do not have an engine. The plane can stay airborne for longer if the glider pilot can find a thermal (a column of rising air).

a Describe three different locations where a glider pilot may find a thermal. In each case explain why a thermal is likely to be located there.
location 1
location 2
location 3

b Suggest how a glider pilot could travel long distances by using different thermals.

B2a The Terrestrial Environment

By the end of this unit you will be able to:

- discuss the factors that influence soil formation.
- compare the types and functions of soils.
- relate soil fertility to the physical and chemical properties of soil.
- identify causes of soil erosion and methods of its prevention.
- compare and contrast methods used in the production of crops.
- discuss food chains and food webs found in an environment.
- describe the oxygen, carbon and nitrogen cycles.

Soil – its formation and components

The soil is one of our most important natural resources. It provides the medium in which plants grow, which is the source from which all animals derive their food. Without the soil and without the growth of healthy plants, animal life would be in jeopardy.

Most soils consist of a fine top soil in which plants grow, a subsoil consisting of coarser grained rocks and stones, and finally the parent rock from which the soil was formed. This is shown in Fig B2.1

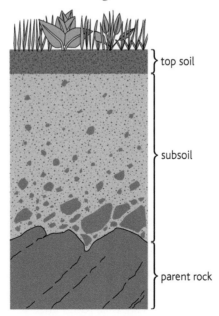

▲ **Fig B2.1** Soils have several distinct layers. The fertile layer is the top soil, derived from weathered particles of parent rock plus organic material from decayed plants and animals, air and water

Factors that influence soil formation

Soil is formed from rocks. The rocks are transformed into finely divided particles by **weathering**. The most important aspects of the weather that affect this process are rain, sun, wind and ice.

When rain falls it beats down onto the surfaces of the rocks. The falling water loosens rock particles, eventually breaking them into smaller pieces. The rain water may also dissolve certain gases in the atmosphere, e.g. carbon dioxide. This results in the formation of weak acids that could help to break down solid rocks. **Acid rain** is a problem in highly industrialised countries. Waste gases like sulphur dioxide dissolve in rain water. The resulting solution defaces buildings and statues, and destroys crops (see Units A3b and C2 for more on acid rain).

The Sun heats the rocks, making them expand. When they cool again at night they contract. This alternate expansion and contraction makes the rock break up into finer particles. The formation of sand in deserts is directly due to intense heat during the day, followed by near freezing temperatures at night.

Wind by itself might not be very effective as an agent of weathering. However, when a gust is 'armed' with grit and small pieces of materials it easily breaks down the rocks. Weathering by wind can be clearly seen on rocks near beaches.

Q1 Explain how soil is formed by weathering.

Components of the soil

Soil is made up of the following components: mineral particles formed by weathering of rocks, organic particles from decaying plant and animal remains, air, water containing dissolved salts, and living organisms.

Mineral particles

The **mineral salts** available for plant growth usually depend upon the nature of the parent rock. Fertile soil contains most, if not all, of the elements necessary for healthy plant growth. These elements have been found by experimentation to include carbon, oxygen, sulphur, potassium, magnesium, calcium, hydrogen, nitrogen, iron and phosphorus. Certain elements are required in small (trace) amounts, e.g. zinc, cobalt, boron, silicon and molybdenum. All of these elements, except carbon, must be taken in by the roots. The nitrogen, absorbed as nitrates, is produced by nitrogen-fixing bacteria in the nitrogen cycle (see page 164).

Organic particles

When **organic particles** like the remains of dead plants and animals decay they form a sticky, colloidal, organic compound called **humus**. The decay is caused by microorganisms such as fungi and bacteria. These are examples of microorganisms that are beneficial to humans. Humus supplies nitrates, which are important in protein synthesis. Because of its fibrous properties it absorbs water easily, and is used, in the form of compost, to improve the crumb structure of sandy soils.

---

weathering – the process of breaking down rocks to make soil

acid rain – rain that has a pH of less than 7

mineral salts – mineral compounds found in the soil that are needed by growing plants

organic particles – particles of materials that were once part of a living organism

humus – a sticky organic compound formed by the breakdown of dead organisms in the soil

Experiment – Estimating the organic content of a soil sample

You will need: some oven-dried soil, a crucible, Bunsen burner, tripod.

1 Put some of the oven-dried soil in the crucible.

2 Weigh the crucible and its contents.

3 Heat the crucible and its contents so that the soil sample glows red. Keep heating for 30 minutes.

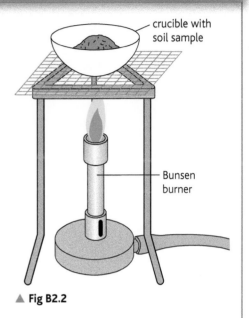

▲ **Fig B2.2**

4 Reweigh the crucible and its contents.

5 Calculate the organic content of the soil using the following equation:

$$\frac{\text{loss in weight}}{\text{weight of original sample}} \times 100 = \text{percentage of organic matter}$$

6 Repeat the experiment using samples of soil from other locations. Is there any difference in the percentage of organic matter found? Which had the greatest? Which had the least?

Q2 In which type of soil do you think plants will grow the best?

Experiment – Showing the solid components of soil

You will need: a measuring cylinder, soil sample, water, salt.

1 Shake the sample of soil in the water in the measuring cylinder.

2 Allow the mixture to stand for some minutes.

3 Measure the width of the various 'bands' or layers which you can see (as in Fig B2.3). Which is broadest? Which is narrowest?

4 Shake the cylinder again, and then add some salt. Does the soil 'settle' more quickly this time? What are the implications of your findings for soil-laden river water entering the sea?

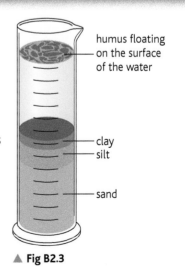

▲ **Fig B2.3**

Soil air

Soil air is found in the spaces between the soil particles. Since different soils have different sized particles (see Table B2.1 below), the amount of air present in a soil sample depends upon the size of its particles. The soil air is necessary for the **aerobic** respiration of the many soil organisms, and for the plant roots themselves. Some clay-type soils can become waterlogged, so there is no longer any air present. Under these conditions only **anaerobic** organisms can flourish.

aerobic – respiration using oxygen

anaerobic – respiration without using oxygen

particle size	soil type
2-0.02 mm	sand
0.02-0.002 mm	silt
0.002 mm	clay

Table B2.1 Size of particles in different soils

Experiment – Showing the presence of air in a soil sample

You will need: a beaker or jam jar, soil samples, water.

1 Put a sample of soil in the beaker and add water as shown in Fig B2.4.

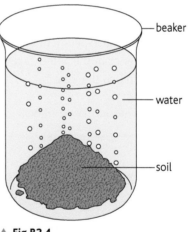

beaker

water

soil

▲ **Fig B2.4**

2 Observe what happens and report.

3 Repeat the experiment using samples of soil from other locations. Which sample has most air?

Soil water

inorganic particles – particles formed from non-living things in the soil such as sand

Soil water forms a thin layer around the **inorganic particles** of the soil. This water is essential for plant nutrition since it provides a medium in which mineral salts can dissolve, and so can be taken up by the plants. The amount of water that a soil holds is a measure of the size of the inorganic particles and also of the crumb structure of the soil.

Experiment – Measuring the water content of a soil sample

You will need: a soil sample, a tin, use of an oven.

1   Record the weight of the tin and then the weight of the tin with the soil sample in it.

2   Heat the tin and its contents in the oven for about six hours at about 100°C.

3   Cool the tin and its contents in a desiccator.

4   Weigh the container and its contents again. Record the weight.

The percentage of water in the soil can be calculated using the following equation:

$$\frac{\text{loss in weight}}{\text{original wet weight}} \times 100 = \text{percentage of water in soil}$$

5   Compare the percentage of water in different soil samples using the above procedure. Which soil has most/least water?

Living organisms

microorganisms – microscopic living organisms

Living organisms are mainly, but not exclusively, **microorganisms**. The following microorganisms are usually found in fertile soil: bacteria, fungi, protozoa and algae. Besides these there are centipedes, millipedes and earthworms. The presence of earthworms is a good indication that soil is fertile. Earthworms help in the mechanical breakdown of humus in the soil. By burrowing, they also increase the aeration (the amount of air) of the soil, as well as bringing fresh subsoil to the surface. The microorganisms help the decomposition of dead and decaying matter, releasing nutrients that are essential to plant growth.

Q3   List the different components of soil.

Physical properties of the soil

sand – large particles usually made from quartz

clay – small particles usually made from silicates

loam – mixture of sand and clay and humus

Soils are usually classified as **sand**, **clay** or **loam**, depending on the size of the mineral particles of which they are formed. This is a product of the nature and structure of the parent rock. Clay is derived from silicates and is rich in plant nutrients; sands are largely made from quartz; loam is a mixture of sand and clay, with the properties of both. Loam is considered the most fertile soil.

Several features could be measured to find out more about the properties of the three soils:

- the feel
- the water retention capability
- permeability
- air content
- capillarity

Texture

The feel or texture can be determined by rubbing the soil between your fingers. This indicates whether the mineral particles are loose, dry or sticky, smooth or gritty. This texture test also gives a fair indication of some of the other properties, such has how wet or dry the sample is.

Water retention

The ability of the soil to hold water is very important if the soil is to support healthy plants. Waterlogged soils encourage anaerobic conditions that could be disastrous for plants. Of course, the ability to retain water is closely related to the nature and size of the mineral particles, as well as to the percentage of organic materials present.

SBA Skills

ORR	D	MM	PD	AI
✓		✓		✓

Experiment – Measuring water retention

You will need: six measuring cylinders, cotton wool, funnel, three soil samples of equal weight – one each of sand, clay and loam.

1 Measure 50 ml of water in a measuring cylinder. Add this quantity of water to each soil sample, as shown in Fig B2.5.

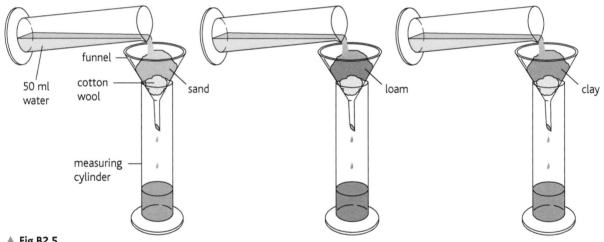

▲ **Fig B2.5**

2 Measure the quantity of water that drains out of each sample.

3 The quantity of water retained by each sample equals the amount poured on (50 ml), less the amount drained out.

Permeability

permeability – how easily water drains through soil

The experiment above could also be used to measure the **permeability** of the soil samples. Permeability is how easily water can drain through different soils. To measure permeability it is necessary to measure the quantity of water that drains through a soil sample over specific time intervals.

Q4 State the structural differences between the three types of soil: sand, clay and loam.

Air content

The quantity of air in the soil is also a measure of the size of the mineral particles. Generally speaking, sandy soils have large particles and large spaces between them. As a result, the air content is high. The size of the air spaces also influences the water retention property of soils, and so influences the drainage.

Capillarity

capillarity – the movement of water through narrow channels and air spaces in the soil

Capillarity is the process by which water moves into very narrow spaces. It takes place in soils that have air spaces. These spaces form the narrow channels through which the water moves. This action takes place especially in sandy soils. In some cases the water brings with it valuable nutrients that are needed for plant growth.

SBA Skills

ORR	D	MM	PD	AI
✓				

To do – Observing capillarity

Place a very narrow glass capillary tube into a beaker of water. A narrow, clear plastic straw will do if you do not have a capillary tube. You will see the water in the beaker drawn up the tube. The narrower your tube, the higher the water will go. This is called capillarity. You may remember in Unit A4b that very narrow blood vessels were called capillaries.

soil	properties				
	size of mineral particles	permeability	retentive ability	air content	capillarity
clay	small	low	high	low	slow
sand	large	very high	low	high	rapid
loam	mixture of large and small	moderate	very high	moderate	moderate

Table B2.2 Properties of different soil types

Q5 Most farmers prefer a loam soil to either sand or clay. Suggest why.

Soil acidity and alkalinity

acid – pH less than 7

alkaline – pH higher than 7

neutral – pH of 7

Soils may be **acid**, **alkaline** or **neutral**, according to their pH value (see page 272). The acidity or alkalinity depends on the nature of the parent rock, and so is linked to some of the physical properties you have already learned about.

Generally speaking, acid soils are those that are waterlogged, while alkaline soils are associated with limestone, and others rocks high in calcium and sodium. The soil's pH can be determined using a pH meter or by using universal indicator.

Plants are usually very specific as to the type of soil on which they grow.

What is soil erosion?

The loose mantle of soil that is found at the top of a soil profile is called the top soil. This is the main medium in which plants grow since it contains the nutrients that are vital to growth. The removal of top soil by wind, rain and running water is called **soil erosion**. Top soil is produced from the parent rock over several hundreds of years, or is transported from one point to another over a long time. However, people's careless and reckless use of the soil could destroy the top soil in a very short period of time.

soil erosion – removal of soil from land caused by wind and rain

SBA Skills

ORR	D	MM	PD	AI
✓				

To do – The Great American Dust Bowl

Use the internet to research 'The Great American Dust Bowl'. It is a terrifying story of how drought and intensive farming destroyed the top soil in the American prairies in the 1930s.

▲ Fig B2.6

Some of the practices that lead to erosion include clearing vast tracts of land for housing and non-agricultural purposes, over-grazing and intensive cropping.

The causes of soil erosion

Wind is a very powerful agent of erosion in some environments. These include areas that are treeless, and those in which the soil is light and loose enough to be carried by the wind.

Rain and running water are perhaps the most widespread causes of soil erosion. They attack almost all exposed areas, whatever the particle size or weight of the soil. The rain drops, by constantly beating on the soil, remove particles by splashing. However, it is running water that causes most damage.

During the seasonal rains that we experience in the Caribbean, floods may be caused. These form large sheets of water that take away the soil when they drain away. This is usually worst on relatively flat or slightly sloping land. Flowing water on steeper slopes tends to form channels that make the water's descent easier. These vary in width and are called rills or gullies. The draining flood waters wash tonnes of fertile soil away and deposit it on the beds of rivers or in the sea.

Q6   List two factors that can cause soil erosion.

Preventing soil erosion

windbreaks – structures that stop wind from damaging crops and blowing away top soil

Erosion by the wind could be prevented by erecting **windbreaks** at the edge of fields. Windbreaks are structures that stop the wind from damaging crops. They can be artificial fences or natural hedges. They are usually placed on the side of the field from which the prevailing wind blows.

Irrigation dampens the loose soil and reduces the amount blown away, while more permanent anchors for the soil can be made by introducing grasses and other plants, which will bind soil particles together with their roots.

contouring – ploughing land flowing contours so that each furrow is at the same level thus preventing water flowing along the furrow and eroding the soil

Erosion by water (rain and running water) can be prevented by various conservation methods. **Contouring** the land involves ploughing across the slope rather than up and down. The channels or furrows block water flowing down the slope.

Strip cropping involves planting row crops, e.g. corn and potatoes, in strips alternating with cover crops like grass. This combines two methods: contouring the land and covering the soil to prevent it washing away.

strip cropping – growing different crops on strips of land

terracing – building walls of stone on hillsides to produce flat growing areas for crops

Terracing erects a series of banks on a slope, giving the appearance of steps. Each bank is held together by plants or rocks.

Each method depends on the terrain and related aspects.

At all costs we must make sure that the mantle of soil is not abused as humans make an impact on the environment.

Q7   Describe three methods a farmer can use to prevent soil erosion.

Crop production

Good crop production requires a balance between getting maximum crop yield while at the same time preventing soil erosion and maintaining soil nutrients and soil structure. Good crop production may also be achieved by terracing, crop rotation, contouring, and greenhouse farming.

Terracing

Many Caribbean farms are found on hillsides. Terracing not only prevents soil erosion by water, but also provides a level surface on the hillside on which to grow crops. This enables the farmer to tend the crops more easily and also provide access for machinery. It is, however, very time consuming to install and relies upon a good local supply of rocks and stones.

▲ **Fig B2.7** Terracing

Crop rotation

Crop rotation involves planting different crops on the same piece of land each year. Because different crops use different nutrients from the soil, crop rotation gives the soil a chance to recover before the same crop is planted again. Some methods of crop rotation involve planting crops such as clover, which can absorb nitrogen from the air and make nitrates. These crops are then ploughed into the soil at the end of the growing season and, as the plants are broken down in the soil, humus is produced and the nitrates enter the soil so that they are available for next season's crops. This technique is called leaving the soil **fallow**. However, it does mean that no crops are harvested for a whole growing season.

fallow – not growing crops on the land for at least a year

Contouring

Contouring not only prevents soil erosion by water but also provides a simple method of growing crops on hillsides without going to all the trouble and hard work of building terraces. It is a good way to grow crops on newly cultivated soil before terraces have been built or if the land is only on a gentle slope. However, extreme wind and rain can destroy the contours and allow erosion to take place.

▲ **Fig B2.8** Contouring

Greenhouse farming

Farming in a greenhouse is expensive and is usually reserved for expensive crops produced on a small scale. It has the advantage of being able to control all the environmental factors such as watering, light and temperature while at the same time protecting the crops from weather and pests.

When farmers are deciding which method they are going to use, they need to take into account which method is going to give them the maximum yield and profit with the minimum depletion of soil nutrients and damage to the soil.

Q8 State the advantages and disadvantages of using greenhouses for crop production.

▲ **Fig B2.9** Caribbean greenhouse

Ecology

Ecology is the study of the relationships between the plants and animals and non-living factors such as wind, rain, soil and light in their environment.

ecology – a study of the relationships between the plants and animals with non-living factors

Ecosystems

An **ecosystem** is all the living and non-living things, such as air, water, soil and light interacting with each other. Ecosystems may be as big as the whole world or as small as a garden pond.

ecosystem – all the living and non-living things interacting with each other

The place in which things live is called a **habitat**. Different organisms live in different habitats. Some habitats near your school may include rainforest, rivers, lakes, wetland or even coastal beaches.

habitat – a place where organisms live

Habitats are places where different **populations** live. A population is a breeding group of the same type of organisms living in the same place at the same time.

population – a breeding group of the same type of organisms living in the same place at the same time

All the different populations in a habitat form a **community**. Communities may consist of only a few different populations or very many of them. The community interacts with the non-living world such as the soil, air, water and light to form the ecosystem.

community – different populations living in the same place at the same time

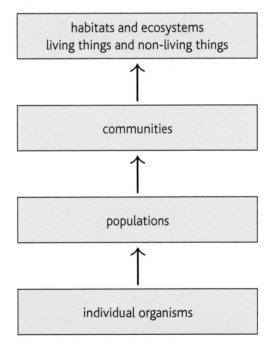

habitats and ecosystems living things and non-living things
↑
communities
↑
populations
↑
individual organisms

SBA Skills

ORR	D	MM	PD	AI
✓				

To do – Looking at an ecosystem

1 Look at the area around your school. Identify a habitat, a community, and a population of things living in the habitat.

2 Make a list of the non-living things in the ecosystem that may affect the survival of the population.

Food chains and food webs

All living organisms are **interdependent**. This means that they all depend upon one another for their survival. All living things need energy. Green plants get this energy from the Sun. They trap the Sun's energy by the process of photosynthesis. Animals get their energy from their food. They either eat plants or other animals. The energy from the Sun is thus passed on from one organism to the next. This is called a food chain.

Sun

Green plant

Grasshopper

Lizard

Snake

▲ **Fig B2.10** A food chain

In reality it is more complicated than this because each animal eats lots of different kinds of food. This means that lots of different food chains are all linked together to make a food web.

The food web in Fig B2.11 shows all the different ways that energy can flow from one organism to the next.

Q9 Choose a food chain to draw from the food web shown in Fig B2.11.

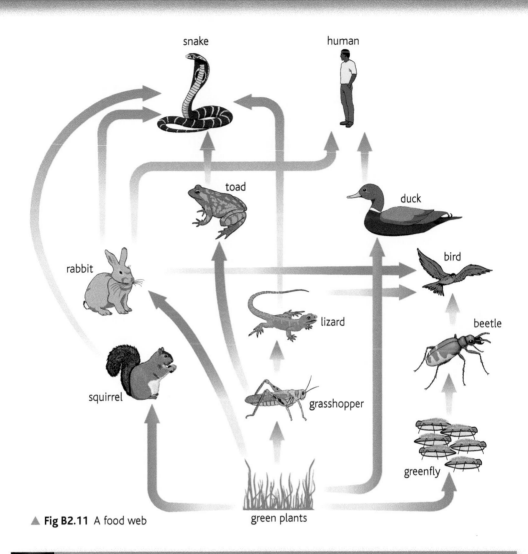

▲ **Fig B2.11** A food web

To do – Looking at habitats

1 Look at the habitats around your school.

2 Write down a food chain to show some of the organisms that live there.

SBA Skills

ORR	D	MM	PD	AI
✓				✓

---

trophic level – the same position in a food chain

pyramid of biomass – an estimate of the mass of living organisms in a food chain

Each step in the food chain is called a **trophic level**. The food chain can be represented as a **pyramid of biomass**, as shown in Fig B2.12. Each step in the pyramid is a trophic level and represents the amount of living material at each level.

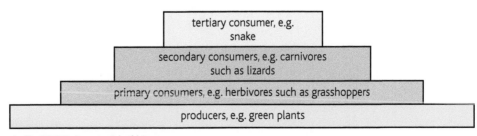

▲ **Fig B2.12** A pyramid of biomass

producers – plants that produce food for a food web

herbivores – animals that eat plants

primary consumers – animals that eat plants

carnivores – animals that eat other animals

secondary consumers – animals that eat animals

tertiary or third-order consumers – animals at the top of the food chain that eat other animals

omnivores – animals that eat both plants and animals

decomposers – organisms that break down dead material

Green plants are called **producers** because they produce the food for all of the food chain.

Animals that eat plants are called **herbivores**. Herbivores are also called **primary consumers** because they are the first in the food chain to consume food from another organism.

Animals that eat herbivores are called **carnivores**. Carnivores are called **secondary consumers** because they are the second link in the food chain that consumes other organisms.

Sometimes there is a further link in the food chain and these are called **tertiary or third order consumers**.

Q10 Suggest why tertiary consumers are so called.

Some organisms in the food web are called **omnivores**. Humans are examples of omnivores as we can eat both plants and animals for food.

Some organisms are **decomposers**. Decomposers are all those organisms such as bacteria and fungi that decompose all the other organisms when they die. Decomposers are very important in a food web because they return all the minerals and nutrients back into the soil so that they can be used again by other organisms such as plants.

This is shown in Fig B2.13.

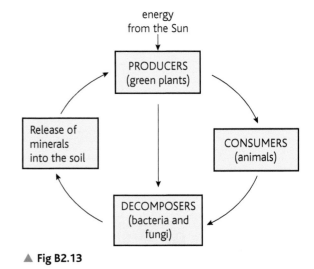

▲ **Fig B2.13**

To do – The local environment

SBA Skills

ORR	D	MM	PD	AI
✓				

1 Look at your local environment.

2 Find examples of producers, consumers, herbivores, carnivores, omnivores and decomposers.

To do – Drawing a food web

SBA Skills

ORR	D	MM	PD	AI
				✓

1 Copy out the diagram of the food web in Fig B2.11.

2 Draw in an extra arrow to show that a human is an omnivore.

3 Draw a new box labelled 'decomposers' and draw arrows to show how decomposers fit into the food web.

Oxygen, carbon and nitrogen cycles

The oxygen cycle

You have already seen in Unit A2a and Unit A3a that animals absorb oxygen from the air for the process of respiration, and that plants release oxygen into the air by the process of photosynthesis.

oxygen cycle – the cycling of oxygen between animals and plants

These two processes together form the **oxygen cycle**, where oxygen is continually being recycled in the environment, as shown in Fig B2.14.

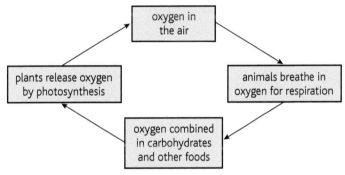

▲ **Fig B2.14** The oxygen cycle

carbon cycle – the cycling of carbon through the ecosystem

nitrogen cycle – the cycling of nitrogen through the ecosystem

Two of the elements that decomposers release from dead, decaying organisms are carbon and nitrogen. The movement of carbon and nitrogen through the ecosystem are called the **carbon cycle** and **nitrogen cycle**.

The carbon cycle

Like oxygen, carbon in the form of carbon dioxide is also passed between plants and animals. Animals breathe out carbon dioxide from respiration, and plants absorb it for photosynthesis. The carbon cycle is slightly more complicated because it also includes the burning of fossil fuels where carbon reacts with oxygen to form carbon dioxide, as shown in Fig B2.15.

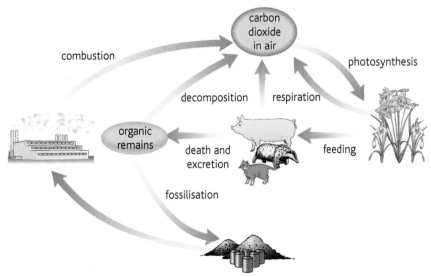

▲ **Fig B2.15** The carbon cycle

The nitrogen cycle

Unlike carbon, which readily reacts with oxygen, nitrogen is very inert and does not like to react with other elements. This could be a problem as nitrogen is required by all living things to make protein. However, there is a type of bacteria called **nitrogen-fixing bacteria** that can live in the soil or in root nodules in plants such as peas, beans and clover. These bacteria can convert atmospheric nitrogen gas into nitrate compounds that plants can absorb and use to make proteins. Unfortunately, some bacteria do this process in reverse and break nitrates down into nitrogen gas. These are called **denitrifying bacteria**. Fortunately, nitrogen-fixing bacteria provide sufficient nitrates for the plants to absorb.

Other bacteria called nitrifying bacteria are responsible for converting ammonium compounds produced by decay into nitrates in the soil.

nitrogen-fixing bacteria – bacteria that convert atmospheric nitrogen gas into nitrates that can be used by plants

denitrifying bacteria – bacteria that convert nitrates into nitrogen gas

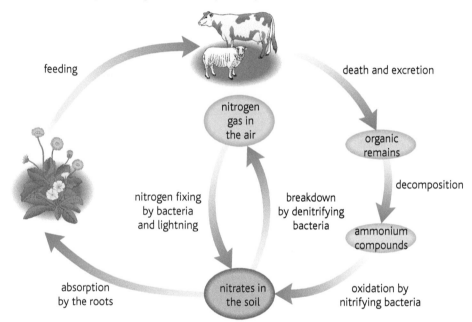

▲ **Fig B2.16** The nitrogen cycle

Q11 Explain the difference between nitrogen-fixing and denitrifying bacteria.

End-of-unit questions

1 Which of the equations would you use for calculating the percentage of water in soil?

A $\dfrac{\text{loss in weight}}{\text{original wet weight}} \times 100$ = percentage of water in soil

B $\dfrac{\text{original wet weight}}{\text{loss in weight}} \times 100$ = percentage of water in soil

C $\dfrac{\text{dry weight}}{\text{original wet weight}} \times 100$ = percentage of water in soil

D $\dfrac{\text{loss in weight}}{\text{dry weight}} \times 100$ = percentage of water in soil

2 Look at the diagram of the nitrogen cycle in Fig B2.17.

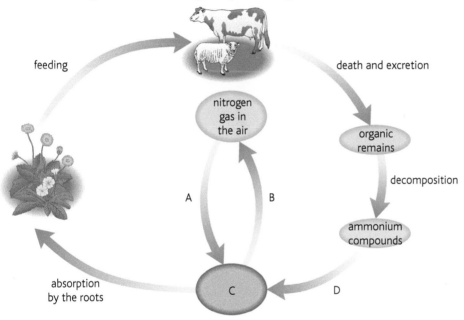

feeding

death and excretion

nitrogen gas in the air

organic remains

decomposition

A B

ammonium compounds

absorption by the roots

C D

▲ **Fig B2.17**

a Which part, A, B, C or D, are nitrates in the soil?

b Which part, A, B, C or D, are nitrogen-fixing bacteria?

c Which part, A, B, C or D, are nitrifying bacteria?

d Which part, A, B, C or D, are denitrifying bacteria?

3 Complete the table to show the names of different soil types. Choose words from the following list: clay   minerals   mud   rock   stone   sand   silt

particle size	soil type
2–0.02 mm	
0.02–0.002 mm	
<0.002 mm	

4 Look at the food web in Fig B2.18.

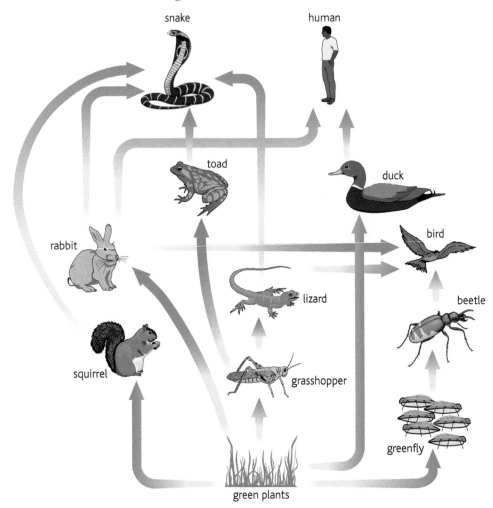

▲ **Fig B2.18**

Use the food web to describe and define examples of the following terms:

a habitat

b population

c community

5 Pesticides used in farming can enter the food chain and get into our food.
   a Explain using food chains how a pesticide such as DDT could get into our food.
   b Very small amounts of pesticide in the environment can build up to lethal concentrations in top carnivores. Use the idea of pyramid of number to explain why.
   You may use a diagram to help explain your answer.

6 You have studied the oxygen, carbon and nitrogen cycles. Sulphur also cycles through nature. Produce a diagram to show the sulphur cycle.
   You should include in your diagram ideas of burning fossil fuels, acid rain, protein and decomposition.

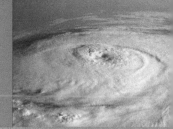

B2b The Weather and Geological Events

By the end of this unit you will be able to:

- describe the various types of air masses.
- distinguish between the four types of local fronts.
- describe the characteristics of a cyclonic storm, particularly a hurricane.
- describe tidal waves and how they are formed.
- explain the causes of the different types of volcanic eruptions.
- discuss the relationship between earthquakes and volcanoes.
- describe how tides are formed.

Weather

Air masses

Fortunately for us in the Caribbean, the weather is nearly always predictable: bright and sunny in the mornings, humid with some rain in the afternoon and then dry and sunny evenings. The reason our weather is so predictable is that we live in the tropics. Air masses from other parts of the world rarely enter the tropics and so the air masses that move over the Caribbean are more consistent than in other parts of the world. Because air masses from the Northern and Southern hemispheres tend not to move into the tropics, this means that areas that are away from the sea such as the Sahara desert are very dry and areas that are near the sea such as the Caribbean get plenty of rainfall.

▲ **Fig B2.19** Global air currents

maritime tropical air mass – a mass of air found over the tropics near the sea

You can see by looking at Fig B2.19 how the air masses in the Northern and Southern hemispheres tend not to move into the tropics. The type of air mass that forms over the Caribbean is called a **maritime tropical air mass**. It tends to be hot and humid. This is because it forms mainly over water in a tropical area. When this mass moves it tends to move northwards over the south eastern United States and is responsible for their weather too.

Q1 Suggest why the Sahara does not have much rainfall.

Moving air masses are often responsible for spreading pollution. Air masses that form over the Sahara desert often contain wind-blown dust that can be spread to different parts of the world as the air mass moves. When it rains, the Sahara dust is washed out of the air mass onto the ground below.

Other types of pollution that can be spread by moving air masses include radioactive fallout, industrial waste from chimneys, landfill fumes and volcanic dust.

Dust from volcanoes can rise high into the atmosphere and be responsible for causing spectacular sunsets all round the world.

▲ **Fig B2.20** A sunset caused by volcanic dust in the atmosphere

Weather fronts

weather front – a mass of warm or cold air moving towards another mass of air

A **weather front** happens when two different air masses meet, especially if one is warm air and one is cold. Warm air tends to hold more water vapour than colder air, which is usually drier. Instead of mixing, a weather front forms between the two masses as the warm, moist air rises above the cold, drier air.

There are four different kinds of weather fronts:

1 cold fronts

2 warm fronts

3 occluded fronts

4 stationary fronts

Cold fronts

cold front – a mass of cold air moving towards a mass of warm air

Cold air is heavier (denser) than warm air. As a mass of cold air moves towards a mass of warm air, the warm air rises up as the denser cold air pushes in and under the mass of warm air. This is called a **cold front**. As the warm air cools it can no longer hold all the water vapour it contains and it begins to rain. With a cold front, the rain always happens after the front edge of the cold air mass has passed over. In other words, it rains behind the cold front.

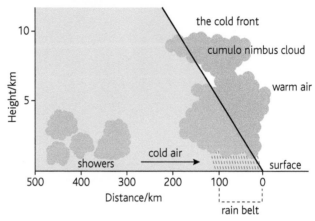

▲ **Fig B2.21** A cold front

Q2 Explain why it rains after a cold front has passed.

Warm fronts

warm front – a mass of warm air moving towards a mass of cold air

A **warm front** happens when a mass of warm air is moving towards a mass of cold air. The warm air rises up and over the trailing edge of the cold front and as the warm air cools it begins to rain. This time, however, the rain always happens ahead of the leading edge of the warm front and when it stops raining you know the warm front has arrived.

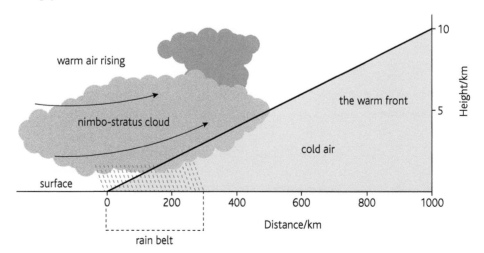

▲ **Fig B2.22** A warm front

Q3 Explain why it rains before a warm front arrives.

Occluded fronts

occluded front – a cold front pushing warm air upwards and meeting another mass of cold air

An **occluded front** happens when a cold front catches up with a warm front. At ground level the two masses of cold air meet and the warm air is pushed upwards above ground level. This can cause a prolonged period of rain. See Fig B2.23.

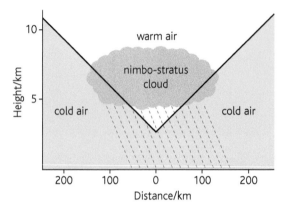

▲ **Fig B2.23** An occluded front

Stationary fronts

stationary front – when weather fronts are not moving

Sometimes neither of the air masses is capable of moving towards the other mass and displacing it. When this happens the weather fronts do not move and are called **stationary fronts**. When there is a stationary front the weather can remain unchanged for a long period of time.

Weather maps

On a weather map warm fronts are shown as

...cold fronts are shown as

...and occluded fronts as

isobar – a line of equal pressure on a weather map

Weather maps also show lines of equal pressure called **isobars**, which show areas of high pressure and low pressure. When two fronts meet the warm air rises above the cold air. More cold air moves in to replace the warmer rising air. Just like water going down a plug hole, the cold air swirls around the warm air in an anticlockwise direction. This mass of swirling air is called a **depression**.

depression – a mass of low pressure air moving in an anticlockwise direction

To do – Feeling the wind

Go outside and stand with your back to the wind. If you are just in the Northern hemisphere, the low pressure area will be to your left and the high pressure area will be to your right.

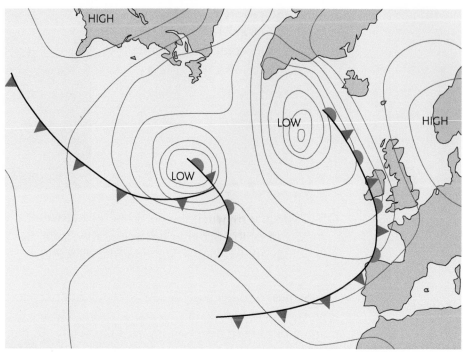

▲ **Fig B2.24** A weather map

Q4 Make a sketch of the weather map in Fig B2.24. Label a cold front, a warm front and an occluded front.

Storms

cyclone – an intense depression

typhoon – a cyclone formed in the China Sea

hurricane – a cyclone formed in the Caribbean

Another word for a depression is a **cyclone**. A cyclone is when winds spiral inwards towards an area of low pressure. The lower the pressure, the faster the winds spiral inwards. Cyclones are known by different names all over the world. In the China Sea they are called **typhoons**. In the Caribbean they are called **hurricanes**.

Think about water going down the plug hole. The water swirls around the plug hole very quickly, but at the centre of the plug hole not much seems to be happening. It's the same with a hurricane. Very high winds spiral in towards the centre of the hurricane but at its centre all is still and calm. It is often called the **eye of the storm**.

eye of storm – the centre of a hurricane where wind speeds are very low

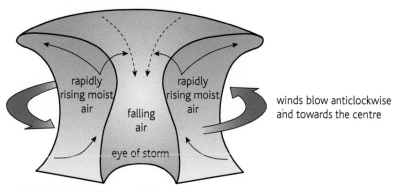

▲ **Fig B2.25** A hurricane and its 'eye'

Look at the picture of a hurricane taken from a satellite. You can see the eye of the storm at the centre.

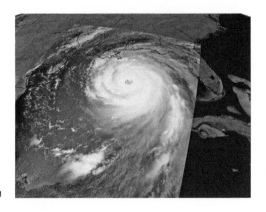

▶ **Fig B2.26** The eye of a storm

Q5 Explain why when the hurricane first arrives the winds are blowing in one direction but after the 'eye of the storm' has passed overhead the winds are blowing in the opposite direction.

Hurricanes form because the air above warm seas holds lots of water vapour and, being less dense, this air starts to rise. The air mass starts to rotate in a counter-clockwise direction towards the centre. In the centre of the hurricane, cooler air falls. Because this falling air is not rotating, the winds at the eye of the storm are very low. Energy from the warm seas feeds the hurricane, which gets bigger and more powerful. It is only when the hurricane reaches land and it is starved of its energy input from the warm seas that it loses its energy and just becomes another depression.

The hurricane season is during the summer when the seas are at their warmest and usually lasts from July to September. This old rhyme reminds us of the danger time.

June too soon
July stand by
August come it must
September remember
October all over

vortex – the rotating mass of air that spirals in towards the centre of a hurricane

The rotating winds are called a **vortex** and winds speeds can be greater than 200 km per hour. Winds of this speed can cause serious damage as they pass over the Caribbean islands. Another problem caused by hurricanes is that the low pressure at the centre of the hurricane can cause sea levels to rise by up to 10 metres. When this high tide reaches land it can cause serious flooding and damage to coastal properties.

To do – Hurricane research

SBA Skills

ORR	D	MM	PD	AI
✓				✓

Research records and list the names of all the hurricanes that have reached your island over the last ten years. Make a note of how powerful each hurricane was. Do the hurricanes seem to be getting stronger, weaker or staying about the same?

Q6 Describe two ways that hurricanes can cause damage to island properties.

Volcanoes, earthquakes and tidal waves

Volcanoes, earthquakes and tidal waves are all caused by effects beneath the Earth's crust. The Earth is not a solid ball of rock. It consists of a **core**, a **mantle** and an outer **crust**, as shown in Fig B2.27. Molten rock, called **magma**, exists beneath the crust, which floats and moves across the mantle.

The crust consists of a set of **tectonic plates** that all move independently of one another. Some are moving towards one another, some apart and some are sliding past one another. It is this movement of tectonic plates that is responsible for volcanoes, earthquakes and tidal waves.

core – the inner part of the Earth

mantle – the layer of the Earth just below the crust

crust – the outer surface of the Earth

magma – hot molten rocks

tectonic plates – plates of the Earth's crust that move against one another

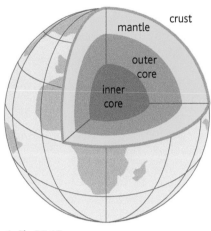

▲ **Fig B2.27**

Volcanoes

Volcanoes do not exist all over the world, but mainly in areas where two tectonic plates meet.

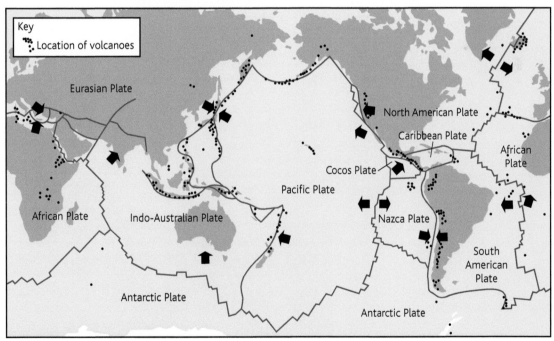

▲ **Fig B2.28**

You can see from the map in Fig B2.28 that several plate boundaries occur near the Caribbean. This explains why there are so many active volcanoes in and around the Caribbean. One of the most active volcanic areas is found to the west of the Caribbean where the Cocos and Caribbean Plates meet. A string of islands have been formed from active volcanoes at the eastern boundary of the Caribbean plate. See Fig B2.29

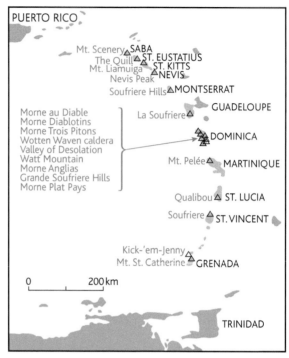

▲ **Fig B2.29** The volcanic islands in the Caribbean

subduction zones – the place where two plates collide and one goes underneath the other

lava – molten rock

Most volcanoes occur where two tectonic plates are colliding. As the plates meet, one of them is pushed underneath the other one, down into the mantle. This is called a **subduction zone**. The intense heat and pressure generated melts the rocks, some of which are forced to the surface as **lava** to form a volcano.

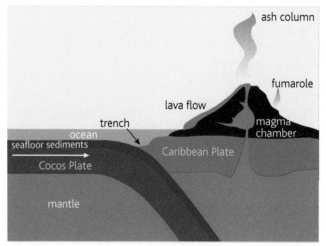

▲ **Fig B2.30** A subduction zone

To do – Modelling a volcano

Use different colours of Plasticine to make a model of a subduction zone and a volcano as shown in Fig B2.30.

SBA Skills

ORR	D	MM	PD	AI
				✓

fumarole – a volcanic vent that releases steam and gases

As the volcano erupts, it is not just lava that spews out at the surface. Ash and gases such as hydrogen sulphide, carbon dioxide and steam are all ejected from the main vent. Smaller vents called **fumaroles** emit gases and steam. The main vent can even eject rocks and boulders weighing many tonnes.

Kick-'em-Jenny is an active submarine volcano. It rises from the sea floor 8 km north of the island of Grenada and rises 1300 metres, where the North American tectonic plate is subducting the Caribbean tectonic plate.

On 23rd July 1939 an eruption occurred, sending debris high into the air and caused tsunamis up to 2 metres high. The summit of this volcano still remains below the surface of the sea.

shield volcano – a flat shield-shaped volcano produced by runny lava, low in silicates

The type of volcano depends largely on the type of lava that is being ejected. Lava that contains a lot of silica tends to be much thicker (viscous) than lava that is low in silicates. Thin lava flows easily and tends to produce flatter volcanoes like the **shield volcanoes** of Hawaii. Viscous lava does not flow easily and produces the typical dome shape of some of the Caribbean volcanoes.

▲ **Fig B2.31** Smoke is seen after an underwater volcano erupts – Tonga

▲ **Fig B2.32** Mauna Loa volcano – Hawaii

▲ **Fig B2.33** Soufriere volcano – Caribbean

Q7 Volcanoes around the Caribbean tend to be a tall, conical shape rather than a flatter, shield shape. Suggest why.

The ecological consequences of volcanoes

Volcanoes are not all bad news. It is true that they can cause untold damage to property and life during a violent eruption, but they also bring prosperity to the surrounding area. It is not by accident that so many people choose to live close to an active volcano. When a volcano erupts, the ashes and dust that are released are rich in nutrients. This makes the local soils very fertile, allowing the local farmers to grow extra crops. Underground heat is also a valuable source of energy and the steam released from deep underground can power turbines to generate electricity.

▲ **Fig B2.34** The start of work on the Nevis geothermal power plant that opened in 2009

Tourism is another industry that benefits from active volcanoes as people visit the islands to see the power of the volcanoes at first hand.

Q8 Explain why people live near volcanoes despite the dangers from eruptions.

Earthquakes

Like volcanic eruptions, most earthquakes happen close to tectonic plates. The force of the plates colliding or moving against each other causes crack or **faults** to form in the plates. When rocks in the fault move against each other, earthquakes happen. Rocks do not move smoothly and continuously past each other. Tension may build up over many years until suddenly the rocks in the fault move. This sudden release of energy causes earthquakes.

fault – a crack in the tectonic plate usually formed close to where two plates meet

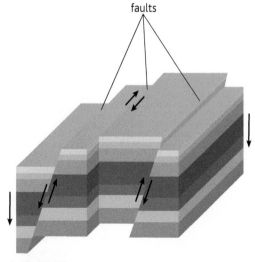

faults

▲ **Fig B2.35**

Richter scale – a scale for measuring the intensity of an earthquake

Earthquakes are measured using the **Richter scale**, which ranges from 0–9.

- Richter scale >2   too small to be felt
- Richter scale 3–4   felt but no damage caused
- Richter scale 4–5   shaking and rattling of items in the house
- Richter scale 5–6   light damage to well designed buildings, major damage to poorly designed ones
- Richter scale 6–7   major destruction to buildings
- Richter scale 7–8   major damage over large area
- Richter scale 8–9   major damage over an area hundreds of kilometres across

Each step from 0 to 9 increases by a factor of ten with each step. Thus, a measurement of 5 is ten times greater than 4, and a measurement of 6 is one hundred times greater than 4.

Q9 How much greater is a Richter scale of 7 greater than a Richter scale of 3?

focus – the point inside the Earth's crust where an earthquake happens

epicentre – the point on the Earth's surface directly above the epicentre

Earthquakes often happen deep down in the Earth's crust called the earthquake's **focus**. The point on the surface directly above the focus is called the **epicentre**.

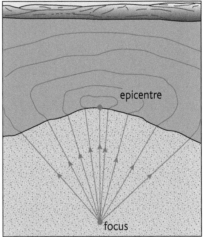

epicentre

focus

▲ **Fig B2.36** The focus and epicentre of an earthquake

seismograph – a device for measuring the intensity of an earthquake

Earthquakes are measured using a **seismograph**. A heavy weight is suspended from a spring. Attached to the weight is a pen, which rests on a moving strip of paper, as shown in Fig B2.37. When the earthquake happens, the heavy weight hanging on the spring does not move very much compared to the moving strip of paper and the pen traces out the record of the earthquake.

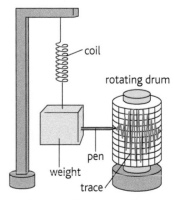

coil

rotating drum

pen

weight

trace

▲ **Fig B2.37** A seismograph

SBA Skills

ORR	D	MM	PD	AI
		✓	✓	

 To do – A home-made seismograph

You can make your own simple seismograph.

1 Attach a pen to a heavy weight suspended from a spring.

2 Rest the pen against a sheet of paper and leave it. If an earthquake were to happen, it would not show the waves but it would make a vertical mark on the paper indicating just how strong the earthquake was.

Q10 Suggest how you could improve your seismograph.

Tidal waves or tsunamis

tsunami – a tidal wave

Tidal waves or **tsunamis** are the result of earthquakes, volcanic eruptions or landslides that occur on or near the sea bed.

▲ **Fig B2.38** The Boxing Day Tsunami of 2004

On 26th December 2004, the Boxing Day Tsunami occurred. Its epicentre was off the west coast of Sumatra. It was caused when part of the ocean bed dropped by 10 metres and caused an earthquake that registered 8.9 on the Richter scale. As the sea bed dropped, it caused waves over 30 metres high when they arrived at land. More than 225,000 people died in 11 different countries.

Waves are very effective at transferring energy, which is why so much damage and loss of life occurred so far away from the epicentre of the earthquake. The waves travelled at speeds of up to 800 kph, which is why so little time was available to give people warning to evacuate coastal areas.

Tides

Tides occur because of the gravitational pull of the Moon. Because the oceans are liquid and not rock like the rest of the Earth, they can move as they are pulled by the gravitational effect of the Moon.

The Earth rotates once every day. As it does so, a bulge of water (the high tide) is being pulled towards the Moon. Because the Earth is rotating, the high tide appears to move around the Earth about once every 24 hours.

However, the Earth is also being pulled towards the Moon. This leaves behind another bulge of water (another high tide) on the side of the Earth facing away from the Moon. This means that there are in fact two high tides roughly every 24 hours, one on either side of the Earth. Because this leaves less water elsewhere on the Earth there are also two low tides every 24 hours. See Fig B2.39.

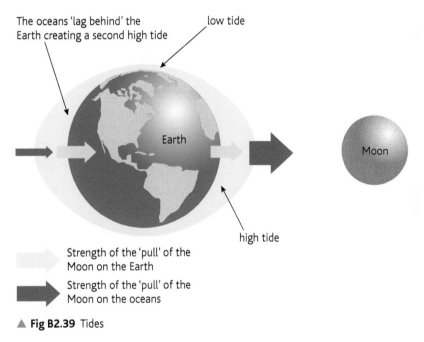

▲ **Fig B2.39** Tides

The times of the high tide and low tide are not the same every day. This is because the Moon orbits around the Earth every 28 days and so each day the Moon is in a slightly different position.

spring tides – very high and low tides caused when the gravitational pull of the Sun is aligned with that of the Moon, either in the same direction or in opposite directions

neap tides – not very high or low tides caused when the gravitational pull of the Moon and the Sun are at right angles to each other

You may also have noticed at the beach that sometimes the tides come in and go out much further than at other times. This is because the Sun also has a gravitational effect on the Earth. Although it is bigger than the Moon, it is much further away so its gravitational effect on the Earth is less than that of the Moon. However, when the Moon and the Sun are both in the same direction from the Earth or directly opposite each other their combined effect is much greater and high tides are even higher and low tides even lower. These are called **spring tides**. When the Sun and the Moon are at right angles to each other the Sun cancels out some of the effect of the Moon's gravitational pull and the high tides are not so high and the low tides not so low. These are called **neap tides**.

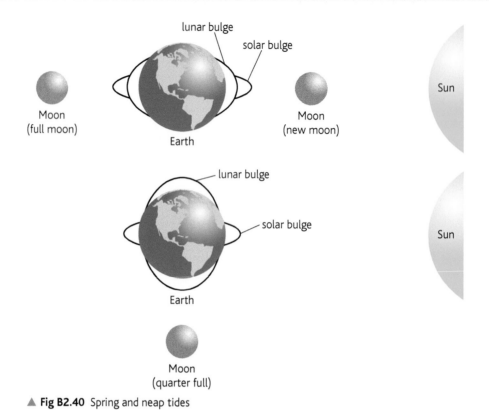

▲ **Fig B2.40** Spring and neap tides

SBA Skills

ORR	D	MM	PD	AI
				✓

To do – Using a tide table

1 Look at the data about tides in the Caribbean in Table B2.3 below.

	day 1	day 2	day 3	day 4	day 5
low tide	0142 hours	0301 hours	0452 hours	0619 hours	0730 hours
high tide	0807 hours	0925 hours	1124 hours	1205 hours	1245 hours
low tide	1428 hours	1527 hours	1732 hours	1832 hours	2008 hours
high tide	2024 hours	2138 hours	?	?	?

▲ **Table B2.3**

2 Predict the times for the high tide on days 3, 4 and 5.

Q11 Suggest what effect increasing the gravitational effect of the Moon would have on the Earth's tides.

Knowledge of the tides is very important, especially for people who live and work by the coast. Knowing and predicting the tides tells us the best times for boats to enter and leave the harbour, when it's safe to harvest shell fish and when it's best to be able to work on your boat.

Q12 Can you think of any other examples of why it is important to know about tides?

End-of-unit questions

1 Look at the diagram of the Earth and the Moon in Fig B2.41.

At which of the following will there be a high tide?

a A only

b A and C only

c C only

d B and D only

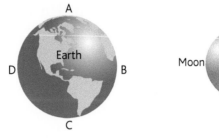

▲ **Fig B2.41**

2 For which of these weather patterns will it rain after the front has arrived?

a a cold front moving towards another cold front

b a cold front moving towards a warm front

c a warm front moving towards another warm front

d a warm front moving towards a cold front

3 Cyclonic winds in the Caribbean are called hurricanes.

a Explain what is meant by cyclonic.

b Describe the conditions required to generate a hurricane.

c Draw a labelled vertical section diagram of a hurricane to show the directions of the air movements.

4 Look at the diagram of a weather front.

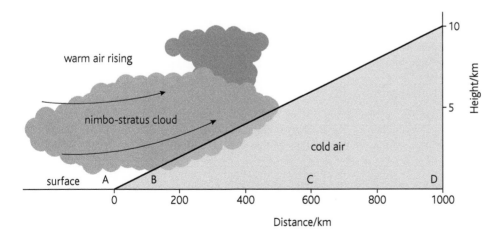

▲ **Fig B2.42**

a What type of weather front is shown in the diagram?

b In which part, A, B, C, or D is rain most likely to fall? Explain your answer.

c Explain what happens when a cold front catches up with another cold front.

5 Tides occur twice each day.

a Explain what causes each of the following:
  i high tides and low tides
  ii spring tides and neap tides

b Explain why there are two tides each day even though there is only one Moon.

c Explain why it is important to be able to predict when high tides and low tides will happen.

6 Weather and weather fronts occur when masses of air at different temperatures and pressures meet.

a Explain why weather is more violent and extreme when the differences in temperature and pressure between the two air masses is greater.

b The type of air mass that forms over the Caribbean is called a maritime tropical air mass. It tends to be hot and humid.
Name three other types of air mass and describe their characteristics.

air mass 1
air mass 2
air mass 3

7 One of the most volcanic active areas is found to the east of the Caribbean where a string of islands have been formed from active volcanoes.

a Explain why the volcanoes have been formed in a string and not dotted randomly through the Caribbean.

b Active volcanoes are not found in the middle of the Sahara desert.
Suggest why.

By the end of this unit you will be able to:

- explain the uses of water.
- describe methods of purifying water.
- explain how the water cycle provides us with fresh water.
- discuss the chemical and physical properties of water.
- state the conditions for flotation.
- discuss the factors affecting the free movement of objects in air and water.
- discuss the effects of water pollution on aquatic life.

Water is the most important substance on the planet. Without water we would all die in just a few days. We often take a clean water supply for granted and just turn on the tap without thinking where all the clean water has come from and just how precious it is. Clean drinking water is a very valuable resource. We should use it sparingly and try to conserve as much as possible.

What we use water for

Most of the fresh water supplied to Caribbean islanders is used for the irrigation of crops and the environment. Only about 20% of the fresh water supply is used for domestic use.

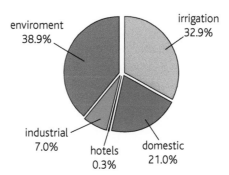

▲ **Fig B3.1** Use of water in Jamaica

Drinking water

The islands of the Caribbean are surrounded by water. In fact most of the Earth's surface is covered in water. Unfortunately for all of this water contains salt. For drinking we need pure, fresh water. Water is important to us because it acts as a solvent and allows all the chemical reactions in our body to take place. It enables substances to be transported around our body, dissolves waste products so we can eliminate them from our body, and it even keeps us cool when we get too hot. We might be able to live for many weeks without food, but without water, we would only survive for a few days.

Water in the home

Many homes are supplied with piped fresh water. We have many uses for this water in our homes.

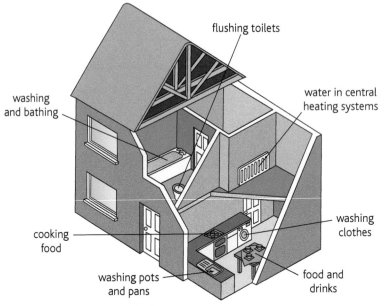

▲ **Fig B3.2** Using water in the home

Water is used in the home for bathing, washing, flushing toilets, cooking food and even for heating our homes in central heating systems.

Here are some facts about water use in our homes.

- A dripping tap can use up to 90 litres of water a week.
- Leaving the tap running while cleaning your teeth can use 9 litres a minute.
- Baths use up to 400 litres of water more each week than showers.
- Putting a water-saving device in your toilet cistern can save 3 litres with each flush.
- Using a hosepipe in the garden uses as much water in one hour as a family of four people use in a day.

To do – Looking at water use in the home

Look at the picture of how we use water in the home (Fig B3.2) and read the list above.

Q1 Suggest three different ways that we waste water in our homes.

Q2 Suggest three different ways by which we could reduce the amount of water that we consume in our home.

Growing crops

Look at the pie chart in Fig B3.1 on page 183.

Q3 How much of the water that we use is used for the irrigation of crops?

Watering our crops is important to ensure that we get a good yield. However, some crops are actually grown with the roots in water and not in soil. This type of farming is called **hydroponics**. Tomatoes grow well by this method. Fertilisers are added to aerated water so that all the roots are supplied with water, nutrients and oxygen. Hydroponics is usually carried out in greenhouses where all the environmental conditions such as temperature and humidity can be controlled. Large amounts of high quality crops can be produced very quickly in quite a small space, making it very profitable for the farmer.

hydroponics – growing plants in water without soil

▲ **Fig B3.3** Growing using hydroponics

Experiment – Growing an onion by hydroponics

Place an onion bulb over a glass container that contains water and plant nutrients. The surface of the water should be just below the base of the onion. Leave for several days. Note what happens.

▲ **Fig B3.4** Growing an onion using hydroponics

Environmental uses of water

Two other uses of water are fire-fighting and generating electricity.

Water is very useful in fire-fighting because it is not flammable. It cuts off the oxygen supply to the flame and it cools the burning material.

Q4 Look at the fire triangle in Fig B3.5. It shows the three things required for a fire to happen. Which two of the three things will water remove when it is directed onto a fire?

▲ **Fig B3.5** A fire triangle

Water can also be used to generate electricity. Many of the Caribbean islands are fortunate in that they have high mountainous regions with lots of rivers and waterfalls. The map shows the locations of hydroelectric power stations in Jamaica.

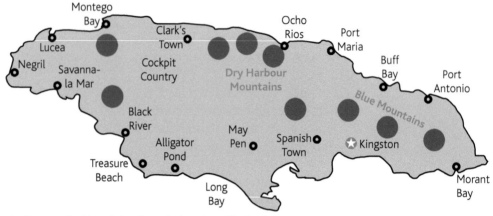

▲ **Fig B3.6** The blue circles show the location of hydroelectric power stations in Jamaica

Falling water has lots of kinetic energy and this energy can be transferred to electrical energy when the falling water is used to turn a turbine attached to a generator.

Most hydroelectric power stations work by damming water in a river to produce a large reservoir. The water is then allowed to fall

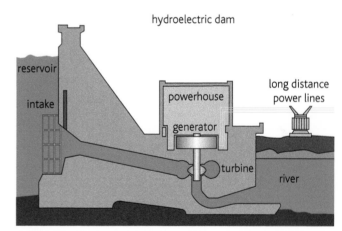

▲ **Fig B3.7** A hydroelectric power station

through tunnels in the dam to turn large turbine blades. As the blades rotate they turn a shaft in a generator that produces the electricity.

Purifying water

Before we drink water, it needs to be cleaned and purified. Water is cleaned at purification plants like the one at the Mona Reservoir in Jamaica. It works by **rapid gravity filtration (RGF)**. Water is taken from the reservoir and chlorine is added to kill off any microorganisms. It then enters a mixing chamber where alum is added to cause any particulate impurities to clump together (**flocculate**) and sink to the bottom of the chamber.

rapid gravity filtration (RGF) – a method of producing clean drinking water

flocculation – the clumping together of particles in water when alum is added

▲ **Fig B3.8** A water purification plant

It is then fed through six filter beds. The filter beds contain sand and anthracite and are regularly cleaned by backwashing them with air and cleaned water. The cleaned filtered water is then disinfected with another dose of chlorine and is stored in a reinforced concrete reservoir.

The cleaned and purified water is then fed into a 90 cm diameter water pipe that carries it into the water distribution system.

SBA Skills

ORR	D	MM	PD	AI
✓				

Experiment – Making a filtration plant

1 Use a large diameter glass tube, a rubber bung and a narrow glass tube to assemble the apparatus as shown in Fig B3.9.

2 Place pebbles at the bottom, small stones next, then gravel and finally sand.

3 Get some dirty water and add a spoonful of alum. Alum is a flocculant, which means it makes all the dirt particles stick together. Shake well, then pour the dirty water into the top of the filter and collect the water as it trickles out of the bottom.

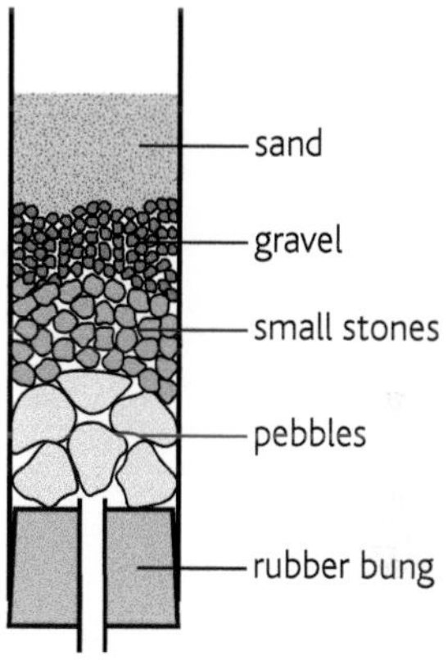

▲ **Fig B3.9**

(labels: sand, gravel, small stones, pebbles, rubber bung)

Q5 What do you notice about the water that comes out of the bottom?

⚠ **SAFETY:** Do *not* drink the water. Even though the water may look cleaner it is still not safe to drink as it may contain bacteria.

Q6 Suggest what could you do to the water to make it safe to drink.

Purifying sea water

The Caribbean islands are surrounded by water. Unfortunately, being sea water we cannot use it for drinking. To make sea water safe for drinking we need to remove the salt. This process is called **desalination**.

desalination – removing salt from sea water to produce drinking water

Distillation

Most desalination plants work by evaporating water from sea water. As the water evaporates, the salt is left behind. The evaporated water is then condensed back into fresh water. This is called **distillation**.

distillation – boiling water and condensing the steam that has been given off

▲ **Fig B3.10** A desalination plant and oil refinery on Sint Maarten, in the Caribbean

Experiment – Distilling water

1 Set up the apparatus as shown in Fig B3.11.

2 Place some sea water in the round-bottomed flask. Heat with a Bunsen burner. Cold water going into the condenser cools down the evaporated water (steam) and condenses it back into water. Collect the condensed water.

Q7 What do you notice about the contents of the round-bottomed flask as the water evaporates?

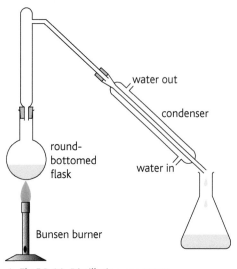

▲ **Fig B3.11** Distillation apparatus

 SAFETY: The water you collect will not contain salt. Even though you may be tempted to taste it to check that it is fresh water, it is safer not to do so.

Chlorination

chlorination – adding small quantities of chlorine to water to kill microorganisms to make it safe to drink

Once water has been filtered it is made safe to drink by the addition of chlorine. This is called **chlorination**. Chlorine is only added in small quantities (about three parts per million) but it kills all the microorganisms that may be present in the water. Hypochlorite and iodine tablets can be used in an emergency to purify river water. Although they will not clean the water, the microorganisms are killed, making the water safe to drink. However it may not look very nice and the tablets may leave an aftertaste in the water.

Experiment – Sterilising river water

1 Get a sample of river water and divide it into two containers.

2 Add a hypochlorite tablet to one sample and allow it to dissolve.

3 Then add one drop of each sample of river water to two different Petri dishes containing sterile nutrient agar. (These will have to be prepared by your teacher.)

4 Seal both Petri dishes with sticky tape and observe what happens over the next few days.

Q8 In which dish do bacteria grow, the one containing untreated river water, or the one containing the river water treated with a hypochlorite tablet? Explain your result.

The water cycle

There are two main sources of fresh water. The first is surface water that collects in rivers and lakes. The second is underground water that is stored in porous rocks underneath the ground.

The average rainfall in the Caribbean is about 2000 mm of rain per year. More than this falls on mountainous regions and less on the coastal areas. It is the responsibility of each territory's water resources authority to control and manage this resource.

water cycle – the cycling of water through the environment

convection – hot air and water vapour rising

condense – evaporated water turning back to liquid water

precipitation – formation of rain drops and rain

Rainfall happens because of the **water cycle**. The Sun heats everything on the Earth. This causes water in the oceans, seas, rivers and ponds to evaporate. In hot environments, like the islands of the Caribbean, the water can evaporate very quickly. The evaporated water rises into the air by **convection**. The higher the water vapour rises, the more it cools. When it is cools enough, the water **condenses** to form rain droplets. When the rain droplets are heavy enough, they fall out of the sky as rain. This is called **precipitation**. Some of the rain runs off the land to form rivers and lakes. Some of it soaks into the land to form ground water. The whole process is repeated over and over again to form the water cycle.

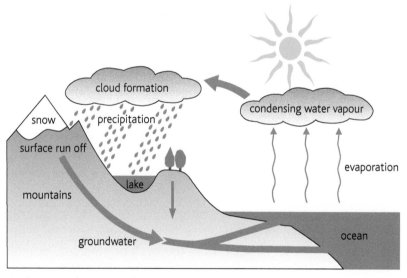

▲ **Fig B3.12** The water cycle

To do – The water cycle

SBA Skills

ORR	D	MM	PD	AI
✓	✓			

Make a drawing of a water cycle to include your own island. Label the names of the mountains, towns, rivers and lakes in your diagram. You could even show the position of your own school.

Q9 Explain why the water cycle provides a constant supply of *fresh* water.

Water and living things

All living things need water and living things are found in both sea water and fresh water. You studied osmosis and the effect of different concentrations of water in living things in Unit A1. The effects of osmosis mean that living things need to be adapted to live in sea water or fresh water.

Fresh water

Organisms that live in fresh water find that water tends to enter them by osmosis. Different plants and animals use different methods to deal with this problem.

Some plants use a stiff cell wall around their cells. This stops the cells from expanding as water enters. Water enters the cell until the pressure inside the cell is the same as the osmotic pressure of water entering the cells.

Some microscopic animals place the water that enters them into a water bubble called a water vacuole. Periodically they push this water vacuole to the outside and thus get rid of the extra water. Other animals like fish simply use their kidneys to produce a lot of dilute urine to get rid of the water.

Sea water

Animals and plants that live in sea water have a different problem. Water tends to *leave* them by osmosis. To counter this problem, some organisms increase the salt content of their cells so that water does not leave them. Other animals, such as fish, solve the problem by drinking extra water. They then excrete all the extra salt that they take in with the water through their gills.

Some mammals have a very efficient kidney and can produce urine that is more concentrated than sea water. This means they can drink sea water and excrete the salt in their urine. Humans are not adapted to living in sea water. If we were to drink sea water our kidneys would not be able to produce urine more concentrated that sea water. This means that we would use up even more of the water in our bodies to get rid of the salt than we actually drank from the sea water. This is why drinking sea water is not safe.

Like some fish, seagulls can excrete salt. You may have noticed a drip of salt water hanging from the beak of a seagull. Seagulls use a special gland in their beaks to excrete very concentrated salt solution that they get from eating salty fish and drinking sea water.

Q10 Explain why humans need a supply of fresh water for drinking.

Water pollution

Water pollution is the presence of unwanted substances in water. Water can be polluted in many ways.

organic matter – matter produced from living organisms such as sewage and from the decay of dead plants and animals

Organic matter

Organic matter is the dead remains of animals or plants or the waste material from animals, such as sewage. The organic matter is a good source of food for bacteria to grow and multiply. As the numbers of bacteria increase, they remove more and more oxygen from the water for respiration. As the oxygen levels drop, animals like fish can no longer survive and start to die of suffocation.

Pesticides and herbicides

pesticides – chemicals used on farms to kill pests

herbicides – chemicals used on farms to kill weeds

Pesticides and **herbicides** are used in farming to control insect pests and reduce the growth of weeds. Because they are soluble in water they sometimes dissolve in rain water and find their way into rivers and underground water sources. Even though they may be present in very low concentrations in the water supply, they can get concentrated to very dangerous proportions as they are passed along a food chain.

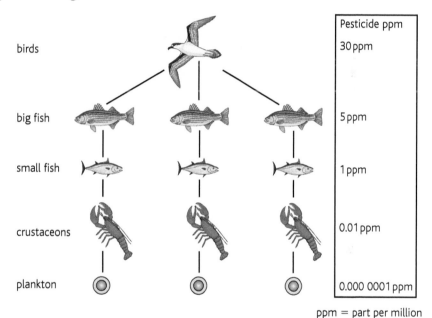

	Pesticide ppm
birds	30 ppm
big fish	5 ppm
small fish	1 ppm
crustaceons	0.01 ppm
plankton	0.000 0001 ppm

ppm = part per million

▲ **Fig B3.13** How pesticides build up to dangerous levels in a food chain

Q11 Suggest why pesticides are much more dangerous to sea birds than they are to plankton.

Look at the photo of the two test tubes in Fig B3.14. One of them looks dirty. It is contaminated with organic matter. The second sample looks clean. But it is contaminated with soluble pesticides.

Q12 The water in the test tube on the right looks safe to drink. Explain why this is not true.

▲ **Fig B3.14**

Fertilisers

fertilisers – chemicals used on farms to increase crop growth

Fertilisers such as nitrates and phosphates can have a devastating effect on the water supply. When fertilisers dissolve in rain water and enter the water supply they cause the growth of green algae. As the water turns to a green algal soup, light can no longer enter the water. Water plants can no longer photosynthesise and begin to die. As they die they rot and large numbers of bacteria grow on the rotting remains of the plants. As the bacteria use up all the oxygen, fish and animals die causing even more bacteria to grow. Soon the water contains nothing but rotting remains and bacteria. This is called **eutrophication**.

eutrophication – process where life in ponds and rivers is killed by rotting algae caused by fertilisers getting into the water supply

Oil spills

Oil is a fuel that is transported around in world in huge ships called tankers. Sometimes these oil tankers get into trouble and are damaged or sink. They can release huge quantities of crude oil into the sea. Oil floats on water to form oil slicks. Sea birds get coated in oil. They can no longer fly and often drown or poison themselves as they try to clean the oil from their bodies.

▲ **Fig B3.15**

To do – Sites of water pollution

SBA Skills

ORR	D	MM	PD	AI
✓				

Identify sites of water pollution in the area where you live. Research how the pollution came about and ways of dealing with the pollution problem.

The chemical and physical properties of water

Water is a most unusual chemical substance. It is an excellent solvent, a liquid at room temperature and, unlike most other chemicals, it expands as it freezes. This is why ice floats instead of sinking. If it were not for this fact alone, all the water on the Earth, apart from the top couple of centimetres of the oceans, would be frozen solid.

Water is colourless and tasteless. It melts/freezes at 0°C and boils/condenses at 100°C. 1 cm³ of water weighs 1 g. This means it has a **density** of 1 g/cm³.

density – mass divided by volume

SBA Skills

ORR	D	MM	PD	AI
✓		✓		✓

Experiment – Checking the density of tap water and sea water

1 Weigh a 250 cm³ measuring cylinder on an accurate balance.

2 Add 250 cm³ of tap water. Re-weigh the measuring cylinder.

3 Determine the mass of the water.

4 Divide the mass by the volume to determine the density of the water. You should find you get an answer that is very close to 1 g/cm³.

5 Now repeat the process using sea water.

Q13 What did you find about the density of sea water? Suggest a reason why the density was different.

Each molecule of water consists of two atoms of hydrogen to one atom of oxygen, giving it the chemical formula H_2O.

It is very difficult to get completely pure drinking water. Water usually contains various types of dissolved chemicals. These dissolve in the water as the water in rivers flows over the river bed or as it passes through porous rocks as ground water.

SBA Skills

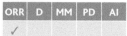

ORR	D	MM	PD	AI
✓				

Experiment – Showing that water contains dissolved chemicals

1 Place some tap water in a shallow container such as a watch-glass.

2 Place the watch-glass over a water bath of boiling water.

3 Continue heating the watch-glass until all of the water has evaporated. Any dissolved chemicals in the water will be left behind on the watch-glass.

Q14 What deposits did you see on the watch-glass?

4 Now repeat the process but this time use sea water.

Q15 What did you notice when sea water was used?

Water not only contains dissolved chemicals. It also contains dissolved air. You may have noticed when you were boiling the beaker of water in the last experiment that bubbles of gas were produced in the water as you started to heat it. Some of this gas was air that had dissolved in the water. This is because cold water can hold more dissolved air than hot water. As the water is heated, the dissolved air comes out of solution to form the bubbles.

Floating and sinking

Some things float in water and some things sink. Objects float in water if they have a density of less than $1\,g/cm^3$. Objects that have a density greater than $1\,g/cm^3$ sink. You have already found out that sea water has a greater density than fresh tap water.

SBA Skills

ORR	D	MM	PD	AI
✓				✓

Experiment – First it floats – then it sinks!

1 Almost fill a $250\,cm^3$ measuring cylinder with sea water. Get a bulb pipette and use the bulb to draw up a small amount of the sea water. Place the pipette into the measuring cylinder. It should float. Continue drawing up small amounts of sea water until the pipette is almost at the point of sinking.

2 Now fill a $250\,cm^3$ measuring cylinder with fresh tap water. Remove the pipette from the sea water taking care not to lose any water from it. Place the pipette into the measuring cylinder of fresh water. Note what happens.

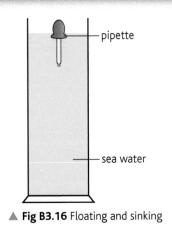

pipette

sea water

▲ **Fig B3.16** Floating and sinking

Q16 Suggest why the pipette floated in the sea water but sank in the fresh water.

Archimedes – a scientist who discovered that an object will float if it displaces its own weight of water

There is an old story about a scientist called **Archimedes**. The story goes that one day he was sitting in the bath wondering why some things floated and some things sank. He suddenly had a 'brainwave' and in his excitement, jumped out of the bath and ran down the street naked shouting "Eureka!" which means "I have found it!". We do not know if the story is true but we do know that Archimedes made a very important discovery.

EUREKA!

▲ **Fig B3.17** Archimedes made an important discovery

Archimedes discovered that if an object displaces its own weight in water it will float. This is why ships that weigh thousands of tonnes can float. As the ship enters the water it starts to push the water out of the way. When the weight of water that has been displaced is equal to the weight of the ship, the ship floats. The force of the **upthrust** of the water is equal to the downward force of the weight of the ship. This is also why ships sink when they have a leak. Water enters the ship, increasing its weight. As its weight increases it has to push or displace more water out of the way so it goes lower in the water. Eventually it cannot displace enough water and the ship sinks. The upthrust of the water can no longer support the weight of the ship.

upthrust – the force of water pushing up on an object in the water

Because an equal volume of sea water weighs more than an equal volume of fresh water, ships do not have to displace as much sea water as fresh water to be able to float. This means a ship will float higher in sea water than in fresh water, just like the experiment you did with the pipette.

Archimedes' principle – that an object will float if it displaces its own weight of water

Archimedes' idea about why things float or sink is called **Archimedes' principle**.

It is important that ship owners and captains know just how their ship will float in different types of water. If a ship is loaded very heavily with cargo when at sea, it could be in danger of sinking when it enters a fresh water river estuary. A scientist called **Plimsoll** came up with a very clever idea to prevent this from happening. He invented something called the **Plimsoll line**. The Plimsoll line is a series of marks placed on the side of a ship to show how deep or high it will float in different types of water. The captain can then see how much cargo he can load onto the ship without any danger of it sinking.

Plimsoll – a scientist who invented the Plimsoll line

Plimsoll line – a line on the side of ships that show how the ship will float in different conditions

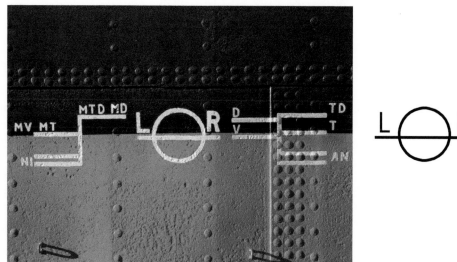
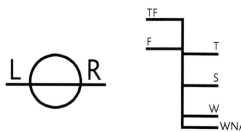

▲ **Fig B3.18** A Plimsoll line on a ship

The original Plimsoll line was just a circle with a line through it. As the cargo was placed on board ship, the ship sank lower and lower into the water. When the water level reached the Plimsoll line, it told the captain that he was now carrying the maximum cargo load. LR on the Plimsoll line stands for Lloyds Register where the ship is insured.

Over the years other load lines have been added to the Plimsoll line. Each line gives the captain additional information about the flotation of the ship in different conditions and places.

The letters on the load line stand for:

- TF – Tropical Fresh Water
- F – Fresh Water
- T – Tropical Sea Water
- S – Summer Temperate Sea Water
- W – Winter Temperate Sea Water
- WNA – Winter North Atlantic

You can see from the Plimsoll line that even temperature affects how high a ship will float in the water. This is because water expands as it heats up. This makes it less dense so the ship has to displace more warm water than cold water to float at the same position.

To do – Plimsoll lines

Different ships may have slightly different Plimsoll lines. Make a note of different Plimsoll lines on ships in a harbour near you. Research what each different line stands for.

SBA Skills

ORR	D	MM	PD	AI
✓				

Q17 In which type of water will a ship float the lowest, Tropical Fresh Water or Winter North Atlantic? Why?

Resistance to movement in air and water

resistance – a force that prevents an object from moving, e.g. air and water resistance

When objects move through air and water they experience **resistance**. This is because they have to push the air or water out of the way to be able to move. Because water is denser than air it is harder to push out of the way and therefore water has more resistance than air.

viscous – a property of liquids – the more viscous a liquid is the less it is able to flow

Some liquids are more **viscous** than other liquids. Treacle is more viscous than water. The more viscous a liquid is the more resistance there is when an object moves through it. Fortunately for us, ships do not have to move through treacle.

friction – a force that resists movement between two surfaces in contact with one another

Friction is another force that stops an object from moving. Friction happens when one object moves against another object. The energy of movement is transferred to heat energy and is lost. This slows the moving object down. When the space shuttle enters the Earth's atmosphere, friction between the hull of the shuttle and the air generates heat. As the shuttle slows, the heat generated is so intense that the shuttle has to be covered with heat resistant ceramic tiles to protect the shuttle from melting.

▲ **Fig B3.19** A space shuttle re-entering the Earth's atmosphere

To do – Friction and heat

Rub your hands together very briskly. You will notice that they soon begin to feel hot. This is because the friction of your hands rubbing together transfers the movement of your hands into heat energy.

We can reduce air and water resistance by streamlining. You may have noticed that ships have a pointed end that moves through the water much more easily than a blunt or flat fronted ship would. Streamlining is also used by fish. One of the fastest fish in the sea is the sailfish, shown in Fig B3.20. It can swim at over 100 kph.

▲ **Fig B3.20** A sailfish

Notice how pointed the front of the sailfish is. This streamlining helps it move through the water at very high speed.

Because boats and ships move through both water and air there are several factors that affect how easily they move. The speed and direction of the wind and the speed and direction of the water currents all affect how the boat moves.

To do – Air resistance and friction

SBA Skills

ORR	D	MM	PD	AI
				✓

Running and cycling both involve movement through air. Tyres and running shoes need to grip the road surface but the human body and the bike need to move through the air with as little resistance as possible. Air resistance and friction both affect running and cycling.

1 Explain the advantages and disadvantages of air resistance and friction to a runner and a cyclist.

Note – sometimes friction can be useful. It is what stops your trainers from slipping when you are running along a track. Think of how friction affects cycling in a positive way.

Like friction, sometimes resistance can be useful. When we are flying a kite air resistance is very useful. It enables the force of the wind to keep the kite in the air.

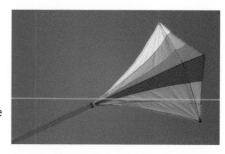

▲ **Fig B3.21** Kites can fly because of their lightweight materials and large surface area

When an athlete throws a javelin or discus they want it to go as far as possible. They have to overcome two forces, air resistance and gravity. To achieve the maximum distance they need to take two factors into account – how hard they throw it and what angle of elevation they use. Obviously the harder they throw it, the further it will go. What is more difficult to know is what angle they should throw it at.

SBA Skills

ORR	D	MM	PD	AI
✓		✓		✓

Experiment – Throwing a javelin

1 Place two nails in a flat piece of wood 10 cm apart. Connect one rubber band to both of the nails to make a simple catapult.

2 Tear a piece of paper into a strip and then fold it in half.

3 Place the folded paper over the rubber band. Pull back and fire!

Care – make sure that you wear safety glasses and that you do not aim the catapult at any other student.

4 Measure how far the paper missile went.

5 Now repeat the process by placing the catapult at different angles. Try angles ranging from flat to the ground to vertically in the air. You must pull the rubber band back by the same amount each time to make it a fair test. Measure the distance the missile travelled in each case.

6 Copy the table below to record your results.

angle of catapult / °	distance travelled / cm
0	
15	
30	
45	
60	
75	
90	

Q18 Which angle was the most effective for getting the missile to go the maximum distance? What advice would you give to a javelin thrower?

End-of-unit questions

1 Which of the following is not a method of conserving clean water?

a Placing a large stone in the toilet siphon.

b Putting less water in the bath.

c Saving the cold water from the shower before its gets hot, to water the garden.

d Installing a garden pond.

2 Which of the following processes is not part of the water cycle?

a evaporation

b distillation

c condensation

d precipitation

3 Which two of the following objects will float in fresh water?

a an object with a density of $1.4\,g/cm^3$

b an object with a density of $2.1\,g/cm^3$

c an object with a density of $0.4\,g/cm^3$

d an object with a density of $0.9\,g/cm^3$

4 Look at the picture of the Plimsoll line in Fig B3.22.

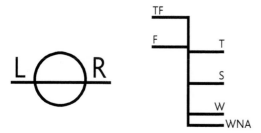

▲ **Fig B3.22**

a What does WNA stand for?

b What will happen to the water level on the Plimsoll markings as the boat is loaded with extra cargo?

c Explain why the boat should not be loaded with cargo such that the TF line is under water.

5 Objects move through air and water. The following all affect how objects move. Explain their meanings.

a resistance

b streamlining

c friction

6 Archimedes discovered how objects float and sink in water.

a State Archimedes' principle.

b A ball of Plasticine sinks. If the shape of the Plasticine is changed into a bowl shape, it floats. Use Archimedes' principle to explain why.

c Explain why the bowl-shaped piece of Plasticine will sink if filled with water.

7 Round the world sailors often use a **solar still** for producing fresh drinking water.

a Explain how a solar still works. You may draw a diagram to help explain your answer.

b Explain why water taken directly from the sea is not safe to drink.

c Explain why water from the solar still is safe to drink.

8 When the Space Shuttle re-enters the Earth's atmosphere, the outer part of the shuttle glows red hot.

a Explain why the Space Shuttle glows red hot when it re-enters the Earth's atmosphere.

b Explain why the Space Shuttle slows down when it re-enters the Earth's atmosphere.

B3b Activities in Water

By the end of this unit you will be able to:

- describe the various methods used locally for fishing.
- describe the various devices used to ensure safety at sea.
- identify water safety devices.
- discuss the hazards associated with scuba diving.

Water for work and recreation

Fish is an important source of food. It is rich in protein and omega oils, which are good for our health. Most of our fish is caught by deep sea fisherfolk. There are three main ways that deep sea fisherfolk catch fish: trawling, dredging and long-line fishing.

Types of fishing

Trawling

Trawlers pull a net through the water. As the fish enter the net, the net gets smaller and smaller until all the fish are trapped at the far end. The net is then pulled in and the fish removed.

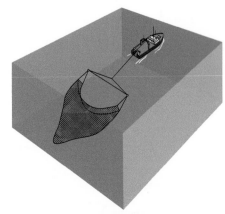

▲ **Fig B3.23** Trawling

Dredging

Dredgers pull a basket net along the sea bed. This catches different kinds of fish that are bottom dwellers. Dredgers have to be careful that they do not damage coral reefs.

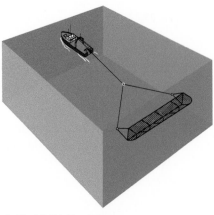

▲ **Fig B3.24** Dredging

Long-line fishing

Long-line fishing uses baited hooks to catch fish. Fishing boats have to be careful not to lose their lines as birds such as albatross can get tangled in lost lines and drown.

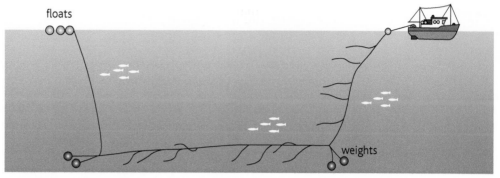

▲ **Fig B3.25** Long-line fishing

Finding fish

Deep-sea fisherfolk may be at sea for several days and travel long distances in search of fish. It is important that they can navigate and find their way safely. One way to do this is by using the position of the Sun and the stars. Navigators can use a piece of equipment called a sextant that measures the angle between the Sun and the horizon. By knowing the time of day, it is then possible to calculate your position. Modern boats use satellite navigation systems to accurately plot their position.

A compass is also used in navigation. It consists of a small bar magnet that is freely suspended so that it can point in any horizontal direction. Magnets have a North and a South pole, as shown in Fig B3.26.

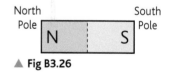

▲ **Fig B3.26**

When two magnets are placed next to each other, opposite poles always attract and like poles always repel. So a North pole is always attracted to a South pole, but two North or two South poles will always repel one another, as shown in Fig B3.27.

▲ **Fig B3.27**

You can use a small magnet, like a magnetised pin, to make a compass if it is balanced on a pivot point. To explain how a compass works, it helps to imagine a huge magnet buried inside the Earth. The South pole of this magnet is actually at the magnetic North pole of the Earth. As opposite poles attract, the North pole of the compass points towards the South end of the buried bar magnet. Hence, the North pole of the compass points towards the magnetic North pole of the Earth.

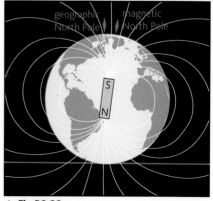

▲ **Fig B3.28**

The buried bar magnet is actually at an angle to the axis that runs from the geographic North Pole to the geographic South Pole. Before satellite navigation systems were invented, navigators would use a compass and the position of the North Star to work out their bearings

SBA Skills

ORR	D	MM	PD	AI
				✓

Experiment – Making a compass

It is quite easy to make a simple compass. All you need is a bowl of water, a piece of tissue paper, a magnet and a steel needle.

1 First you need to turn the steel needle into a magnet. You can do this by stroking a real bar magnet along the length of the needle several times. You can check that it is magnetised by picking up another steel needle with it.

2 You then need to float the magnetised needle onto a bowl of water.

Normally a needle sinks in water because it is much denser than water. However, there is a way to get around this because a needle is very light and water has a remarkable property called **surface tension**. Surface tension forms something like a skin on the surface of water. It is why little drops of water stay bead-shaped and do not simple flow into a flat little puddle. Place the needle on the tissue paper and lay the tissue paper onto the surface of the water in the bowl. The tissue paper will get waterlogged and sink leaving the needle supported by the surface tension of the water. Take care to remove the tissue paper without disturbing the needle. You may need to try this several times before you can get it to work.

surface tension – formed on the surface of a liquid by an increased attraction of molecules resulting from forces of attraction on fewer sides of the molecules

3 The 'floating' needle will now act like a compass and will swing round to face North/South. You can check its accuracy by comparing it to a real compass.

▲ **Fig B3.29**

4 Get a map of the Caribbean. Use the compass to orientate your map so the top of the map is facing North. Then use the map to identify the direction of your nearest neighbouring island.

Other navigation devices

Other navigation devices include sonar, radar and GPS.

sonar – an underwater detection system

Sonar is an underwater detection system. A series of sound waves are sent out as pulses that reflect off underwater objects such as shoals of fish. It is rather like a form of underwater radar. The reflected echoes are displayed on a screen. The sonar can calculate the distance away from the object by recording the time taken for the echo to return.

radar – a detection system that works in the atmosphere

Radar is similar to sonar but uses microwave radiation instead of sound and works above the sea rather than below it.

GPS – a Global Positioning System

GPS or Global Positioning System uses a number of satellites and a computer to calculate the ship's latitude and longitude and display the data on a screen showing a map of the surrounding area. It is very accurate and has the advantages of working in bad weather.

Safety at sea

Safety standards for fishing are set by the Regional Boards and Fishery Departments.

Fishing at sea can be a very dangerous activity. Weather conditions can change rapidly and violent storms are common. It is important that regulations are in place to ensure that all reasonable safety precautions are taken to protect the lives and the safety of fisherfolk.

To do – Safety standards for fishing

Use the internet to search for the websites of the Regional Boards and Fishery Departments for your island. Look at the safety standards set by the different regional boards.

SBA Skills

ORR	D	MM	PD	AI
✓				

Fisherfolk and sailors use various devices to ensure their safety while at sea.

Life rafts

Life rafts may be rigid small boats or inflatables. Large ships carry sufficient life rafts for all their passengers. Smaller boats may use inflatables to save space.

▲ **Fig B3.30** Life boats and an inflatable life raft

Life jackets

Life jackets are flotation devices designed to keep someone from drowning when they are in water. They are brightly coloured so that they can easily be seen. They also have location devices such as a flashing light, whistles and radio detection. Once the life jacket is being correctly worn, it can be inflated automatically with a gas canister of compressed air, or by blowing it up with an inflatable tube.

▲ **Fig B3.31** A life jacket

Scuba diving

Scuba diving is diving underwater with a tank full of breathable air. It can be a very enjoyable activity. But it does have its dangers. You should never undertake scuba diving without first being properly trained by an expert and always make sure you have a 'buddy' to go with you in case of an incident.

Some of the dangers associated with scuba diving are;

▲ **Fig B3.32** A scuba diver

- *Injuries due to changes in air pressure* – Pressure increases rapidly in water as you descend. Pressure on the eardrum and sinuses increases as you descend and needs to be equalised as you go deeper. Failure to do so can result in severe pain and damage to the eardrum and sinuses.
- *Decompression sickness* – As the diver goes deeper and the pressure increases, more and more air dissolves into the blood. This is not harmful until the

▲ **Fig B3.33** A decompression chamber

  diver decides to ascend to the surface. If the ascent is too quick, the effect is like undoing a bottle of lemonade. All the gas dissolved in the solution fizzes out. Bubbles of air can appear in the blood causing extreme pain and blocking blood vessels. This is called an **embolism**. The diver needs to return to the same depth to make the gases re-dissolve into the blood, or go into a decompression chamber where the pressure is gradually reduced to normal atmospheric pressure, so that the dissolved air is breathed out without forming bubbles in the blood.
- *Nitrogen narcosis (rapture of the deep)* – Air is mostly nitrogen. As more and more nitrogen dissolves into the blood stream it can cause a state similar to drunkenness and even cause hallucinations. Divers have been known to take off their masks and drown.
- *Oxygen toxicity* – This is caused when too much oxygen dissolves into the blood. It can be prevented by making sure that the dive limit for breathing air is not exceeded. To dive any deeper the diver must use a different air mixture that often includes the gas helium.
- Other hazards include losing too much body heat, and cuts and grazes from sharp corals, or stings from poisonous fish.

embolism – an obstruction of a blood vessel

End-of-unit questions

1 Which of the following is not a type of fishing?

   a trawling

   b collecting

   c dredging

   d long-line

2 Other than using a compass, describe two ways that fisherfolk can navigate to find fish at sea.

3 Fisherfolk use a compass to find their way at sea. The compass needle is a magnet.

   a Explain what happens when two magnets are placed close to one another.

   b The North pole of a freely suspended compass always points North. Use the magnetic properties of magnets and the Earth to explain why this happens.

   c Explain how you could make your own compass.

4 Safety at sea is important.

   a Explain why safety at sea is regulated by Regional Boards and Fishery Departments.

   b Describe three ways that safety at sea is ensured.

5 Explain the safety hazards associated with scuba diving.

6 You go to spend a day on the beach with your friends.

   a Write down three ways that you and your friends can stay safe on the beach.

     1.

     2.

     3.

   b Write down three ways you and your friends can stay safe if you go out fishing on a boat.

     1.

     2.

     3.

7 You have studied the hazards associated with scuba diving.

   a Write down three hazards associated with snorkelling.

     1.

     2.

     3.

   b Explain how you should do a risk assessment before you go snorkelling or scuba diving.

B4 Pests, Parasites and Sanitation

By the end of this unit you will be able to:

- describe the conditions that promote the growth of microorganisms.
- discuss the principles of food preservation.
- discuss the conditions that encourage the breeding of household pests.
- discuss the different types of waste.
- discuss the need for personal and community hygiene.

Microorganisms

Microorganisms are living microscopic organisms. They consist of bacteria, viruses, fungi and protozoa.

Types of microorganism

bacteria – a single-celled microorganism that does not have a nucleus

Bacteria are single-celled organisms. But unlike normal cells they do not contain a nucleus and the DNA is just found inside the cell. Some bacteria can cause disease in humans. Inside our body they rapidly multiply and release poisonous toxins that make us feel ill. Most bacteria are harmless to humans and some are even useful. We use bacteria for making yoghurt and cheese.

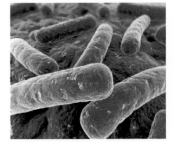

▲ **Fig B4.1** Bacteria

virus – a microorganism that is smaller than bacteria and reproduces inside the cells of living things

Viruses are much smaller than bacteria and cannot even be seen with the most powerful light microscope. Viruses invade cells in living organisms where they multiply before being released to invade even more cells. Some viruses invade the cells of humans and can give us diseases such as influenza, measles and AIDS.

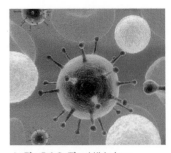

▲ **Fig B4.2** The HIV virus

fungi – moulds, mushrooms and toadstools that live on the dead or decaying remains of living things

Fungi are a group or organisms that are more like plants than animals, but they do not contain chlorophyll. This means they have to grow on other living things or dead remains of plants and animals. Athlete's foot is a fungus that grows on humans.

▶ **Fig B4.3** Athlete's foot is caused by a fungus

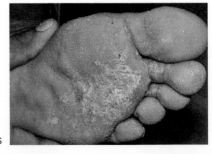

mucor – the biological name for a mould that can grow on foods such as bread

Bread mould (**mucor**) is a fungus that grows on bread.

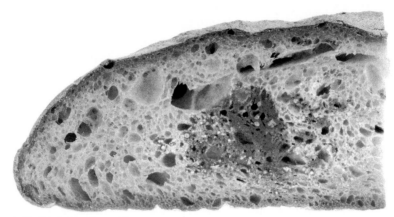

▲ **Fig B4.4** Bread mould

protozoa – a single-celled microorganisms, such as plasmodium that causes malaria

Protozoa are another group of single-celled organisms. Unlike bacteria, however, they do have a nucleus that contains their DNA. They tend to be larger than bacteria and can cause diseases in humans, such as malaria and sleeping sickness.

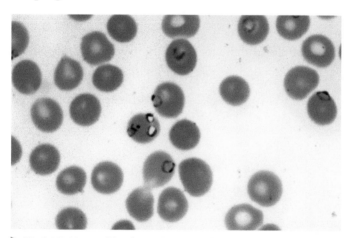

▶ **Fig B4.5** Human blood containing a malarial parasite

Preserving food from microorganisms

All living organisms require certain conditions to be able to survive. Just like us, microorganisms need warmth, food and water for their survival. We can use this knowledge to our advantage when we want to preserve food.

When microorganisms land on food they may use the food in order to grow. Bread mould (mucor) releases enzymes into the food to digest it before the cells of the mould absorb the digested bread in order to grow. Bacteria also multiply on food and can release poisonous toxins that can give us food poisoning when we eat the contaminated food. Our bodies are quite capable of destroying just a few bacteria on our food, but when left, the bacteria quickly multiply to numbers that can overwhelm our body's defences when we eat the food. The bacteria can then grow in our gut and this can also give us food poisoning.

SBA Skills

ORR	D	MM	PD	AI
				✓

> ### To do – Multiplication of bacteria
>
> Bacteria can divide into two about every 20 minutes. Imagine a bacterium landed on some food. If the bacterium divided into two every 20 minutes calculate how many bacteria there would be on the food after eight hours. Remember the bacterium starts as 1, after 20 minutes there will be 2, after 40 minutes there will be 4, after 1 hour there will be 8 etc.
>
> Would you really want to eat the food after just eight hours? How many bacteria would there be?

Fortunately, there are several ways of preserving food and preventing it from being spoiled by microorganisms. These involve either removing microorganisms from the food and sealing it, or making the conditions within the food very unfavourable for microorganisms to grow and reproduce.

Salting

salting – adding salt to food to preserve it

Adding salt to food (**salting**) has been used for centuries as a means of preserving food. In Unit A1 you studied osmosis and how water moves from a less concentrated to a more concentrated solution. When salt is added to food, water moves out of the food into the salt solution. When a microorganism lands on the food they also lose their water as it leaves their cells by osmosis and enters the salt solution. This dehydrates and kills the microorganism.

Q1 Suggest why salt is not used to preserve fruit such as bananas.

Adding sugar

sugaring – adding sugar to food to preserve it

Adding sugar to food (**sugaring**) works in the same way as adding salt. Just as with salting, when a microorganism lands on the food it dies because water leaves the microorganism by osmosis. The process is used to make jam and marmalade.

Q2 Suggest why sugar is not used to preserve fish.

Pickling

pickling – adding vinegar to food to preserve it

Pickling is adding vinegar to food. Vinegar contains an acid called ethanoic acid. This is why it tastes so sharp. Microorganisms do not like acidic conditions and are unable to grow and reproduce in them. This is why foods such as pickled onions stay fresh for such a long time.

Adding other chemical preservatives

preservatives – chemicals added to food to preserve it

Preservatives are chemicals that are added to food that are harmless to humans, but prevent the growth of microorganisms. Technically, salt, sugar and vinegar are preservatives, but the term is usually used to mean chemicals that are added in smaller quantities. Some of them such as anti-oxidants preserve the texture and flavour of the food. Others, such as sodium metabisulphite and sulphur dioxide, kill any microorganisms in the food.

SBA Skills

ORR	D	MM	PD	AI
✓				

To do – Preservatives in food

1 Look at the food label in Fig B4.6. Name the preservative that has been added to the food.

2 Collect labels from other foods and make a list of all the chemicals that are added to them.

▲ **Fig B4.6** A food label

Sometimes we want to preserve food without adding anything extra to it.

Q3 Suggest why we may want to preserve food without adding anything extra to it.

Canning

canning – a method of preserving food by sealing it away from microorganisms

When food is heated above 100°C, microorganisms are killed. The food can then be sealed in a tin can (**canning**) and it will stay fresh for a very long period of time. This is because no microorganisms can get to the fresh food. The reason why canned food has a 'use by' date is not that it will be unsafe to eat, but that after a long period of time in the can, chemical reactions within the food may slightly change the food's colour or flavour.

Refrigeration and freezing

refrigeration and freezing – methods of preserving food by cooling it

Because microorganisms are living things, they need warmth to be able to grow and reproduce. If the food is cooled down (**refrigeration**), the rate of reproduction of microorganisms is also slowed down. If the food is frozen, the growth of microorganism comes to a stop. **Freezing** does not kill microorganisms like heating does. Once the food is defrosted and allowed to warm, the growth of the microorganisms continues. This is why we are told that once food has been defrosted, it should not be refrozen.

Q4 Explain why once food has been defrosted, it is not a good idea to refreeze it.

Heating

heating – a method of preserving food by killing microorganisms through heat

We have already said that **heating** food to above 100°C will kill all the microorganisms. However, heating food to such a high temperature also changes the way the food tastes. This is fine when we are heating food such as meat that is being stored in a tin can. However, some food such as milk tastes quite different when it has been boiled. Fortunately, when milk is heated to 63°C for 15 seconds and then rapidly cooled, all the dangerous bacteria in the milk are killed. This process is called **pasteurisation** and is now used worldwide to make milk safe to drink.

pasteurisation – a way of making milk safe to drink through heating

Drying

drying – a method pf preserving food by removing its water

Drying food is also an effective way of preserving it. Microorganisms need water in order to live. Some food, such as rice, is naturally very dry and stays fresh for a long time provided it does not get damp. Other foods, such as some breakfast cereals, are manufactured to be dry. They also have a very long shelf life, even once the packet is opened. Other foods, such as fruit, can be eaten dried, for example grapes are dried to make sultanas.

Q5 Suggest why sun-dried tomatoes stay fresh for a long period of time.

SBA Skills

ORR	D	MM	PD	AI
✓				✓

Experiment – How to stop mould from growing on bread

1 Take five pieces of bread. Place one piece of bread in each of the following five conditions and leave for at least one week.

- Leave one piece of bread in a plastic bag at room temperature.
- Place one piece of bread in a refrigerator.
- Place one piece of bread in a freezer.
- Toast another piece of bread until it is dry and crispy and leave in a dry place.
- Sprinkle one piece of bread with salt.

2 Observe which of the samples of bread grow mould, and which do not.

Did you know? Salt is added to bread when it is being made to help keep the bread fresher for longer.

Pests

pest – an animal that is a nuisance to humans

A **pest** is an organism that is a nuisance to human beings. It may live in places where we do not want it to live, such as our homes. Or it may cause damage to our belongings or eat our crops and spoil our food. Sometimes they even act as a **vector** (carrier) of disease.

vector – an animal that can carry disease-causing microorganisms

Unlike microorganisms, pests do not cause disease, but are often capable of spreading it by carrying the disease-causing microorganisms from one place to another. The bubonic plague (Black Death) that killed millions of people in the 14th century was caused by a bacteria spread by fleas that lived on black rats. When the flea bit a human they were injected with the bacteria that caused the plague.

parasite – an organism that lives on or in another organism at the expense of that organism

host – an organism on which a parasite lives

The flea is an example of a **parasite**. A parasite is an organism that lives on or in another organism called the **host**. Parasites obtain shelter and food from their hosts.

Rubbish tips are ideal breeding sites for pests. This is because there is plenty of food and water for them to feed on.

▲ **Fig B4.7** A rubbish tip

Many pests are attracted to houses where humans live. This is because they find ideal conditions in which they can grow, breed and find food. Pests that are attracted to our homes include rats, mosquitoes, flies and cockroaches. The best way to keep pests out of our homes is to clean regularly and make sure that we do not leave waste food lying around the house or kitchen.

Q6 Suggest three things that we could do in our homes to prevent pests from living there.

Rats

Brown rats are a serious nuisance. There are more rats living on the planet than humans, so even though we may not see them, we are never very far away from a rat. Because, just like humans, rats are mammals, they can carry lots of diseases that humans can catch. Their urine and faeces can easily contaminate our food and drinking water passing on diseases such as **leptospirosis**. Leptospirosis can be fatal if it is not treated and causes damage to our brain and other organs of our body.

▲ **Fig B4.8** A brown rat

leptospirosis – a disease spread by rats

There are several ways that we can control rats. Common methods include using poisoned bait, traps and even using cats to catch them. However, rats breed very quickly and rat numbers can soon recover if we are not vigilant in keeping our environment clean and free of waste food.

Mosquitoes

Mosquitoes are blood-sucking insects. They are also carriers (vectors) of disease. This makes them extremely dangerous because as they suck our blood they can transmit disease-causing microorganisms into our blood. Diseases that are transmitted by mosquitoes include malaria and sleeping sickness.

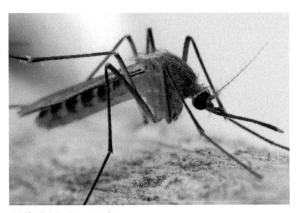

▲ **Fig B4.9** A mosquito

Mosquitoes are insects and therefore have the usual insect life cycle (see Fig B4.10). The female mosquito lays her eggs in stagnant water. Within a few days, the eggs hatch into wriggling **larvae** that hang just beneath the surface of the water. The larvae breathe through a breathing tube that they push through the surface of the water. They feed on microscopic life in the water and grow quickly. They then turn into **pupae** before hatching out into the adult mosquito.

larvae – part of the life cycle of an insect

pupae – part of the life cycle of an insect

It is the female mosquito that sucks blood and after hatching she flies away looking for her first blood meal. Mosquitoes use special mouth parts that pierce the skin of their victim, inserting a narrow tube through which they can suck the blood. First, however, they inject some **anticoagulant** to stop the blood from clotting. This is the point where an infected mosquito also injects the victim with the malarial parasite.

anticoagulant – a chemical that stops blood from clotting

Knowing the life history of the mosquito is important if we want to know how to control them and stop them spreading malaria. Not giving them places to breed is a good start. Anywhere that water can collect is a potential breeding site for mosquitoes, even an old tin can filled with rain water. We can spray their breeding sites with insecticides to kill them.

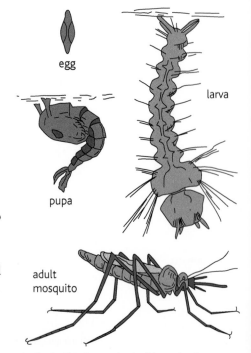

▲ **Fig B4.10** Stages in the life cycle of a mosquito

Because mosquitoes tend to feed at night, using mosquito nets when we sleep can be very effective in stopping the spread of malaria.

▲ **Fig B4.11** Using a mosquito net

Flies

There are many different types of flies but house-flies are a common household pest. Like the mosquito, the house-fly is an insect. However, it lays its eggs in faeces or rotting organic matter. The eggs hatch after about two weeks into wriggling larvae or maggots that feed on the rotting remains. After about a week the larvae turn into pupae. Pupa have a hard brown case and inside the larva

▲ **Fig B4.12** A house-fly

changes into an adult fly and emerges about five days later.

Although the fly does not spread disease-causing microorganisms like the mosquito, because it feeds on dead rotting waste, it spreads disease-causing microorganisms when it lands on our food. Flies feed by releasing a liquid that contains digestive enzymes. These enzymes start to digest the food, which the fly then sucks up. If a fly has been feeding on dog faeces and then lands on your food, just imagine all the microorganisms that can contaminate your food.

Cockroaches

Cockroaches are very well adapted to living in our homes. They can cause a lot of damage to our food, clothing and even household objects like books and curtains.

The best way to ensure that cockroaches do not breed is to keep all food enclosed in containers. They can also be controlled by spraying insecticide into cracks and under ledges where the insects can hide and lay their eggs.

▲ **Fig B4.13** A cockroach

Control of pests

Pests can be controlled in a number of different ways.

biological control – using another organism is used to control a pest

Biological control – is when another organism is used to control the pest. An example of this is putting tilapia fish into lakes. The fish eat the mosquito larva so fewer larvae hatch out into mosquitoes.

mechanical control – using a physical structure to control a pest such as a mosquito net

Mechanical control – Mosquito nets are an example of mechanical control. They are a physical barrier between the sleeping person and the mosquito.

chemical control – using insecticide to control a pest

Chemical control – An example of this is the spraying of insecticide on the places where mosquitoes can lay their eggs and breed.

sanitary control – keeping the environment clean to prevent the spread of pests

Sanitary control – This is something we can all do. Not leaving tin cans lying about that can fill with rain water and act as a reservoir for breeding mosquitoes. Not leaving food lying around that can attract rats and cockroaches.

Waste

domestic waste – waste produced form our homes

industrial waste – waste produced from factories and places of work

biological waste – sewage

All of us produce waste. This waste may be **domestic waste** from our homes, **industrial waste** from factories and the workplace, or even **biological waste**, such as sewage.

It is important that we reduce our waste as much as possible and that what waste we do produce is recycled.

To do – Waste diary

Keep a diary of every single thing that you throw away in one day. You will be surprised! Imagine this multiplied by 365 and this is the amount of waste that you throw away each year.

SBA Skills

ORR	D	MM	PD	AI
✓				

Biodegradable and non-biodegradable waste

biodegradable – breaks down in the environment

non-biodegradable – is not broken down in the environment

Some waste is **biodegradabl**e and will be quickly broken down by sunlight and microorganisms in the environment. This waste is less of a problem. Other waste is **non-biodegradable**. This includes waste like most plastics that will remain in the environment for hundreds of years without breaking down. It is important that we recycle as much of this waste as possible.

Disposal of waste

Domestic or household waste is all the waste that we throw away from our homes. It includes the remains of food, packaging from things that we buy, glass and plastic containers, paper, card, old clothes and even sewage.

sanitation – ensuring our living environment is clean and free from pests, parasites and disease-causing microorganisms

It is important that this waste is disposed of properly and either taken to proper collection sites or, in the case of sewage, flushed down the toilet. **Sanitation** ensures that our living and working environment is clean, tidy and free from pests and parasites that could cause disease. To achieve this we need good personal hygiene and good methods of disposing of sewage and waste from our homes and workplaces, such as garbage collection.

There are three ways of dealing with waste other than sewage: landfill, incineration and recycling.

Landfill

landfill – burying waste material to dispose of it

Landfill is where waste is buried beneath the ground. Old abandoned quarries are often used. The waste is then covered with a layer of earth and allowed to rot. The rotting produces methane, which can be collected and use as a fuel.

The problem with landfill is that it is often difficult to find a piece of land that can be used. It can also attract pests such as rats and even contaminate local water supplies. To avoid problems with pests and smells, the landfill is often done in a series of stages where each stage is covered with earth before starting on the next stage.

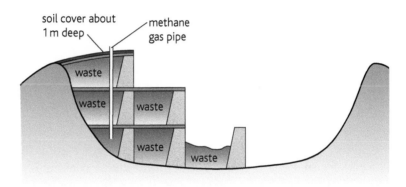

▲ **Fig B4.14** Landfill

Incineration

incineration – burning waste material to dispose of it

Waste can also be burned, or **incinerated**. The disadvantages of incineration are that the incineration plant is expensive to build and it is important that toxic gases are not released into the atmosphere. The advantages of incineration is that the waste is permanently disposed of and the heat generated can be used to heat local buildings. It also does not use up land, attract pests or contaminate local water supplies.

Recycling

recycling – reusing materials so that we do not keep consuming a limited supply of raw materials

Many of the things that we throw away can be **recycled**. Paper can be recycled to produce recycled paper. This avoids cutting down trees to make the wood-pulp for new paper. Glass bottles can be melted down to make new glassware and metals such as copper and aluminium can be recycled rather than destroying our environment by mining them from the ground. Recycling not only preserves our environment but it is also often cheaper than using new materials.

To do – Recycling plastic

Plastic bottles can also be recycled. Use the internet to research all the uses that recycled plastic can be put to.

SBA Skills

ORR	D	MM	PD	AI
✓				

Sewage

It is important that sewage is disposed of properly. Failure to do so can have serious consequences for our health. Unsanitary conditions can rapidly lead to the spread of microorganisms and lead to diseases such as cholera and typhoid. Sewage should be disposed of properly either by a flushing toilet or an earth latrine that is properly looked after. It is important that sufficient toilets are provided for our community. It is also important that we observe personal hygiene by washing our hands after we use the toilet to avoid food contamination.

To do – Researching the discovery of the cause of cholera

Years ago people did not realise that diseases such as cholera were caused by poor sanitation. Use the internet to research how Dr Snow discovered the cause of cholera and what he did to prevent it. You could create a short presentation to give to the rest of the class.

SBA Skills

ORR	D	MM	PD	AI
✓				

If a sewage system has been provided for the community, the sewage is taken by a series of pipes to the sewage station. Most sewage will flow through the pipes by gravity but sometimes it may have to be pumped to the sewage station.

When the sewage reaches the sewage station there are four ways that it can be treated. Most sewage stations use a combination of at least two of them.

Filtration

The sewage first goes through a grid to remove any large debris that has got into the sewage. You would be amazed at what some people flush down the toilet. The sewage is then allowed to stand in a collection tank where large particles separate out by falling to the bottom of the tank. The liquid is then sprayed through a rotating sprinkler onto a filter bed of stones and grit. Microorganisms grow on the grit and break down the organic matter in the sewage as it passes through the filter. By the time the water leaves the filter it is clean enough to be fed back into streams and rivers. The sludge left in the tanks may then be treated by the activated sludge method.

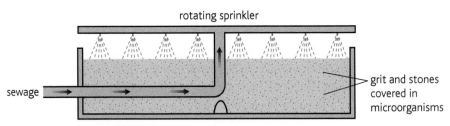

▲ **Fig B4.15** A sewage filter bed

Activated sludge method

activated sludge – sludge produce from sewage works

In the **activated sludge** method the sludge is placed in large tanks and air is bubbled through the liquid. The oxygen in the air encourages the growth of microorganisms that break down the organic matter in the sludge. The liquid is then clean enough to be fed back into streams and rivers and any solid that remains can be placed in digesting tank to release methane, which can be used as a fuel. Any solid remains can then be dried and used as a fertiliser on farms and gardens.

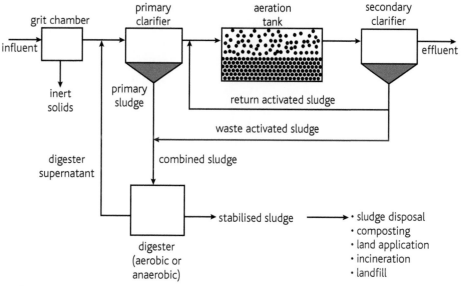

▲ **Fig B4.16**

Water stabilisation ponds

water stabilisation ponds – a way of treating sewage to make it safe

Small-scale sewage treatment can be done using sewage **water stabilisation ponds**. The liquid sewage is allowed enter a reed pond. As the liquid filters through the pond, microorganisms break down the organic remains. The liquid overspills into second and even third ponds where the breakdown of organic remains continues. Finally the clean water is allowed back into rivers and streams.

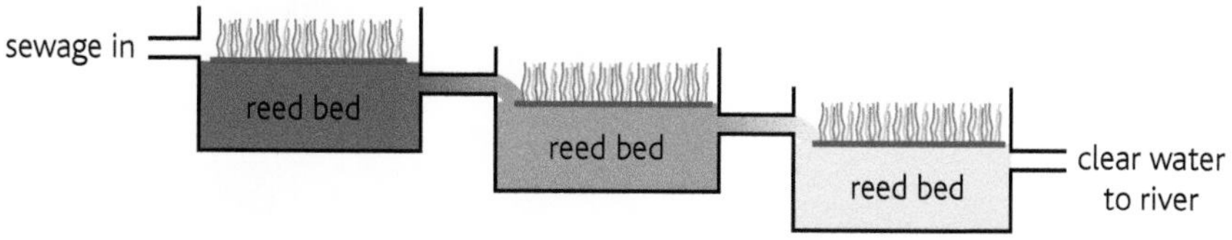

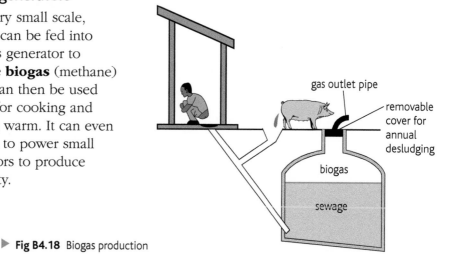

▲ **Fig B4.17** Water stabilisation ponds

Biogas generators

biogas – a gas called methane that is produced when organic waste material decays

On a very small scale, sewage can be fed into a biogas generator to produce **biogas** (methane) which can then be used as fuel for cooking and keeping warm. It can even be used to power small generators to produce electricity.

▶ **Fig B4.18** Biogas production

Consequences of lack of personal and community hygiene.

We sometimes take clean drinking water and disposal of our waste for granted. We turn on the tap and clean water is there. We flush the toilet or put our waste in bins, and then it is gone. The consequences for both our personal health, and that of the community, depend upon these things happening. We only have to look at those parts of the world where war or famine prevent these things from taking place, to see what the consequences are.

One of the first effects of poor sanitation is the spread of diseases such as cholera. You have already investigated how Dr Snow stopped the spread of cholera just by providing clean drinking water and good sanitation. Cholera is a water-borne disease. Once our water supplies become contaminated with sewage, it is not long before the first cases of cholera appear. Other diseases such as typhoid also become commonplace.

Other parasites that quickly spread when sanitary conditions break down are parasitic roundworms. These live in the gut and reproduce and grow to large numbers. Some parasitic worms such as *Ascaris* can grow to over a foot in length and partially block the intestine.

▲ **Fig B4.19** The *Ascaris* worm can be big enough to partially block the intestine

SBA Skills

ORR	D	MM	PD	AI
				✓

To do – What if?

Imagine that in the area that you live, the water supplies became contaminated with sewage. Predict what the specific consequences would be for your area. Think about what would happen to schools and hospitals. How would people get clean water?

Discuss your thoughts as a class.

Next time you turn on the tap or flush the toilet, just think 'What if?'

End-of-unit questions

1 Which of these types of microorganism are responsible for making bread go mouldy?

   a bacteria

   b viruses

   c fungi

   d protozoa

2 Food can be preserved by adding salt. Which of the statements best explains why?

   a Salt is poisonous to microorganisms.

   b Salt makes the food taste unpleasant to microorganisms.

   c Salt dries out microorganisms.

   d Salt acts as a barrier to microorganisms.

3 Which of the following diseases is spread by an insect pest?

   a influenza

   b measles

   c scurvy

   d malaria

4 Waste needs to be disposed of properly.

   a Describe three different ways of disposing of waste.

   b Explain the consequences of not correctly disposing of human sewage.

5 Sewage is disposed of at a sewage treatment works.

   a Explain how a sewage treatment works disposes of sewage.

   b Explain how a biogas generator works.

6 Food can be preserved in many different ways.
One way is to preserve food in alcohol.

   a Suggest how preserving food in alcohol works.

   b Write down the name of three foods that could be preserved in alcohol.

   c Explain why you would not preserve bread in alcohol.

7 Some people live in villages with no fresh water, no mains sewage and no rubbish collection system.
Suggest how the villager should:

   a dispose of their sewage.

   b obtain fresh drinking water.

   c dispose of their rubbish.

B5 Safety Hazards

Safety hazards and risks

One of the most dangerous places we can be is in our own home. More accidents happen there than anywhere else. Most accidents happen because someone has been careless and done something silly.

Did you know that people even have accidents while putting on their socks? How often have you stood on one leg and toppled over when putting socks on? Getting up in the morning can be a dangerous business.

▲ **Fig B5.1** Even putting socks on can be dangerous

hazard – something that can cause you harm

risk – the possibility of harm being caused by a hazard

A **hazard** is something that can cause you harm. It can be anything from a sharp pair of scissors, to crossing the road when you leave your home. A **risk** is the possibility of harm being done to you or someone else by a particular hazard.

It is important to identify hazards because then we can do something to reduce the risk. We need to say why a particular hazard is dangerous and then say what we should do to reduce the risk.

For example, when you use a Bunsen burner at school, the Bunsen burner is a hazard. If you do not take care there is a risk that you will burn yourself. Fortunately there are ways of reducing the risk of getting burned.

- First of all, when using a Bunsen burner, if you have long hair it should be tied back. That reduces the risk of your hair falling into the flame and catching light.

▲ Fig B5.2

- Also, the flame of a Bunsen can be controlled by turning the gas tap on or off and by adjusting the air hole at the base of the Bunsen burner.

- When the Bunsen is being used the air hole should be gradually opened until all trace of the yellow flame disappears. This produces a hotter but less luminous flame as the gas is burned with the oxygen in the air. However, when the Bunsen is not being used the Bunsen should be turned off, or if not being used for just a short time, the air hole should be closed. This

▲ **Fig B5.3** A Bunsen burner with air hole

reduces the oxygen reacting with the gas and the flame burns with a much more visible yellow flame, due to less efficient **combustion**.

combustion – burning when a fuel and oxygen combine to release heat and light

- As you can see in Fig B5.4, it is much easier to see the yellow flame than the pale blue flame. Making it easier to see means we are less likely to burn ourselves or set fire to our hair. So once we identify the hazard, we can do things to reduce the risk.

▲ **Fig B5.4** A pale flame (hard to see), and a yellow flame easy to see)

Humans are not very good at assessing risk. We will quite happily get in a car and drive at high speeds without any worries at all. Then we will worry about whether our drinking water should have fluoride in it or not.

Q1 Which do you think is more dangerous, travelling at high speed in a car or drinking fluoridated water? Explain your answer.

Safety in the home

Unlike most places where we work, there are very few regulations that tell us how we should behave to keep our homes safe to live in. We are often our own worst enemy when it comes to accidents in the home.

SBA Skills

ORR	D	MM	PD	AI
✓				✓

To do – Hazards in the home

▲ **Fig B5.5**

1 Look at the pictures of possible hazards in our home in Fig B5.5.

2 Make a list of as many hazards as you can see.

Some things in the home are more dangerous than others. Gas and electricity are two hazards found in most homes.

Gas

SBA Skills

ORR	D	MM	PD	AI
				✓

Experiment – Exploding tin can

Your teacher will do this for you as a demonstration behind a safety screen.

1 Get a tin can with a removable lid. Make a small hole in the centre of the lid. Cut a larger hole in the base of the can so that it will fit over the top of a Bunsen burner.

2 Place the can with the lid in place on top of a Bunsen burner. Turn on the gas and fill the can with gas. The can must be full of gas with no air present.

3 Turn off the gas and light the gas coming out of the hole in the top of the can, as shown in Fig B5.6.

After a short time the flame burns down. Air is drawn into the bottom of the can. When the gas/air mixture reaches the correct proportions – BANG! Off flies the lid.

Q2 What are the hazards associated with this experiment and what do you need to do to reduce the risk?

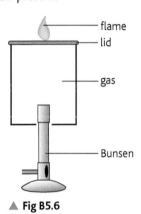

▲ **Fig B5.6**

The experiment above is just a small version of what would happen if there was a gas leak in our home; the consequences would be horrendous. It also teaches us not to delay lighting the gas when we turn the cooker on. Fortunately, a nasty smell is added to the gas in our homes so that if there is a leak, we can smell it.

Electricity

How often have you seen a situation looking like this?

When we put too many plugs into one socket, the wires are in danger of overheating. This can lead to fires. We should also check that we use the correct fuses for each appliance (see Unit C1b). Electric fires and heaters need bigger fuses up to 13 amp. Electric radios need much smaller fuses of about 3 amp. We should always check the manufacturer's instructions and make sure we use the correct size fuse for each appliance.

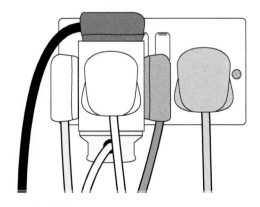

▲ **Fig B5.7**

Faulty electrical equipment can give us an electric shock. Electric shocks over 100 volts can kill. This is why good maintenance is very important and why electrical appliances should be checked regularly to ensure that they are working safely.

It's not even safe watching TV

Even TVs and radios can be dangerous in the home. How often have you rested a cup of coffee on top of the TV or radio? Just imagine what would happen if you spilled it.

radiation – a type of energy released from a source

Old types of TV that use a cathode ray tube also give out low level X-ray **radiation**. These are very low dose, but it is still not advisable to sit up close to a TV for long periods of time. Fortunately, modern flat-screen TVs do not have this problem.

▲ **Fig B5.8** Watching TV can be a risky business

To do – Electrical hazards and risks

SBA Skills

ORR	D	MM	PD	AI
✓				

In groups, discuss the hazards of careless handling of radios, television sets and other equipment that operate from the mains electrical supply. Then discuss how the risk can be minimised.

Food

We do not normally think of food as being a hazard. But, as you learned in Unit B4, food that is left out in a warm place can grow bacteria. Bacteria produce toxins on the food as they grow so that by eating the food we can get food poisoning. Food should be eaten shortly after preparation or stored in a cool place such as a freezer or refrigerator. This slows down the rate at which the bacteria reproduce and keeps the food fresher for longer.

▲ **Fig B5.9** Kitchens can be dangerous places

SBA Skills

ORR	D	MM	PD	AI
				✓

To do – How quickly can food become contaminated by bacteria?

1 Bacteria can divide once every 20 minutes in ideal conditions. Imagine a single bacterium lands on your food. In 20 minutes there will be 2, in 40 minutes there will be 4 and in an hour there will be 8. Calculate how many bacteria will be on the food in 24 hours.

2 Cooling slows down the rate at which bacteria multiply. Now do the same calculations but assume that the bacteria only reproduce every two hours. Compare your two answers.

Q3 How can you reduce the risk from food in your home?

Chemicals

Our homes contain a surprising number of dangerous chemicals. Some of the chemicals we use in our homes would not be allowed to be left around in a school classroom. Some of the chemicals are **caustic**. This means they will burn our skin. Some are **toxic**. This means they are poisonous.

caustic – chemicals that will burn our skin

toxic – chemicals that are poisonous

▲ **Fig B5.10** Hazardous chemicals in the home

225

To do – Chemicals in the home

SBA Skills

ORR	D	MM	PD	AI
✓				✓

1 Make a list of all the hazardous chemicals found in your home.

2 Say which are caustic and which are toxic.

3 Explain how you could reduce the risk of these hazardous chemicals.

Safety in the workplace

The workplace is the place where we work. For us this will be at school. When we leave school it will be where we get a job.

The school laboratory is usually a very safe place because teachers are trained in safety, but there are lots of potential hazards if you do not act carefully.

To do – Hazards in the laboratory or classroom

SBA Skills

ORR	D	MM	PD	AI
				✓

▲ Fig B5.11

1 Look at the picture of a school laboratory in Fig B5.11. It is not such a safe place as yours.

2 Copy and then complete Table B5.1. The first hazard has been done for you.

hazard	risk – why is it dangerous?	what should be done to reduce the risk?
pupils running	they could trip over and hurt themselves	no running in the laboratory

Table B5.1

Rules for practical work
Do not enter the room until your teacher gives you permission.
Wear eye protection when the teacher tells you.
Do not touch equipment unless your teacher says you can.
Put your bag under the bench and keep your bench clean and tidy.
Do not run in the room.
Report any spills, breakages or accidents to the teacher.
Do not eat or drink anything in the lab.
Follow instructions carefully.

Table B5.2

To help us identify hazards at school or the workplace, hazardous chemicals and equipments have hazard signs.

Look at the two hazard signs in Fig B5.12. They tell us when a substance is toxic and corrosive.

TOXIC

CORROSIVE

▲ **Fig B5.12** Hazard signs

SBA Skills

ORR	D	MM	PD	AI
✓				

To do – Hazard signs

1 Collect different hazards signs from your school laboratory. Make a drawing of each sign and write next to it what hazard it is warning you of.

2 Visit other departments in your school such as Industrial Arts, Home Economics or Visual Arts. Identify hazards and collect any hazard signs that you see.

General hazards

Hazards are all around us. We need to be able to identify these hazards so that we can reduce the risk to ourselves and others. Sometimes hazards are not objects. Hazards also include a lack of light. Many accidents can happen in poorly lit places when people cannot see what they are doing. Loud noise is also a hazard. It may just be distracting, but if it is too loud it can damage our ears, resulting in permanent hearing loss.

Sometimes hazards can be invisible, such as when we get food poisoning from eating invisible bacteria or toxins. In 1986 next to Lake Nyos in the Cameroon, West Africa, 1746 people died because a cloud of carbon dioxide had been released from a nearby volcanic vent. The gas cloud was heavier than air and rolled down the hillside, suffocating the villagers while they slept.

SBA Skills

ORR	D	MM	PD	AI
				✓

To do – Spotting hazards around us

▲ **Fig B5.13**

Look at the pictures in Fig B5.13. Identify the hazard in each one. Say how the risk could be reduced.

Fires

Most of us think that fire is just fire. But in fact there are six different types of fire:

Class A: Solids – such as wood, paper and plastic
Class B: Liquids – such as petrol and oil
Class C: Gases – such as methane, propane and butane
Class D: Metals – such as magnesium and aluminium
Class E: Electrical fires – these are where electricity is involved
Class F: Cooking fat – these are fires involving cooking fat and oil

It is essential that the correct type of fire extinguisher is used with each type of fire. For example, if you used water to put out an electrical fire you could end up being electrocuted. Also, if you use water to put out a chip pan fire, the water will sink to the bottom of the pan, reach boiling point, and then explode burning fat all over the kitchen making matters much, much worse.

There are four different types of fire extinguisher that are available to the public, as shown in Fig B5.14.

▲ **Fig B5.14** A water, foam, dry powder and carbon dioxide fire extinguisher

- Water: Identified by having a red label. Used on Class A fires but NOT where electricity is involved.
- Foam: Identified by having a cream label. Used on Class A and B fires but not recommended for fires where electricity is involved.
- Dry powder: Identified by having a blue label. Used on Class A, B and C fires. But any gas supply should be turned off before being used.
- Carbon dioxide: Identified by having a black label. Used on Class B and E fires.

SBA Skills

ORR	D	MM	PD	AI
✓		✓		✓

Experiment – Making a home-made foam fire extinguisher

You will need: safety glasses, 250 ml glass flask, washing up liquid, dilute hydrochloric acid (bench strength or vinegar will do instead), sodium hydrogencarbonate (sodium bicarbonate or baking powder).

1 Put your on safety glasses.

2 Place $1\,cm^3$ of washing up liquid and two teaspoons of sodium hydrogencarbonate into the flask.

3 Add 10 ml of acid and observe what happens.

4 State what types of fires you could use your foam fire extinguisher on.

When fire fighters are putting out fires they need to think about the fire triangle.

AIR   FIRE   HEAT

FUEL

▲ **Fig B5.15** The fire triangle

A fire needs all three sides of the triangle to be able to burn. Remove any one side and the fire goes out.

To do – Fighting fires

Suggest how a fire fighter could put out the following fires.

a a fire inside a metal box

b a bonfire partly burning

c smouldering furniture next to a switched on electric fire

SBA Skills

ORR	D	MM	PD	AI
				✓

Bush fires

Bush fires are different from most other fires in that they may cover a much larger area and do far more damage.

Q4 What types of substance would you use to extinguish a bush fire – water, foam, dry powder or carbon dioxide? Explain your answer.

Bush fires do not only burn bush. They also threaten the homes and lives of local residents.

▲ **Fig B5.16** Bush fires can spread quickly and be difficult to control

The Caribbean Association of Fire Fighters was set up on 26th October 2000 in Castries, St Lucia. It provides an organisation for all the fire fighters in the Caribbean with an aim to improve conditions in all fire fighting departments through training and support.

▲ **Fig B5.17** The badge of the Caribbean Association of Fire Fighters

Bush fire prevention

fire break – a strip of land from which vegetation is removed to stop bush fires from spreading

One method of fighting bush fires is to use **fire breaks**. Fire breaks are strips of bare land that are intended to halt the spread of the fire. They also provide quick access routes for fire fighters to get to the fire. However, they do not always work. In windy conditions, bits of burning debris can be blown by the wind, allowing the fire to jump across fire breaks.

▲ **Fig B5.18** A fire break

Protective clothing

Protective clothing is important for our safety. We even wear protective clothing at school.

▲ **Fig B5.19** A student wearing safety clothes

Q5 What type of protective clothing is the girl wearing in Fig B5.19? Suggest what she may be doing at school.

SBA Skills

ORR	D	MM	PD	AI
				✓

To do – The right clothing for the job

1 Look at the people in Fig B5.20. They are all wearing protective clothing.

▲ **Fig B5.20**

2 Make a list of the protective clothing that each one is wearing.

Q6 The surgeon is missing one vital piece of protective clothing. Can you think what it is?

First Aid

Everyone should know basic First Aid. Accidents are not planned. They happen when we least expect them. Even when we are taking care and obeying all the rules, accidents can still happen. Knowing some basic First Aid may save someone's life.

Electric shock

Electric shocks can be fatal. This is what you should do.

1 DO NOT touch the patient until you have assessed the situation and carried out Steps 2 and 3, ONLY IF it is safe for you to do so.
2 If the electric shock has come from equipment plugged into the mains, switch off the mains switch.
3 Remove the contact between the patient and the source of the electric shock by using a non-conductive stick such as a wooden broom handle.
4 Call for help or phone for an ambulance.
5 Check to see if the patient is unconscious and breathing.
6 If breathing has stopped, begin CPR (Cardio Pulmonary Resuscitation) – see page 234.
7 If the patient is breathing but unconscious, place them in the recovery position.

▲ **Fig B5.21** The recovery position

8 Treat the patient for shock by loosening tight clothing and reassuring them. Treat any burns as described below.

Q7 Suggest why you should always switch off and isolate the electrical supply before handling the patient.

Burns

Burns can be caused by heat, chemicals or electric shock. This is what you should do.

1 Place the burned area under clean, cold running water for at least 10 minutes, longer if the burn is caused by chemicals. This dissipates the heat and prevents further burns. It also helps to reduce pain. If the burns are severe, call for help while the washing proceeds.
2 Remove any jewellery such as rings. Swelling may prevent their removal later.
3 Lightly cover the burned area with sterile lint-free dressing. Clean cling film will do if a proper dressing is not available.
4 Seek medical help if necessary.

DO NOT
1 Use creams.
2 Burst any blisters or remove skin.
3 Give the patient food or drink. This will hinder any operation that the patient may need.
4 Waste time. The sooner you can get help and get cold water on the burn, the better.

Q8 Explain why rings and jewellery should be removed from burns patients.

Mouth to mouth resuscitation or CPR

Cardiopulmonary resuscitation (CPR) is used whenever a patient is unconscious and has stopped breathing. This is what you should do.

1 Check to see if the patient is unconscious. If they are, **call the emergency services**. This means saying who you are, where you are, your telephone number, and the nature of the emergency.
2 Tilt the patient's head back and listen for breathing. If they are not breathing, pinch their nose and cover their mouth with yours and blow until you see their chest rise. Give two breaths, one each second.
3 Place the palms of your hands between their nipples and push down hard 30 times, about two pushes per second.
4 Give two more breaths and 30 pushes. Repeat until help arrives.

This process is shown in Fig B5.22.

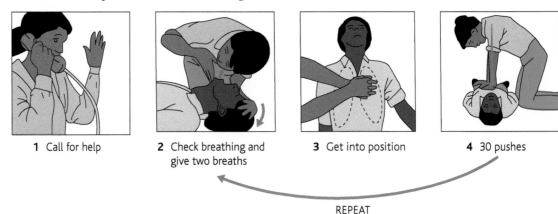

| 1 Call for help | 2 Check breathing and give two breaths | 3 Get into position | 4 30 pushes |

REPEAT

▲ **Fig B5.22**

Q9 Suggest why the time taken to start resuscitation is of crucial importance.

As you can see, hazards and risks are all around us. It is up to us to identify these hazards and then minimise the risk. That way we should stay safe.

End-of-unit questions

1 Which of the follows describes a hazard?

a an accident

b an emergency

c an object that can harm you

d the chance that you might get hurt

2 Which of the following describes a risk?

a an accident

b an emergency

c an object that can harm you

d the chance that you might get hurt

3 Describe how a Bunsen burner should be safely used in a school laboratory.

4 Look at the fire extinguisher in Fig B5.23.

▲ **Fig B5.23**

Explain when you would use it and when you would not.

5 The following pictures show CPR (cardiopulmonary resuscitation).

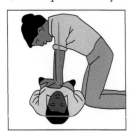

A

B

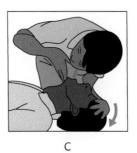

C

D

▲ **Fig B5.24**

a They are in the wrong order. Write out the correct order.

b For each picture write down exactly what you should doing.

6 Imagine you are being taught in a school that is not as safe as it should be.
Design and draw a hazard for each of the following situations.

a Very steep stairs

b A balcony with no hand rail

c A deep hole next to a path.

7 You go on work experience to a garage that spray paints cars.

a Write down three hazards that you might encounter

i

ii

iii

b Write down the nature of the risk (why each hazard is dangerous).

i

ii

iii

c Write down a way that you could reduce the risk from each hazard.

i

ii

iii

B6 Materials

By the end of this unit you will be able to:

- relate the uses of metals and non-metals to their properties.
- discuss the advantages and disadvantages of using plastics.
- describe the reactions of metals with oxygen, acid, alkali, water and steam.
- discuss the advantages and disadvantages of using cooking or canning utensils made of aluminium.
- discuss methods of cleaning household appliances.
- discuss the benefits of using alloys to make household items.
- discuss the conditions that cause rusting.
- identify the factors that affect the rate of rusting.
- discuss the methods used to reduce or prevent rusting of iron or steel.

Metals

Life without metals would be difficult to imagine. We find their properties very useful when making a variety objects to meet the needs of modern society. Here is a list of the general properties of metals. Metals:

- are good conductors of electricity
- are good conductors of heat
- are hard and strong
- have high melting points
- have high densities
- are shiny
- are **ductile** (can be drawn out into wire by applying forces)
- are **malleable** (can be hammered into shapes without smashing)
- are **sonorous** (they ring when struck).

ductile – describes a material that can be drawn out into wire by applying forces in opposite directions

malleable – describes a material that can be hammered into shapes without smashing

sonorous – describes a material that rings when struck with a hard object

▲ **Fig B6.1** Metals have a wide variety of uses

So, for example, we use copper metal for electrical wiring because it is a good conductor of electricity and it is ductile.

Q1 Choose from the list of properties above to explain why we use metals for:

a pans

b musical steel drums

c chicken-wire fencing

d jewellery

The metals iron, cobalt and nickel are also magnetic but most metals are not.

Alloys

alloy – a mixture of a metal with one or more other metals or non-metals

Steel is also magnetic but that is because steel is made up mainly of iron. Steel is an example of an **alloy**. An alloy is a mixture of a metal with one or more other metals or non-metals. We make metal alloys by melting a metal, then adding in the other metals and stirring them together. The metallic elements do not react with each other. That is why we say that an alloy is a mixture and *not* a compound. For more on mixtures and compounds, see Unit B7a.

We make alloys to improve the properties of metallic elements to make them better suited for particular uses. Pure metals are made up of atoms that are all the same size. The metal atoms pack together closely in layers. Therefore, if we apply a large enough force the layers of atoms will slide over each other quite easily. However, we can make pure metals stronger (and harder) by adding atoms of a different size. This makes it more difficult for the layers to slip past each other. It resists stress forces, such as stretching or impact, more effectively. We say that the alloy has a greater **tensile strength** than the pure metal. Look at Fig B6.2:

tensile strength – the ability of a material to resist pulling forces and to keep its shape

Pure metal

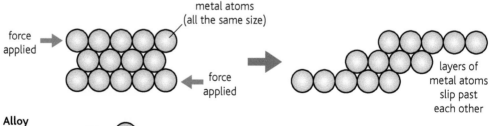

Alloy

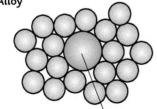

some atoms of a different size make it harder for layers to slip

▲ **Fig B6.2** Alloying a metal can make it stronger and harder

To do – Making springs

1 Cut a piece of copper wire 30 cm long. Wrap it tightly around a pencil so it forms a spring.

2 Do the same to a piece of wire alloy of the same diameter (e.g. steel or constantan).

3 Make a hook-shape at each end of the wires.

4 Collect some slotted masses (10 g masses should be sufficient if the wire is not too thick).

Now plan an investigation to find out which metal would make the better spring – copper or the alloy. Use Fig B6.3 to guide you. Remember to take any measurements as precisely as you can to back up your decision.

Explain your conclusion in terms of the data you have collected and using what you know about pure metals and alloys.

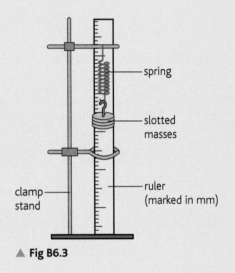

▲ Fig B6.3

Q2 Explain why 'copper' coins are actually made of an alloy of copper with other metals such as zinc and tin.

Common examples of alloys

steel – an alloy made mainly of iron with varying traces of carbon, and sometimes other metals added to modify its properties

Steel is an alloy made mainly of iron (usually over 95%) with varying traces of carbon, and sometimes other metals added to further modify its properties. For example, a very hard and tough steel is made by adding a little tungsten. Stainless steel, which does not rust, is made by adding nickel and chromium. Table B6.1 shows some carbon steels and their uses.

▲ Fig B6.4 The high-carbon steel of the cutting tool can be used to shape softer types of steel

type of steel	amount of carbon	hardness	uses
mild steel	0.2%	can be easily shaped	car bodies, wires, pipes, bicycles
medium steel	0.3–0.6%	hard	girders, springs
high-carbon steel	0.6–1.5%	very hard	drills, hammers, other tools

Table B6.1

brass – an alloy of copper and zinc

Brass is an alloy of copper and zinc. Musical instruments made of brass make a pleasant, sonorous sound. The alloy also allows the intricate shapes of the brass instruments to be stamped or pressed out of the sheet metal.

Brass is an attractive gold colour. It is also much harder than copper or zinc. This has led to its use in making door fittings that need to withstand a lot of wear and tear and also look good. The alloy has a lower melting point than either of the metals from which it is made so it is easier to cast into shapes in a mould.

▲ **Fig B6.5** Brass is a shiny, hard alloy

solder – an alloy of lead and tin

Solder is usually an alloy of lead and tin. It is also known as soft solder. It is used to connect the different parts of an electrical circuit. You use a soldering iron to do this job. The hot iron easily melts the solder, which is bought as a thick wire. It solidifies to form the connection.

Other similar alloys melt at even lower temperatures, which makes them ideal for use in automatic fire sprinkler systems. The alloy is located in each sprinkler. If the temperature of the sprinkler reaches about 70°C, the alloy melts and lets water in the pipes above it spray down onto the fire.

▲ **Fig B6.6** Solder melts at a low temperature

To do – Using solder

SBA Skills

ORR	D	MM	PD	AI
	✓			

1 Use a soldering iron and some solder to join two pieces of copper wire. Take care not to touch the hot iron – remember that it will stay hot for a while after you switch it off.

2 Test the connection you have made with a battery and a light bulb.

3 Draw a diagram of the circuit you used.

Q3 Explain why the head of a hammer is made of steel and not pure iron.

Q4 Which type of steel would you use to make a sink? Why?

Q5 Why is brass used to make door knobs?

Q6 Which metals are used to make soft solder?

Chemical reactivity of metals

It is important for materials scientists, as well as builders, to know about the chemical reactivity of metals so that they can choose the best metal for a particular job. For example, copper is not only used in electrical wiring in homes. The water pipes are also often made of copper. Copper is suitable for use by plumbers because they can easily bend the copper pipes to run around corners and awkward shapes. However, even more important than that is the fact that copper metal does not react with water.

To find out about the reactivity of some common metals we can compare their reactions with substances they are likely to come into contact with. For example, oxygen (the reactive gas in the air), water, dilute acid and alkali, which are found around the home and in the workplace. We can start by looking at the reaction of a fairly reactive metal, magnesium, with dilute acid.

SBA Skills

ORR	D	MM	PD	AI
✓		✓		✓

Experiment – Metals plus acid

1 Add a small piece of magnesium ribbon to a 2 cm depth of dilute sulphuric acid in a test tube.

2 Hold a boiling tube upside down over the mouth of the first tube. Hydrogen gas is less dense than air so will displace air from the boiling tube.

3 Remove the boiling tube and test the gas collected by putting a lighted splint near its mouth. What happens?

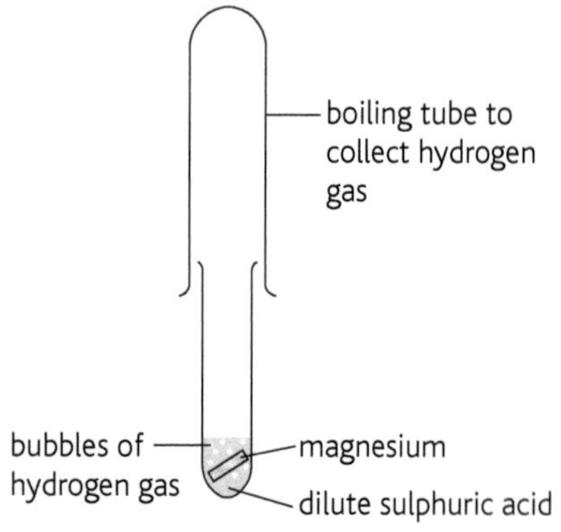

boiling tube to collect hydrogen gas

bubbles of hydrogen gas

magnesium

dilute sulphuric acid

▲ **Fig B6.7**

4 Now repeat steps **1** and **2**, but instead of using magnesium use iron, aluminium, zinc, and copper. Use sandpaper to clean the surface of each metal before adding it to the acid.

5 You should observe and record how quickly any bubbles of gas are given off to judge how quickly the metals react. You do not need to test for the gas – if a gas is given off it will be hydrogen.

6 What is the order of reactivity, starting with the most reactive metal?

If a metal reacts with a dilute acid it gives off hydrogen – a flammable gas. Here is the general word equation:

$$\text{metal} + \text{acid} \longrightarrow \text{a salt} + \text{hydrogen}$$

A salt is a compound formed when the hydrogen in an acid is wholly, or partially, replaced by a metal. Different acids will form different salts. Sulphuric acid forms salts called sulphates, hydrochloric acid forms salts called chlorides and nitric acid forms salts called nitrates.

For example, in the reaction between zinc and sulphuric acid:

$$\text{zinc} + \text{sulphuric acid} \longrightarrow \text{zinc sulphate} + \text{hydrogen}$$

Q7 Write the word equation to show the reaction between:

a iron and sulphuric acid

b magnesium and hydrochloric acid

Table B6.2 is a summary of the reactions you need to know. The metals are arranged in order of reactivity, with aluminium the most reactive.

metals (in order of reactivity)	reaction with oxygen	reaction with water (or steam)	reaction with dilute acid	reaction with dilute alkali
aluminium (Al)	Rapidly tarnishes in the cold. A thin coating of aluminium oxide forms.	If the oxide coating is removed, Al reacts with water, giving off hydrogen and a lot of heat.	Gives off hydrogen gas, and a salt is formed.	Dissolves, giving off hydrogen.
zinc (Zn)	Zinc oxide forms, slowly in the cold, but more rapidly when heated.	Displaces hydrogen from steam.	Gives off hydrogen gas, and a salt is formed.	Dissolves, releasing hydrogen.
iron (Fe)	Forms an oxide if heated. Rusts in the cold, if moisture is present.	Reacts with steam to produce hydrogen.	Gives off hydrogen gas, and a salt is formed.	No reaction.
tin (Sn)	Forms an oxide if heated.	Only a slight reaction with steam.	Reacts slowly when the acid is warm.	Reacts slowly, giving off a few bubbles of hydrogen.
copper (Cu)	Forms black copper oxide if strongly heated.	No reaction.	No reaction.	No reaction.
silver (Ag)	Forms silver oxide (tarnishes).	No reaction.	No reaction.	No reaction.

Table B6.2 The reactions of some common metals

tarnish – when a metal gets covered by a layer of its oxide on its surface when left in air

oxidation – a chemical reaction in which oxygen is added chemically to a substance, for example in the tarnishing of metals

Metals react with oxygen in the air at different rates, if at all. Gold is so unreactive that it doesn't **tarnish** (i.e. get covered by a surface layer of its oxide). It remains shiny in air.

When the metals react with oxygen, they form a metal oxide. For example:

$$\text{aluminium} + \text{oxygen} \longrightarrow \text{aluminium oxide}$$

This is called an **oxidation** reaction. We say that the aluminium has been oxidised.

Q8 Lead metal is more reactive than copper but less reactive than tin. Use Table B6.2 to predict what would happen in reactions with oxygen and with cold water. Write a word equation for any reactions you predict would occur.

Aluminium

Aluminium is a very useful metal, mainly because it has a low density for a metal and it is resistant to corrosion so it keeps its silvery appearance. Its lack of reactivity with oxygen and water might seem strange at first sight. After all, aluminium lies just below magnesium and above zinc in order of reactivity. However, aluminium is protected on its surface by a *tough layer of aluminium oxide*.

Once the initial outer layer of aluminium oxide has covered the metal, the aluminium atoms beneath do not come into contact with water or dilute acids. This explains its apparent lack of reactivity. That's why this fairly reactive metal can be used outdoors, for example in ladders, patio doors and window frames, without corroding.

▲ **Fig B6.8** Aluminium is extracted from its ore, which is called bauxite. Bauxite is dug from the ground in Jamaican open-cast mines that can scar the landscape

Aluminium's high melting point and malleability makes it useful as cooking foil. Its ductility and good electrical conductivity means that it is also one of the metals used in overhead power cables. To improve its strength, we often mix aluminium with other metals to make useful light-weight alloys. For example, aluminium alloys are used extensively in aircraft manufacture.

▲ **Fig B6.9** Aluminium alloys are strong enough to withstand the stresses put on an aeroplane during flights

Q9 Duralumin is an alloy made from aluminium with 4% copper and a little magnesium added. It is used to make aeroplanes. Why are aeroplanes made mainly from aluminium alloys rather than steel, which is much cheaper?

Many pots and pans are also made from aluminium, as well as drinks cans. However, dissolved aluminium in the body has been linked in some studies to Alzheimer's disease, although this has not been proved conclusively. Alzheimer's disease affects the brain and is the most common form of dementia.

The protective aluminium oxide layer that forms naturally on the pans reacts with both acids and alkalis. This exposes fresh aluminium metal inside the pan to react further and dissolve in the food. So there is a potential hazard when highly acidic foods, such as tomatoes, rhubarb, cabbage or soft fruits, are boiled in aluminium pans. The same applies to highly alkaline mixtures, for example those containing baking soda. It is also best not to store acidic food in uncoated aluminium cookware as dissolved aluminium can build up over time. Again, scientific research differs in how much and how dangerous dissolved aluminium is.

Coated (non-stick) or sealed (anodised) pans are perfectly safe. The acid or alkali cannot get through a non-stick coating or through the thicker protective aluminium oxide layer formed on an anodised pan.

Cleaning metals

You can clean the inside of aluminium pans by boiling some water with vinegar added in the pan for 10 minutes. However, cleaning uncoated aluminium pans can also increase the dissolved aluminium in food. The acid will react with the aluminium oxide layer, leaving newly exposed metal, which is more likely to dissolve into food. The use of abrasive cleaners is not recommended as these will scratch the surface of the relatively soft aluminium.

Table B6.3 gives some methods suggested for cleaning other household metals.

metal	method to clean the tarnished metal surface
copper	Make a paste out of lemon juice and salt; rub with a soft cloth then rinse with water and dry.
iron	Any rust can be removed by scouring with steel wool or a scouring powder. If an uncoated cooking utensil is scoured, the surface will have to be protected again by rubbing on a thin layer of oil or fat.
tin	Only a thin layer of tin is used to coat objects. If the coating is scratched by using abrasives, the iron or steel beneath will rust, in which case, treat as iron above. Ensure a very fine steel wool is used to avoid further damage to the decorative tin coating.
brass (alloy of copper and zinc)	Use a metal polish that will contain solvents and detergents to remove the tarnish, mild abrasives to polish the metal, and oils to act as a barrier between the exposed metal and air.
silver	Sprinkle baking soda on a damp cloth and rub the tarnish off; then rinse and dry.

Table B6.3 Cleaning tarnished metals

In the methods given in Table B6.3, the use of vinegar and lemon juice, which are both weakly acidic, is an example of removing the basic metal oxide tarnish by chemical means. Some metal oxides are amphoteric, meaning they can react with acids or bases. This explains the use of baking soda, which is a weak base, on silver that is tarnished.

Acids and bases react together in neutralisation reactions (see page 276). This principle can also be applied in removing rust stains from clothing. Rust is a form of iron oxide (see page 245) which will react and dissolve in acidic solutions. So rubbing the stain with lemon juice or white vinegar, mixed with salt to make a paste, will help dissolve the rust marks, which can then be rinsed away.

Q10 The chemical reaction between an acid and a basic metal oxide produces a salt plus water. Write a word equation for the reaction between copper oxide and sulphuric acid.

abrasive – a hard material used to scrape away softer materials from surfaces

The use of **abrasives** is a physical (or mechanical) means of removing any substances from the metal surface.

▲ **Fig B6.10** Metal polishes contain a fine suspension of fine abrasive particles to remove tarnishing, plus oils to protect the shiny metal surface

Problems with abrasives

Grinding tools may be used to remove serious corrosion from metal surfaces before reapplying a protective coating of some kind, for example a lacquer (which forms a hard varnish-like coating) or paint. Before the protection can be applied, any grease must be removed from the metal surface. Solvents called degreasers are used for this purpose although there are concerns about their toxicity if disposed of irresponsibly in the environment.

Abrasives are also used for 'blasting' metals clean. Care must be taken as abrasives, such as sand, are blasted at high pressure. They produce a fine dust made up of the abrasive and particles of the metal. The dust can damage the lungs, as can any tiny particles of solid we breathe in. Also particles of some metals inhaled, such as copper, are toxic. Therefore blasting should be carried out in well-ventilated areas and face masks should be worn.

Q11 Why is it a bad idea to remove the tarnish from a silver ornament using steel wool?

Rusting

rust – a form of hydrated iron oxide that forms on the surface of iron in contact with both air (oxygen) and water

The **rusting** of iron and steel (which contains a very high proportion of iron) costs millions of dollars each year. In the following activity you can find out what causes the corrosion of iron.

SBA Skills

ORR	D	MM	PD	AI
✓				✓

Experiment – What is needed for iron to rust?

Set up the test tubes as shown in Fig B6.11 below and leave for several days.

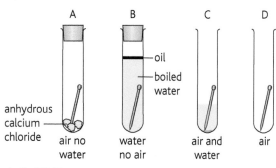

▲ **Fig B6.11**

Then answer the following questions:

1 Why is the water boiled in test tube B?

2 Why is a layer of oil and a stopper added to test tube B?

3 Why is anhydrous calcium chloride added to test tube A?

4 What is the purpose of test tube D?

5 What conditions are needed for the nails to rust?

Iron needs both air (oxygen) and water in order to rust. The iron corrodes to form a layer of hydrated iron oxide, known as rust. We can think of the hydrated iron oxide formed as iron oxide with water bound in its structure. The rusting of iron is an example of oxidation.

$$\text{iron} + \text{oxygen} + \text{water} \longrightarrow \text{hydrated iron oxide}$$
$$\text{(rust)}$$

This rust is a crumbly substance that flakes away and exposes fresh iron to attack so that the iron can corrode completely. This is a problem because iron, often in the form of steel, is the most widely used metal in the construction industry.

▲ **Fig B6.12** Structures containing iron are weakened by rusting

Q12 Which substance is oxidised in the process of rusting?

Q13 Explain the differences you would see if an aluminium rod and an iron rod were left outside for a year.

Factors affecting rusting

People who live near the sea know that bicycles and cars seem to rust more quickly at the seaside than in areas inland. So why is this and what else might affect how quickly iron rusts?

SBA Skills

ORR	D	MM	PD	AI
✓				

To do – Speeding up rusting

1 Place an iron nail in each of two containers with lids, one containing only water and the other containing water with salt added.

2 Leave the nails in contact with the water and salt solution for a week. What do you observe?

Scientists have also found that rusting takes place faster in tropical climates where it is warm and humid, with lots of water vapour in the air. As well as near the sea, other areas prone to excessive rusting are found around industrial plants that give off acidic gases, such as power stations and metal smelting plants. Areas affected by acid rain will also suffer from rapid rusting. From their observations, scientists conclude that rusting is speeded up by:

- high temperatures
- salt (sodium chloride)
- acid.

Preventing rust

The obvious way to protect iron and steel from rusting is to keep air and water away from the metal. We can form a barrier on the surface of the iron by:

- covering with oil or grease
- painting
- coating in plastic
- coating in tin
- coating in zinc.

Q14 For each of the methods listed above, name an object it is used to protect against rusting.

To do – Rusting survey

SBA Skills

ORR	D	MM	PD	AI
✓				✓

1 Carry out a survey into rusting.

2 Record the name of the object, how it has been protected and its condition.

3 Try to explain why different methods of protection are used on different objects.

electroplating – the coating of one metal by another in an electrolytic cell

Tin cans used in the food industry are actually steel cans coated by a very thin layer of tin. The tin is applied to the steel by a process called **electroplating**. The tin layer keeps air and water from the iron. But when the tin is scratched the steel underneath will start rusting. People can get food poisoning if a can rusts and lets bacteria get into the food.

In the following experiment you can see how electroplating works by coating a piece of copper with nickel.

To do – Electroplating copper

1 Ensure the surface of the copper foil to be electroplated is clean and shiny to start with. You can rub it with an abrasive such as emery paper.

2 Using Fig B6.13, set up the circuit as shown to electroplate the copper.

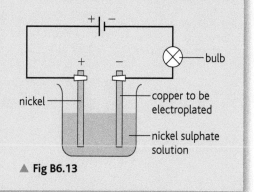

▲ **Fig B6.13**

In electroplating, the object to be plated is always attached to the negative pole of the cell or battery. A block of the plating metal is attached to the positive pole. The object to be plated and the block of plating metal are called electrodes. They are placed in a solution of a compound of the plating metal, which conducts electricity (called an electrolyte).

galvanising – coating an iron or steel object with a layer of zinc to protect it from rusting

A more effective way to protect iron or steel than plating them with tin is to coat them with a more reactive metal, such as zinc. This is called **galvanising**. We use zinc to coat steel bins because they are likely to get knocked about when they get emptied. But, unlike other methods, even when the layer of zinc gets scratched, the zinc still protects the exposed steel underneath. Remember that zinc is more reactive than iron (see Table B6.2 on page 241). So the air and water will attack the zinc rather than the iron. We call this **sacrificial protection**.

sacrificial protection – a method of rust prevention in which iron or steel is protected by a more reactive metal, often zinc

Magnesium is also more reactive than iron so it, too, can be used as sacrificial protection to prevent rust. It is used where extreme conditions could cause rapid rusting, for example in the sea or in underground pipes.

▲ **Fig B6.14** This ship has magnesium blocks bolted to its hull for sacrificial protection

stainless steel – a steel alloy that does not rust (resists corrosion), which is formed by mixing nickel and chromium with the steel.

Another good way to stop iron corroding is to make objects out of **stainless steel**. We looked at steel alloys on page 238. Adding small amounts of nickel and chromium to molten steel form this rust-proof alloy. However, it is expensive. That's why we still use cheaper, but less effective, methods like painting.

To do – Investigating the prevention of rust

Plan an investigation under controlled conditions to find out which method of preventing rusting is most effective. If possible, carry out your plan and evaluate each method.

SBA Skills

ORR	D	MM	PD	AI
			✓	

Non-metallic materials

There are an enormous number of useful non-metallic substances that people use in their everyday lives. It is difficult to generalise their properties because of their great variety. When comparing metallic and non-metallic properties the main difference is in electrical conductivity. All metals are good conductors of electricity but most non-metallic materials are not. Graphite (a form of carbon) is the only non-metallic element that conducts electricity well. Therefore, non-metallic materials find uses as electrical, as well as thermal, insulators.

Some non-metallic materials occur naturally but many are processed or synthesised from raw materials.

▲ **Fig B6.15** Ceramic insulators are used on electricity pylons. Ceramics also have high melting points and are weather resistant

Wood

Wood is an example of a useful naturally occurring material. There are many types of wood. They range from soft, low density balsa wood to hard, dense mahogany. Wood is made of long fibres of cellulose that are lined up along the length of a tree's trunk. The wood can be split along its grain, as when cutting firewood. However, it is much tougher across its grain.

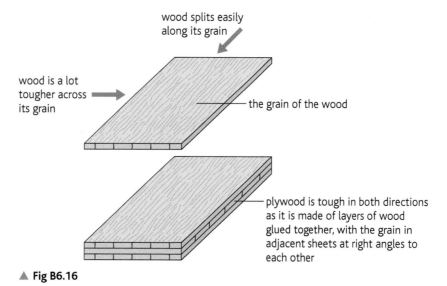

wood splits easily along its grain

wood is a lot tougher across its grain

the grain of the wood

plywood is tough in both directions as it is made of layers of wood glued together, with the grain in adjacent sheets at right angles to each other

▲ **Fig B6.16**

To do – Splitting wood

1 Try splitting a wooden ice lolly stick along its length and across its width.

2 Explain your observations.

SBA Skills

ORR	D	MM	PD	AI
✓				✓

Plastics

The main group of synthetic non-metallic materials to be developed in the last century are the plastics. This section will teach you about their advantages and disadvantages.

Many different types of plastics have been developed by chemists. Their properties depend on the starting materials, called monomers, the conditions in which the monomers react together and how the plastic is processed. For example, yoghurt pots can be made of polystyrene. But when gas is blown into the plastic during the setting stage, it makes expanded polystyrene foam, often used in packaging electrical goods.

Some plastics are called thermoplastics. These are flexible and melt at low temperatures. They can be recycled and remoulded by heating to make new objects. Other plastics are called thermosets. These are hard, rigid and are heat resistant. Very high temperatures will eventually cause this type of plastic to burn and char, but not melt.

Q15 Explain whether you would use a thermoplastic or a thermoset to make the following objects:

    a the cover for an electric plug

    b a carrier bag

    c a rope

    d a pan handle

So plastics can be made for a wide variety of uses. Table B6.4 shows some common plastics and some of their uses.

plastic	some uses
polythene	washing-up bowls, carrier bags, toys, bins
polystyrene	packaging, insulation, disposable cups, yoghurt pots
PVC (polyvinylchloride)	waterproof clothing, window frames, gutters, insulation around wires
polypropene	carpets, milk crates, fishing nets, ropes
nylon	toothbrushes, sports clothing, shirts, ropes, tights
melamine	table tops, ash trays, kitchen work surfaces, picnic ware
phenolic resins	light switches, plug covers, pan handles

Table B6.4

To do – Plastics survey

Conduct a survey of the plastics you use at school. For each plastic say what it is used for and which of the plastic's properties make it suitable for that particular use.

SBA Skills

ORR	D	MM	PD	AI
✓				

Q16 What are the main useful properties of the plastic used to make these items?

    a washing-up bowls

    b carpets

    c ash trays

    d light switches

    e ropes

Q17 Discuss the advantages and disadvantages of using:

    a paper carrier bags and polythene carrier bags

    b PVC gutters and iron gutters

There are few things we use that don't have some plastic components in them. However, in the future we will have to think of new ways to make plastics. That's because the majority of our plastics use crude oil as the main raw material. Crude oil is a fossil fuel (see page 317). As our supplies of crude oil are running out, new raw materials will be needed.

Another problem with plastics is what to do with them when we've finished with them. Most waste plastic ends up as rubbish in landfill tips. Other rubbish in the tips rots away quite quickly as microbes in the soil break it down. But what was a useful property during the working life of the plastic (its lack of reactivity) becomes a disadvantage when we throw it away. Many plastics last for hundreds of years before they are broken down completely. So they take up valuable space in our landfill sites.

▲ **Fig B6.17** About 10% of household waste is made up of plastics

Q18 Why is waste plastic proving to be a problem for society?

Degradable plastics

However, scientists are working to help solve the problems of plastic waste. We are now making more plastics that do rot away in the soil when we dump them. These plastics that can be broken down by microbes are called **biodegradable**. Scientists have found different ways to speed up their decomposition. One way uses granules of starch built into the plastic. The microbes in soil feed on the starch, breaking the plastic into small bits that will rot more quickly.

biodegradable – plastics that rot away by the action of microorganisms in the soil when we dump them

We also have some plastics that are broken down by sunlight. Chemists have made long polymer molecules with groups of atoms along the chain that absorb the energy from sunlight. This splits the chain down into smaller bits. These are called **photodegradable** plastics.

photodegradable – describes plastics that are broken down by sunlight

Finally, we have plastics that are soluble in water – not a property we normally associate with plastics! However this plastic, called polyethenol, can be designed to dissolve at a variety of temperatures. It is used in hospitals where dirty laundry collected in these soluble plastic bags can be put straight into washing machines without staff having to touch it.

Burning plastics

Some countries reclaim the energy stored in plastics, instead of burying them in the ground. For example, Switzerland burns about three-quarters of its household plastic waste and uses the energy given out to generate electricity, saving fossil fuels.

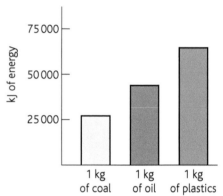

▲ **Fig B6.18** Waste plastics are potentially a useful source of energy

However, there are issues to consider when burning plastics. Some plastics, such as polythene, are hydrocarbons so should burn to give carbon dioxide and water if enough oxygen is present. However, these conditions are difficult to achieve and toxic carbon monoxide gas and particles of carbon are also produced, which causes air pollution. Many other plastics have other types of atom in their molecules, besides carbon and hydrogen. These can cause extra problems when we burn them. For example, chlorine in plastics such as PVC produces acidic hydrogen chloride gas. Nitrogen in plastics makes toxic hydrogen cyanide gas when it burns.

However, research on better ways to burn plastics continues. The incinerators must operate at high temperatures in order to break down other poisons, called dioxins, which can be produced. Chimneys can be fitted with 'scrubbers' to remove the acidic gases given off.

Q19 What are the advantages and disadvantages of burning plastic waste?

Recycling plastics

We can recycle some plastics rather than throwing them away. You might have seen plastics with these symbols on them:

▲ **Fig B6.19** Recycling plastics will conserve crude oil, which is the raw material for most plastics

These plastics can be melted down and remoulded to make new objects. However, we need more people to recycle their plastic waste for this to have a greater impact on the problem.

Q20 What do the letters LDPE, PP and PS stand for on the recycling symbols shown here?

Q21 Make a table to summarise the advantages and disadvantages of using plastics.

Plastics in sport

As in most areas of life, plastics have helped make significant improvements in sporting equipment.

We can look at one plastic to show the impact on sport. Kevlar® was discovered about 40 years ago by an American scientist called Stephanie Kwolek. She could barely believe the strength of the new plastic for such a light-weight material. Here are some of its useful properties:

- high tensile strength
- low density
- high toughness and durability
- abrasion and cut resistant
- flame resistant

▲ **Fig B6.20** How are plastics used in basketball?

This has led to its use in making kayaks and oars for canoeing, tennis racquet strings, volleyball nets, fencing suits, fire-proof suits for racing drivers and 'leathers' for motorcycle racers.

▲ **Fig B6.21** Kevlar can save a rider who skids across the track from serious injury

▲ **Fig B6.22** Kevlar is used in the strings in tennis raquets. It is also used to strengthen the frame of some raquets

Q22 How could Kevlar save the life of a grand prix motor racing driver?

Q23 What is your favourite sport? Think of how plastics are used and explain their advantages over traditional materials that were used for the same job.

End-of-unit questions

1 Which one of the following properties would be best for distinguishing a metallic material from a non-metallic material?

A high melting point

B good electrical conductivity

C high density

D shiny appearance

2 Which of the following is the best explanation for aluminium's corrosion resistance?

A Aluminium is a very unreactive metal.

B Aluminium reacts with water but not with oxygen.

C Aluminium reacts with oxygen but not water.

D Aluminium is protected by a layer of aluminium oxide on its surface.

3 Which **two** of these statements about rusting are correct?

A Iron only needs a supply of oxygen to rust.

B Iron only needs water to rust.

C Iron only needs both oxygen and water to rust.

D Iron is oxidised in the process of rusting.

4 a Explain how blocks of magnesium attached by wires to an underground iron pipeline can stop the pipes rusting.

b Magnesium is quite an expensive metal. Discuss why the owners of the pipeline do not use a different method to protect the pipes against rusting.

5 A research chemist was testing five new forms of photodegradable plastic. The plastic was being developed to make environmentally friendly carrier bags. The chemist tested how long samples of the different plastics took to decompose to half their original mass when left in light. Here are her results:

plastic sample	number of days to decompose to half its original mass
A	17
B	25
C	36
D	13
E	10

a Name two things the chemist would have to do to make this a fair test.

b Draw a bar chart to show the chemist's results.

c Which sample decomposed most quickly?

d Why are the carrier bags described as 'environmentally friendly'?

e Another sample of the plastic was tested and was found to decompose in normal light in a few hours. Why was this not suitable for the new carrier bags?

f Besides degradability, state two other properties that the new plastic should have.

6 Use the data in this table to answer the following question:

Metal	Reaction with oxygen	Reaction with dilute acids	Relative cost	Tensile strength (MN/m)	Density (g/cm³)	Melting point (°C)
aluminium	forms non-porous layer; reforms if scratched	Reacts very slowly	2.5	140–400	2.7	659
stainless steel	forms non-porous layer; reforms if scratched	No reaction	4	500–1000	7.9	1440
carbon steel	forms porous layer; flakes away	Reacts slowly	1	250–400	7.9	1539

Discuss the advantages and disadvantages of using aluminium, stainless steel and carbon steel to make: **a** car bodies **b** car exhaust pipes.

B7a Mixtures

By the end of this unit you will be able to:

- explain the differences between elements, compounds and mixtures.

- identify solutions, suspensions and colloids as mixtures and differentiate among them.

- discuss some consequences of the solvent property of water.

- explain the difference between hard and soft water.

- discuss ways of softening water.

- describe the effects of soaps on hard and soft water.

- discuss the use of solvents in stain removal.

Have you ever wondered why a small amount of salt added to a large pot of water adds flavour to whatever is being cooked in the pot? How is it possible to identify some spices used in foods that you eat without seeing them? It is because all matter is made of small particles invisible to the naked eye. The particles of salt and the spices are dispersed throughout the food and so you cannot see them.

▲ **Fig B7.1** Why can we taste things we cannot see?

atom – the simplest part of an element

molecule – the simplest unit that makes up an element or compound, made from two or more atoms chemically combined

element – a substance that cannot be broken down into a simpler substance by chemical means

Elements, compounds and mixtures

These tiny building blocks of matter are known as **atoms** and **molecules**. Atoms are the smallest particle. When two or more atoms are chemically combined, they form a molecule. If the atoms are of the same type then the molecules make up an **element**. The element oxygen, for example, is made from oxygen molecules. Each oxygen molecule is made of two oxygen atoms.

compound – a pure substance that is made up of at least two different elements that are chemically combined

mixture – impure matter made up of two or more types of elements or compounds that are not chemically combined

fixed composition – having set ratio of atoms or molecules throughout the substance

variable composition – having no set ratio of atoms or molecules throughout the substance

If the atoms are different, then the molecule formed makes up a **compound**. Compounds are made up of at least two different elements that are chemically combined.

Elements and compounds are described as pure matter. The word 'pure' as used by the scientist means *not* mixed with anything *else*. Impure matter is called a **mixture**. Mixtures are also made up of two or more types of elements or compounds but they *are* not chemically *combined*. Compounds should not be confused with mixtures.

A compound is always made up of the same elements in the same ratios every time. We say that compounds have a **fixed composition**. A molecule of the compound water is always made up of one atom of hydrogen combined with two atoms of oxygen. Mixtures have a **variable composition** as the ratios of the component substances can change. A mixture of salt and water can be made from any amount of salt mixed with water.

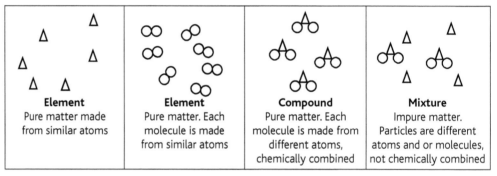

Element	**Element**	**Compound**	**Mixture**
Pure matter made from similar atoms	Pure matter. Each molecule is made from similar atoms	Pure matter. Each molecule is made from different atoms, chemically combined	Impure matter. Particles are different atoms and or molecules, not chemically combined

▲ **Fig B7.2** Diagrammatic representation of particles of elements, compounds and mixtures

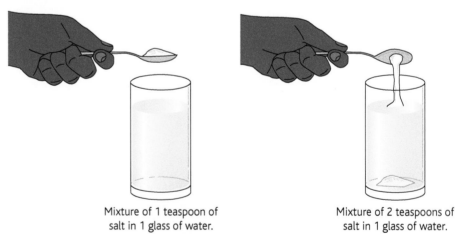

Mixture of 1 teaspoon of salt in 1 glass of water.

Mixture of 2 teaspoons of salt in 1 glass of water.

▲ **Fig B7.3** The composition of a mixture of salt and water can vary

▶ **Fig B7.4** The composition of water is fixed – two hydrogen atoms and one oxygen atom combine to form one molecule of water

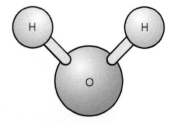

The properties and separation of elements, compounds and mixtures

Scientists use the properties of substances to predict their behaviour and identify them. Physical properties refer to those that can be measured or perceived without changing the chemical make-up of the substance. These include colour, melting point, boiling point, mass and density. Chemical properties refer to those that determine how substances behave in chemical reactions such as their reaction with water or when heated. For example, if you put a teaspoon of baking powder in a glass of water, you will notice that it fizzes as a gas is produced. Heat also causes substances to change chemically. Cooking an egg is an example of a chemical process.

Because of their fixed composition, compounds have unique physical and chemical properties. For example, at a pressure of 1 atmosphere, the boiling point of water is 100°C. However, impure water does not boil at a fixed temperature but will boil over a temperature range slightly above 100°C.

The properties of compounds are generally different from the elements that make them up. Hydrogen and oxygen (the elements that make up water) are gases at room temperature, but water is a liquid.

Mixtures have the properties of the substances that make them up. Because the components are not chemically combined it is relatively easy to separate them to produce the component substances. On the other hand, separating compounds to form the component elements requires a lot of energy and is very difficult to do.

Table B7.1 summarises some important differences between compounds and mixtures

points of comparison	compounds	mixtures
composition	two or more elements in fixed composition	two or more substances with variable composition
property	properties are different from the component elements	properties are similar to the substances making up the mixture
method of separation	can be separated in to elements by chemical processes – but it is difficult	can be separated in to components by physical process – usually an easy process

Table B7.1 Comparison of the properties of mixtures and compounds

Types of mixtures

Most of the common chemicals and juices you will find in a household are mixtures. Examine the labels of some of the cleaning products around the home. You will find that most of them contain different chemical substances.

Mixtures can be classified according to their appearance and the size of the particles that make them up.

SBA Skills

ORR	D	MM	PD	AI
✓		✓		✓

To do – Making mixtures

You will need: water, flour, sugar, syrup, oil, milk (liquid or powdered), five colourless drinking glasses or transparent plastic cups, spatula or spoon, measuring cylinder.

1 Measure approximately 100 cm³ of water and pour into each glass.

2 Add one teaspoon or spatula of flour to the water in the first glass. Stir with a spoon or glass rod and leave for a few minutes.

3 Repeat step 2 with each of the remaining materials – sugar, syrup, oil and milk.

4 Observe each mixture and record your observations of:

   a the colour

   b whether the mixture is clear (transparent) or cloudy

   c whether or not particles settle at the bottom when left standing for a few minutes.

5 Create a suitable table to display your results. Classify the mixtures made as solutions, suspensions or colloids. If you are unsure the following section will help you.

solution – a transparent, homogeneous mixture formed when a solute dissolves in a solvent

solute – a substance that dissolves in a solvent

solvent – a substance that dissolves a solute

aqueous solution – a solution in which the solvent is water

non-aqueous solution – a solution made from a liquid other than water

soluble – will dissolve in a solvent

insoluble – will not dissolve in a solvent

dilute – a relatively small mass of solute dissolved in a large volume of solvent

concentrated – a relatively large amount of solute dissolved in a small volume of solvent

homogeneous – uniform in structure or composition throughout

phase – any form or state in which matter can exist at a specific temperature and pressure

solubility – the amount of solute that can be dissolved in a specified amount of solvent for a given temperature

Solutions

A **solution** is a transparent mixture formed when a substance known as a **solute** dissolves in a **solvent**. Solutes may be solids, liquids or gases. Solvents are mostly liquids. Water is described as the 'universal solvent' because of the wide range of substances that can dissolve in it. However, there are some that do not, like oil. Solutions that have water as the solvent are called **aqueous** solutions. Those in which other solvents are used instead of water are called **non-aqueous** solutions. Common non-aqueous solvents are ethanol, propanone (acetone) and methyl benzene.

Substances that dissolve in a solvent are described as being **soluble**, while those that will not dissolve are **insoluble**. When solutions are made, the solvent is always present in larger quantities than the solute. **Dilute** solutions are those in which a relatively small mass of solute is dissolved in a large volume of solvent. Solutions can be made less dilute and more **concentrated** by adding more solute. The particles of a solution do not settle on standing and will pass through a filter paper. Solutions are classified as **homogeneous** mixtures as they are the same all the way through. We say they consist of only one **phase**. This means that there is one uniform state throughout the mixture.

The **solubility** of a substance is the mass that will dissolve in a given volume of solvent at a stated temperature. The solubility of most solids increases with increasing temperature, while the solubility of gases decreases with increasing temperature.

When one liquid dissolves in another liquid, the liquids are said to be **miscible**. For example, lime juice and water are miscible liquids which are combined when preparing lemonade. Liquids that do not dissolve in each other are said to be **immiscible**. Oil and water are immiscible liquids. Oil will float on the water because it is the less dense of the two liquids.

miscible – a liquid that is able to dissolve in a solvent

immiscible – a liquid that is insoluble in a solvent

colloid – a solid, liquid or gaseous mixture in which particles (the dispersed phase) are suspended in a dispersion medium

Colloids

A **colloid** is a solid, liquid or gaseous mixture in which one type of substance, called the dispersed phase, is suspended in another substance, called the dispersion medium. The dispersed phase and the dispersion medium may be solids, liquids or gases. The particles in the dispersed phase remain suspended and do not settle on standing, giving some colloids a cloudy appearance. This occurs because the particles of a colloid are larger than those of a solution but not large enough to settle and so remain dispersed in the mixture.

Colloids can be distinguished from solutions by the effect they have on light when it passes through the mixture. Solutions are transparent because light will pass through them. Many colloids appear to be translucent due to the scattering of the light rays by the suspended particles as they pass through. Some colloids are opaque.

Common examples of colloids are given in Table B7.2.

dispersion medium	dispersed phase		
	gas	**liquid**	**solid**
solid	glass	gels, e.g. gelatine, silica gel, jelly	styrofoam
liquid	whipped cream	emulsions, e.g. milk, mayonnaise, hand cream	ink, blood
gas	none	fog, mist, cloud	smoke, soot

Table B7.2 Examples of colloids

▲ **Fig B7.5** We use many colloids in our daily activities

When a mixture of oil and water is shaken vigorously the oil breaks up into tiny droplets that become suspended in the water but will soon settle to form two layers again if left undisturbed. Mixtures of immiscible liquids can be treated with special substances called **emulsifiers** that will allow the droplets to remain suspended. Mixtures of liquids suspended in liquids are called **emulsions**. Mayonnaise is an example of an emulsion. Examine the label of a bottle of mayonnaise and you will see that oil and vinegar are two of the main ingredients. These are the immiscible liquids. Egg yolk is normally the emulsifier used in making mayonnaise.

emulsifier – a substance used to keep particles suspended in a fluid

emulsion – a suspension of small globules of one liquid in another

SBA Skills

ORR	D	MM	PD	AI
		✓		

To do – Making mayonnaise

Use the recipe below to make home-made mayonnaise.

You will need:

1 egg yolk

2 teaspoons cider vinegar

2 teaspoons lemon juice

$\frac{1}{4}$ teaspoon salt

$\frac{1}{8}$ teaspoon sugar

$\frac{1}{4}$ teaspoon white pepper

$\frac{1}{2}$ teaspoon dry mustard

$\frac{1}{2}$ cup vegetable oil

1 Mix the egg yolk, vinegar, lemon juice, salt, sugar, pepper and mustard in a small bowl with a rotary beater until well blended.

2 Add the oil slowly in stages and mix thoroughly after each stage until all the oil has been added. Store in a covered container in a refrigerator.

Suspensions

suspension – a heterogeneous mixture of a solid in a liquid in which the solid particles fall to the bottom on standing

A **suspension** is a mixture of a solid in a liquid in which the solid particles settle on standing. The particles of the suspension are larger than those in solutions and colloids, and cannot pass through a filter paper. The large particles in suspensions make them opaque. Examples include sand in water, chalk in water, and milk of magnesia.

heterogeneous – consisting of dissimilar components or parts

Colloids and suspensions are examples of **heterogeneous** mixtures as they consist of more than one state of matter. You might have read the labels of common household products and seen the instruction 'Shake well before using'. This is because these products are suspensions and they settle out on standing.

Directions: Shake well before each use. DANGER: harmful or fatal if swallowed. Harmful if inhaled. Eye and skin irritant.
Suggested uses: brass, chrome, copper, stainless steel, pewter.
Contains petroleum distillates and silica.

▲ **Fig B7.6** You may find this information on a can of metal cleaner

SBA Skills

ORR	D	MM	PD	AI
✓		✓		✓

To do – Properties of solutions, suspensions and colloids

You will need: mixtures of salt, chalk and milk in water, three conical flasks, filter funnel, beakers, measuring cylinder, supply of filter paper, flashlight with narrow beam.

1 Measure out approximately 100 ml of the following mixtures and pour into the beakers:

   a salt and water (solution)

   b chalk and water (suspension)

   c milk and water (colloid)

2 Turn on the flashlight and focus the beam of light through the liquid mixture. Hold a piece of paper on the other side of the beaker in line with the beam of light. Determine whether the light passes through the mixture, is diffused by the mixture or does not pass through.

3 Fold the filter paper, insert it into the filter funnel provided and place the filter funnel over the conical flask. Pour the solution into the funnel. Note what happens.

4 Repeat steps **2** and **3** using the suspension followed by the colloid. Compare the appearance of the liquid in the three conical flasks (**filtrate**) and any solid (**residue**) on the filter paper.

5 Write a summary of the differences between suspensions, solutions and colloids.

filtrate – the product that passes through a filter during filtration

residue – the product that remains on the filter after filtration

Summary of the properties of solutions, suspensions and colloids

property	solution e.g. salt dissolved in water	colloid e.g. a few drops of milk in a glass of water	suspension e.g. flour in water
appearance	clear, transparent and homogeneous	cloudy – heterogeneous but generally uniform	cloudy – heterogeneous, at least two substances visible
effect of light	none – light passes through	light is dispersed by particles in dispersed phase	generally opaque depending on the quantity of suspended particles
relative particle size	molecule size	larger than molecule, visible but smaller than that of suspension	visible – larger than that of colloid
tendency to settle on standing	none	none	particles will eventually settle out

Table B7.3

The role of solvents in stain removal

Knowledge of solvents can be used for removal of stains from fabrics. Many stains are not water soluble and so alternative solvents are required to remove them. Oil-based paint, for example, cannot be removed with water. A non-aqueous solvent is required to remove it. Common non-aqueous solvents such as turpentine (paint thinner), methylated spirit and acetone can be used to remove various types of stains. The following experiment will help you to determine which of these solvents are most effective for removing some common stains.

SBA Skills

ORR	D	MM	PD	AI
✓				✓

Experiment – Stain removal

You will need: pieces of cotton fabric cut into 5 cm squares, solvents (acetone, turpentine and methylated spirit), three Petri dishes, forceps.

 SAFETY: These chemicals are flammable so you should keep them away from any naked flame.

1 On each piece of fabric make stains with nail polish, grass, and waxed crayons. Allow the stain to dry.

2 Pour approximately 10 ml of each solvent into different Petri dishes.

3 Dip one piece of the stained fabric into the solvent in each Petri dish. Use the forceps to keep the stained fabric submerged in the solvent for 5 minutes.

4 Remove the fabric and blot dry using a hand towel or tissue paper.

5 Observe the changes to each stain made on the fabric for each solvent.

6 Identify the solvents that dissolve each stain.

Dry cleaning

Dry-cleaning dates back to the mid 19th century and was accidentally discovered by a French dye worker Jean Baptiste Jolly. Jolly discovered that when kerosene was accidentally spilled on his table cloth by his maid, the table cloth became cleaner. He capitalised on the discovery and today dry cleaning is widely used.

In dry cleaning, an alternative solvent is used instead of water for cleaning fabric. Different non-aqueous solvents have been used over the years as the industry has developed. These include gasoline, kerosene and carbon tetrachloride. Nowadays the solvent most widely used in tetrachloroethylene also known as perchloroethylene (perc) in the industry.

Water

The solvent properties of water

The fact that water is such a good solvent is important for supporting life. Water serves as a transport medium and as a solvent for nutrients needed by plants (see Unit A4a). It contains dissolved oxygen needed for the process of respiration in aquatic plants and animals (see Unit A3a). Many biological activities in animals require water. For example, water serves as a solvent for oxygen, nutrients and hormones and also for the removal of various toxins, excretory waste and dead cells from the body (see Unit A4a).

Hard and soft water

The excellent solvent properties of water can sometimes be a problem. Rain leaves the clouds as chemically pure water, but as it falls to the ground it mixes with and dissolves carbon dioxide and other acidic gases, forming weak dilute acids such as carbonic acid. In highly industrialised areas, rain water will dissolve other acidic gases to form stronger acids such as sulphuric and nitric acids.

$$\text{carbon dioxide} + \text{water} \longrightarrow \text{carbonic acid}$$
$$\text{(acidic gas)}$$

$$\text{nitrogen dioxide} + \text{water} \longrightarrow \text{nitric acid}$$
$$\text{(acidic gas)}$$

$$\text{sulphur trioxide} + \text{water} \longrightarrow \text{sulphuric acid}$$
$$\text{(acidic gas)}$$

When the rain water runs over soil and rocks that contain calcium and magnesium salts (limestone, chalk, dolomite, gypsum and magnesite), they are dissolved and soluble compounds of calcium and magnesium hydrogencarbonate are formed. The calcium and magnesium compounds find their way into our water supplies and make it difficult for the water to form lather with soap – far more soap is required.

hard water – water that does not lather easily with soap due to the presence of calcium and magnesium salts

soft water – water that lathers easily with soap

Such water is known as **hard water** and is not ideal for washing purposes. Water that does not contain dissolved calcium and magnesium salts and easily lathers with soap is called **soft water**.

Soaps are compounds made by combining sodium hydroxide with animal fat or oils from plants. Hard water is less effective in forming a lather with soap because the calcium and magnesium salts in hard water react with the soap to form insoluble salts, grey in colour, known as scum. This is what produces the ring in bath tubs when soap is used.

$$\text{sodium stearate} + \text{calcium salt} \longrightarrow \text{calcium stearate} + \text{sodium salt}$$
$$\text{(soap)} \qquad \text{(in hard water)} \qquad \text{(scum)}$$

The cleaning action of the soap can begin only when all the calcium or magnesium salts present have been removed as scum.

▶ **Fig B7.7** Scum is the insoluble grey substance formed when soap reacts with the calcium salts in hard water

SBA Skills

ORR	D	MM	PD	AI
✓		✓		✓

To do – Comparing water samples for hardness

You will need: six similar jars (e.g. jam jars) with lids, eyedropper, soap flakes made by grating a bar of soap, samples of tap water, distilled water, boiled water, water in which a little Epsom salts (magnesium sulphate) has been dissolved and water to which a little washing powder has been added.

1 In one of the jars make some soap solution by dissolving a tablespoon of soap flakes in half a cup of hot water and set aside to cool.

2 Measure out half a cup of tap water and place into one of the jars.

3 Using the eye dropper, add five drops of the soap solution to the tap water in the jar.

4 Cover the jar and shake vigorously. Examine the jar to see if any lather appears.

5 Continue adding soap until lather appears. Note the number of drops of soap that it takes for the lather to appear.

6 Repeat steps **2–5** using the different samples of water.

Q1 Order the samples of water starting with the sample with greatest degree of hardness.

Q2 Explain the basis of your answer to **Q1**.

Q3 What is the substance formed on the side of the jar when the lather disappears?

The effects of hard water

Many people enjoy drinking hard water and it is widely believed that it helps to prevent heart disease. However, hard water has several disadvantages as it causes the build-up of scale or fur in pipes, water heaters and tea kettles. Scale is formed when water containing soluble calcium or magnesium salts is heated. The heat decomposes the hydrogencarbonates and the insoluble carbonates are formed.

calcium hydrogencarbonate $\longrightarrow$ calcium carbonate
    (soluble in water)   (insoluble in water)

This tends to clog pipes and reduces the efficient transfer of heat resulting in more energy being required to heat water.

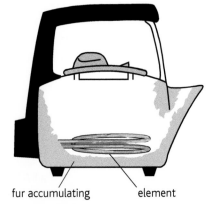

▲ **Fig B7.8** Water has to be heated for a longer period due to the build-up of fur inside the kettle

fur accumulating   element

Methods of softening water

Soft water is preferred to hard water in laundering where fewer soaps and detergents are used, and in boilers so that scale does not build up. Softening water is achieved by removing the calcium or magnesium salts that cause the problem in the first place. Several methods can be used to soften water:

1 *Boiling* – Boiling can remove temporary hardness caused by the presence of calcium or magnesium hydrogencarbonate. Boiling converts the hydrogencarbonates to insoluble carbonates. Boiling will not remove permanent hardness.

2 *Addition of sodium carbonate (washing soda)* – This method removes permanent hardness, which is caused by the presence of calcium or magnesium sulphate. The sodium carbonate is soluble and combines with the calcium and magnesium salts in water to form insoluble calcium and magnesium carbonate. This process is known as precipitation, as a solid (precipitate) is formed from the reaction taking place in a solution.

$$\text{sodium carbonate + calcium sulphate} \longrightarrow \text{calcium carbonate + sodium sulphate}$$

3 *Distillation* – Water can be softened temporarily by distillation.

Distillation process

Distillation is a technique that separates mixtures according to the boiling points of the different components. Simple distillation is used to separate miscible liquids. This is possible because the liquids present in the mixture have different boiling points. For example, the boiling point of ethanol is 60°C while that of water is 100°C. Two physical processes are involved in distillation. These are vaporisation – the conversion of liquid to gas by heating, and condensation – the conversion of gas to liquid (the distillate) by cooling.

During simple distillation, the mixture is heated and the liquid with the lower boiling point will vaporise first. This vapour passes through a special apparatus called a condenser. The Liebig condenser has an outer tube through which water runs and an inner tube through which the vapour is channelled. The running water cools the vapour as it passes through the condenser. As the vapour cools, it changes back to a liquid (Fig B7.19).

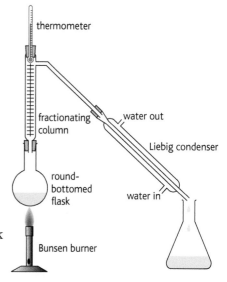

▲ **Fig B7.9** The distillation process

Because the liquids in the mixture have different boiling points, only one liquid will boil, vaporise and condense at a time. In this way the different liquids can be separated from each other. For distillation to be effective, the boiling points of the liquids in the mixture should be fairly far apart – usually separated by about 20°C. In cases where liquid mixtures have components with boiling points closer than 20°C, fractional distillation is used to separate them. For example, fractional distillation is used to separate crude oil into its various parts. With fractional distillation, the vapour passes through a fractionating column before passing through the condenser.

Q4 In separating a mixture of ethanol and water, which liquid do you expect to boil first? Why?

A solution of water containing dissolved solids can also be purified by simple distillation. This is possible because the dissolved solids have a much higher boiling point and will not evaporate with the steam. This process is used to produce distilled water and is also a temporary method for softening water.

SBA Skills

ORR	D	MM	PD	AI
✓				✓

Experiment – Separation of a mixture of copper sulphate and water by simple distillation

This activity should be done in the science laboratory. Your teacher will help you to set up the distillation apparatus. This is very costly equipment, so every effort should be taken to handle it with care.

You will need: solid copper sulphate, tap water, measuring cylinder, distillation apparatus (distillation flask, Liebig condenser, thermometer, conical flask, rubber bung), Bunsen burner.

1 Measure 50 cm³ of water and pour into the distillation flask.

2 Add one spatula of the solid copper sulphate to the water in the distillation flask. Swirl to mix it and form a solution.

3 Add a few anti-bumping granules to the mixture to help to create a smoother boiling action.

4 Set up the distillation apparatus as shown in Fig B7.10. Ensure that the distillation flask is air tight.

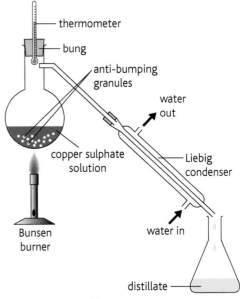

▲ **Fig B7.10** Distillation apparatus

5 Light the Bunsen burner and observe what takes place.

Q5 Why do you think the cold water enters the condenser nearer the end where the distillate is leaving it and not at the other end where the vapour enters?

Q6 Why do you think the condenser is sloped downward?

Q7 What is the distillate?

Q8 What is the temperature of the vapour as the distillate is being collected?

Q9 Why is it necessary for the distillation flask to be air tight?

Detergents

detergent – a cleansing agent similar in action to soap but made of different chemicals

Detergents are substances that improve the cleaning power of water. They are cleaning agents that are used for dishwashing, laundry, in personal care products and in industrial cleaning. There are two types of detergents:

1 soapy detergent (soaps) – made from fats and oils and an alkali (sodium or potassium hydroxide)

2 soap-less detergents – made from chemicals obtained from petroleum

Today we mostly use soap-less detergents as they are able to lather well in hard water. Soap-less detergents include shampoos, dishwashing liquids and bath gels.

scouring powder – a powdered abrasive cleaner

Other cleaning agents include **scouring powders**. These are abrasive powder cleansers that are not usually soluble in water. Various commercial brands exist. However, a simple homemade cleanser can be made from mixing borax, salt and baking soda in equal proportions. When made, this should be stored in an air tight container and used as you would with a regular commercial cleanser.

How do detergents work?

Water is a poor cleansing agent because it has poor wetting properties and it does not dissolve grease. Detergents improve the wetting properties of water and help to disperse grease and dirt. They serve the same function as emulsifying agents by keeping dirt suspended in water. Detergents are able to do this because they can mix with both grease and water. The detergent molecule can be represented as in Fig B7.11.

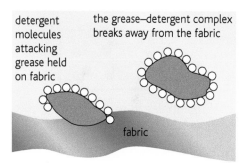

▲ **Fig B7.11** How detergents work

The water-loving head of the molecule attaches itself to the water, while the grease-loving tail attaches itself to the dirt particles in the fabric to be cleaned.

Choosing the right detergent

Most detergents today contain phosphates as water softeners. Calcium salts in the water will react with the phosphate to produce insoluble calcium phosphate, thereby removing the calcium salts from the water. However, calcium phosphate causes much environmental damage. When the phosphates in waste water enter waterways such as lakes, ponds and rivers, the phosphorus-rich environment causes excessive algal growth, reducing the availability of oxygen for aquatic organisms. Many die of oxygen starvation. Detergent manufacturers are now trying to reduce this harmful effect by using alternative softeners that are biodegradable.

End-of-unit questions

1 Which of the following substances is pure?

   A steel

   B brass

   C sodium chloride

   D carbonated drinks

2 Fig B7.12 below shows the types of particles present in three substances. Which of the diagrams represent mixtures?

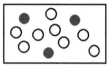

Key
○ = atom 1
● = atom 2

Substance 1

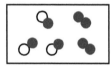

Substance 2

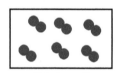

Substance 3

▲ **Fig B7.12**

   A 1 only

   B 1 and 2 only

   C 3 only

   D 1, 2 and 3

3 The salt in sea water is the:

   A solute

   B mixture

   C solvent

   D solution

4 Water is referred to as the 'universal solvent' because:

   A it is made up of small molecules.

   B it is a liquid.

   C it dissolves many substances.

   D it consists of hydrogen and oxygen atoms chemically combined.

5 Scum was formed when soap was added to water. The water is likely to contain particles of:

   A oxygen

   B chlorine

   C calcium salt

   D sulfur

6 A student carried out an experiment in which equal volumes of two different water samples were shaken with the same amount of soap and the height of the lather in the tube measured (see Fig B7.13.)

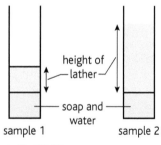

height of lather

soap and water

sample 1   sample 2

▲ **Fig B7.13**

   a Suggest a suitable aim for the experiment that the student conducted.

   b Which sample of water might be more suitable for doing laundry?

   c Suggest one reason for the differences in the height of the lather in the two water samples.

   d Identify one means by which the quality of the water in sample 1 may be improved to produce more lather with soap.

B7b Acids, Bases and Salts

By the end of this unit you will be able to:

- list some common household chemicals and their uses.
- classify common household chemicals as acids, bases and salts.
- describe the pH scale.
- describe the use of indicators to identify acids and bases.
- explain the use of acids and bases in stain removal.
- give applications of neutralisation reactions.
- describe steps for using, storing and disposing of household chemicals.

substance – a special form of matter that has a fixed composition

In our everyday use of the word 'substance' we use it to refer to any form of matter. However, from a chemist's perspective, a **substance** is a special form of matter that has a fixed composition. This means that the component particles of substances are always in the same ratios or proportions, for example, like the compounds and elements you learned about in Unit B7a.

product name	chemical substance	use
acids		
battery acid	sulphuric acid	the electrolyte in car batteries
club soda	carbonic acid	a soft drink
vinegar	ethanoic acid	in dilute solutions used in pickles and food preservation
bases		
magnesia	magnesium oxide	in medicines to neutralise stomach acid
rouge	ferric oxide	make-up
oven cleaner	sodium hydroxide	cleaning ovens
salts		
baking soda	sodium hydrogencarbonate	a raising agent in baked products, stain removal, cleaning jewellery
chalk	calcium carbonate	writing, dressmaking
Epsom salts	magnesium sulphate heptahydrate	reduce swelling
plaster of Paris	calcium sulphate hemihydrate	keep broken limbs in fixed position
table salt	sodium chloride	cooking, food preservation

Table B7.4 Common household chemicals

Chemical substances around the house

If you have a look around the house, you will find lots of chemicals. Water is the most common household chemical, and a very important one because of its wide range of uses such as for drinking and washing. In particular it is used as a solvent for many other chemicals (see Unit B7a).

If you look at some everyday cooking and cleaning products, you will find more chemicals. You will recognise their common names such as vinegar, alum, and borax on their labels. If you carefully examine the labels you should find the chemical names of the substances present in the product.

Table B7.4 shows a list of some common household chemicals according to their product names, their chemical names and their uses.

To do – The wonders of baking soda

You will need: tarnished metal objects (such as aluminium pots, used aluminium foil or bronze, brass or copper ornaments), baking soda, soft pieces of cloth.

1 Make a paste with half a cup of baking soda and water.

2 Apply this paste to the damp cloth and use it to polish the metal.

3 Rinse with warm water and dry with a soft cloth.

4 Observe any changes in the metal.

SBA Skills

ORR	D	MM	PD	AI
✓				

Acids, bases and salts

acid – a chemical with pH less than 7 that react with metals to produce hydrogen

base – a substance that will neutralise an acid

salt – a compound formed as a result of a neutralisation reaction

The chemicals in Table B7.4 have been classified as **acids**, **bases** or **salts** according to their chemical composition.

Salts

Salts are solid compounds derived from reacting acids with substances such as bases and some metals. Salts derive the second part of their names from the acids from which they are produced. The first part of the name of the salt comes from the metal or base.

Table B7.5 gives the names of some common acids, the salts they form, and some examples of common salts and their uses.

acid	name of salt formed from acid	examples	common uses
hydrochloric acid	chlorides	sodium chloride	preservative, cooking
		ammonium chloride	soldering
		calcium chloride	drying agent
sulphuric acid	sulphates	magnesium sulphate (Epsom salts)	soothing sore or swollen muscles
		potassium sulphate	fertilisers
		copper sulphate	pesticides
		calcium sulphate	plaster
nitric acid	nitrates	potassium nitrate	explosives, fertilisers
		sodium nitrate	fertilisers
carbonic acid	carbonates	sodium carbonate	baking powder, manufacture of glass, water softener

Table B7.5 Uses of common acids

Acids and bases

The list in Table B7.5 represents a small fraction of the chemicals that we encounter in our daily lives. Many of the foods we eat contain acids. Acids have a sour taste and are found in fruits such as oranges, grapefruits, lemons, and cherries. These fruits contain ascorbic acid, which is commonly known as vitamin C.

In the school laboratory you will find other common acids such as hydrochloric acid, sulphuric acid and nitric acid. These are called mineral acids as they are prepared from naturally occurring minerals. Unlike the acids found in foods and household chemicals, mineral acids are more harmful and could cause burns if they come in contact with the skin or create holes when spilled on clothing. The mineral acids should be handled with great care.

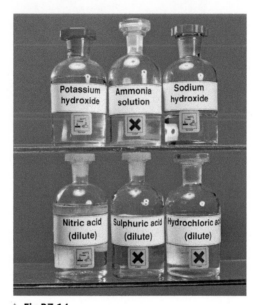

▲ **Fig B7.14**

Household bases are likely to be more harmful than household acids. They are widely found in cleaners as they dissolve grease. Drain cleaners and oven cleaners contain sodium hydroxide, which can cause severe irritation and burns to the skin. Ammonia cleaner will irritate the eyes and nose. Bases are usually bitter in taste. Most bases are insoluble in water but those that are soluble are called **alkalis**. Alkalis feel soapy. If you dip your fingers in ammonia solution and rub your fingers together, you will experience this soapiness.

alkali – a soluble base

While it is important to know these properties of acids and bases, you should not touch or taste chemicals to identify them. More trustworthy and objective results are obtained by the use of instruments or observing the changes that take place when other chemicals are added.

Q1 Conduct research in your school library or on the internet to find the names of the acids found in the following.

a car battery

b the human stomach

c ants

d acid rain

Using pH to identify acids and bases

pH scale – a scale of numbers ranging from 0 to 14 that give the acidity and alkalinity of substances. Substances with pH less than 7 are acids and those with pH greater than 7 are alkaline. A pH of 7 indicates a neutral substance

Scientists have devised a scale called the **pH scale**, which will give a measure of the strength of an acid or alkali. The pH scale consists of numbers ranging from 0 to 14.

▲ **Fig B7.15** A pH scale

neutral – a substance with a pH of 7

The stronger the acid, the lower the pH. A solution of pH 7 is neither acid nor alkaline. It is **neutral**. Solutions with pH greater than 7 are alkaline.

pH	description	colour with universal indicator	household substance	chemical it contains
14 13 12	strongly basic	deep blue	oven cleaner	may contain sodium hydroxide
11 10 9 8	weakly basic	green	household ammonia milk of magnesia and other antacids baking powder	magnesium hydroxide potassium hydrogencarbonate
7	neutral	yellow	blood, milk, water	
6 5 4 3	weakly acidic	orange	vinegar oranges, health salts, lemon juice	ethanoic acid citric acid, ascorbic acid
2 1 0	strongly acidic	red	muriatic acid	hydrochloric acid

Table B7.6 The pH of some common household substances

universal indicator – an indicator that can be used to give the pH of a substance

In the laboratory, pH can be measured using **universal indicator** paper or solution. A universal indicator is accompanied with a colour chart with colours ranging from red through orange, yellow, green, blue, and deep purple. Each colour on the chart is associated with a specific pH.

When the indicator paper is dipped into an acidic or alkaline solution being tested, the colour of the paper will change. The new colour is matched against the colours on the colour chart. The corresponding number on the colour chart gives the pH of the solution being tested.

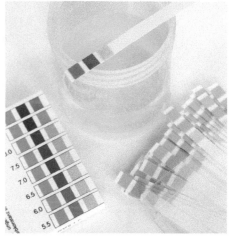

▲ **Fig B7.16** Indicator paper tells you how acidic or alkaline a solution is

A more precise method to measure pH is with a pH meter.

▲ **Fig B7.17** A pH meter

Some of the more common indicators used in school laboratories and their colours in acids and bases are given in Table B7.7. Common acid–base indicators can give information on whether a substance is an acid or a base. They will not give a pH value. However, universal indicator can give the pH value of a substance.

indicator	colour in acids	colour in bases
litmus	red	blue
methyl orange	pink	yellow
phenolphthalein	colourless	pink
screened methyl orange	light red	green

Table B7.7 Colour of common indicators in acids and bases

Experiment – Finding the pH of toothpaste

You will need: four different brands of toothpaste, water, test tubes, universal indicator paper.

1 Add a little water to a spatula load of each brand of toothpaste in a test tube and mix to make a suspension.

2 Use the universal indicator paper to determine the pH of each brand of toothpaste.

3 Record the results obtained in a suitable table. What conclusion can you draw from your results?

Using indicators to identify acids and bases

Indicators are dyes that can be used to identify acids and bases. They change to one colour in acids and another colour in bases. Commercial indicators are derived from a variety of plants, including sorrel, beetroot, red cabbage, poinsettia and rose.

Experiment – Using home-made indicators to identify acids and base

You will need: petals of sorrel flower, droppers, test tubes, beaker, samples of baking soda solution, orange juice, shampoo (dilute), vinegar, ammonia cleaner, soft drinks.

Part a: Making the indicator

1 Crush the sorrel in a beaker and add about 10 cm³ of hot water to extract the colour.

2 Filter to remove the plant and set the filtrate aside to cool.

Part b: Determining the indicator colour in acid and base

1 Pour approximately 3 cm³ of baking soda solution (base) and vinegar (acid) into two separate test tubes.

2 To each test tube add two drops of the sorrel indicator that you have made.

3 Observe the colour changes in the acid and base when the home-made indicator is added. Keep these samples as a reference for completing the rest of the experiment.

Part c: Testing for acids and bases

1 Pour approximately 3 cm³ of shampoo into a test tube.

2 Add two drops of the indicator solution and mix.

3 Observe the colour of the mixture.

4 Record your results in a suitable table.

5 Repeat the process using other solutions, such as orange juice, soft drinks, ammonia cleaner and tea.

6 Based on your results on part **b**, identify the acids and the bases present.

Q2 How does the use of pH paper differ from using litmus paper when testing for acids and bases?

Reactions of acids and bases

Acids and bases undergo several important reactions. Some of these reactions can be used to identify when an acid or a base is present. These reactions are described below.

Reaction of acids with carbonates and hydrogencarbonates

Carbonates and hydrogencarbonates react readily with acids to produce effervescence or fizzing as the gas carbon dioxide is produced. Other products of the reaction are salt and water. This reaction can easily be demonstrated in the home by using vinegar and baking soda – sodium hydrogencarbonate. The chemical name for vinegar is ethanoic acid. The salt formed is sodium ethanoate.

The reaction between vinegar and baking soda is represented below:

$$\text{sodium hydrogencarbonate} + \text{ethanoic acid} \longrightarrow \text{sodium ethanoate} + \text{carbon dioxide} + \text{water}$$

The reaction between acids and carbonates can be used to remove the build-up of calcium carbonate in tea kettles. This is the white deposit produced when kettles are used over time, as you may remember from the discussion on hard water in Unit B7a. Pouring vinegar into the tea kettles and leaving it for a while will cause the calcium carbonate to react with the acid and so remove the build-up of calcium carbonate.

Q3 Why is vinegar used to clean the scale deposit in tea kettles in preference to lime juice?

The reaction between calcium carbonate and ethanoic acid is given below:

$$\text{calcium carbonate} + \text{ethanoic acid} \longrightarrow \text{calcium ethanoate} + \text{carbon dioxide} + \text{water}$$

Egg shells are also made of calcium carbonate. Carry out the following activity at home to show how the egg shell of an uncooked egg can be removed without breaking the egg.

To do – Removing the shell of an uncooked egg

You will need: one uncooked egg, vinegar, a clear drinking glass.

1 Carefully place the egg in the glass.

2 Add enough vinegar to cover the egg.

3 Leave for about 4 hours.

4 Carefully remove the egg from the vinegar.

Reaction of acids with bases – neutralisation

Acids react with bases to produce a salt and water. Heat is also produced. The reaction is described as **neutralisation**. Neutralisation reactions are widely used in the home at school and in industry. Equations of neutralisation reactions involving some common acids and bases are given below:

$$\text{sodium hydroxide} + \text{hydrochloric acid} \longrightarrow \text{sodium chloride} + \text{water}$$
$$\text{(base)} \qquad\qquad \text{(acid)} \qquad\qquad \text{(salt)}$$

$$\text{Calcium hydroxide} + \text{nitric acid} \longrightarrow \text{calcium nitrate} + \text{water}$$
$$\text{(base)} \qquad\qquad \text{(acid)} \qquad\qquad \text{(salt)}$$

If an alkali is gradually added to an acid, a point will be reached when all the acid has reacted with the alkali to produce a salt and water. This point is called the neutralisation point. Acid–base indicators are used to indicate when neutralisation takes place. The following experiment shows how this could be done in the science laboratory.

neutralisation – the reaction between an acid and a base to produce a salt and water only

SBA Skills

ORR	D	MM	PD	AI
✓		✓		✓

Experiment – Neutralisation reaction

You will need: dilute hydrochloric acid, dilute sodium hydroxide solution, methyl orange indicator, measuring cylinder, dropper.

 SAFETY: Solutions should be prepared by the teacher.

1 Using a measuring cylinder, carefully measure out $10\,cm^3$ of dilute hydrochloric acid solution and pour into the beaker.

2 Add two drops of methyl orange indicator to the acid in the beaker. Swirl gently to ensure mixing. Note the colour of the solution.

3 Using a dropper, add dilute sodium hydroxide solution to the acid in the beaker, swirling after the addition of each drop to ensure mixing. Be careful not to spill any of the acid mixture.

4 Continue adding the indicator until one drop of alkali causes the colour to change from pink to light orange. At this point the acid is completely neutralised.

5 Record the number of drops needed to change the colour of the acid solution.

Q4 Write a word equation to show what reaction took place between the acid and the alkali.

Q5 Add another five drops of alkali to the mixture in the beaker and note that the colour changes to yellow. Provide a suitable explanation for what you observe. Use the information in Table B7.7 to assist you.

property	acids	bases (alkalis)
feel	not slippery	slippery
taste	sour	bitter
reaction with litmus	changes blue litmus red	changes red litmus blue
pH	less than 7	greater than 7
reaction with carbonates and hydrogencarbonates	produces carbon dioxide	no visible change

Table B7.8 Summary of the properties of acids and bases

Some other useful neutralisation reactions

1 *Reducing the likelihood of tooth decay* – Toothpaste is alkaline with pH greater than 7. The bacteria present in our mouths convert starch and sugars in the foods we eat to acid with pH less than 7. This acid, if allowed to build up in our mouths, will destroy the teeth and cause cavities to develop. Toothpaste helps to neutralise the acid and reduce the likelihood of cavities being formed.

2 *Reducing indigestion* – Sometimes the foods we eat lead to acid build-up in the stomach, which can be quite uncomfortable. This discomfort can be relieved by drinking an antacid such as milk of magnesia or sodium hydrogencarbonate (baking soda) solution. These antacids are alkaline and they neutralise the acid build-up in the stomach, thereby reducing the discomfort.

3 *Metal cleaning* – Copper ornaments often get tarnished by a thin coat of copper oxide when exposed to the air. This causes the metal to become darker in appearance, losing its sheen and attractiveness over time. Copper objects can be cleaned effectively by rubbing with lemon juice. The acidic lemon juice will neutralise the basic copper oxide.

To do – Getting rid of tarnish

You will need: copper ornament such as a vase that has been tarnished, cotton wool, lemon juice

1 Dip the cotton wool in the lemon juice.

2 Use this to rub the copper ornament so as remove the tarnish.

3 Continue until the entire ornament is clean.

4 Rinse in warm water and blot dry with a soft rag.

4 *Hair and scalp care* – People who use hair crèmes to straighten hair require a neutralising shampoo to be used immediately after processing. The alkaline shampoo neutralises the excess acid from the crème to reduce the risk of scalp burns.

5 *Stain removal* – Some stains can be removed by usind the reactions between acids and bases.

SBA Skills

ORR	D	MM	PD	AI
✓				✓

Experiment – Stain removal

You will need: pieces of cotton fabric cut into 5 cm squares, solutions of borax and baking soda, various fruits, red wine, tea, two beakers, forceps, hand towel or tissue.

1 On two separate pieces of fabric make stains with the fruits provided. Allow the stains to dry. Stain two other pieces of fabric with tea and red wine.

2 Pour approximately 25 cm³ of the borax solution and 25 cm³ of the sodium hydrogencarbonate solution into different beakers.

3 Dip one piece of the stained fabric into the solvent in each beaker. Use the forceps to keep the stained fabric submerged in the solvent for five minutes.

4 Remove the fabric and blot dry using hand towel or tissue.

5 Observe the changes to each stain made on the fabric.

6 Repeat steps **3–5** for the remaining sets of stained fabric.

Q6 Given that baking soda solution is alkaline and borax forms an acidic solution, what can you deduce about the acidic and basic properties of the substances used to stain the pieces of fabric?

Handling hazardous household chemicals

Several of the chemicals used around the house for gardening, car maintenance or housework can be quite harmful to the user as well as to the environment if not handled carefully. The labels of the more dangerous ones warn that the substances may be flammable, explosive, toxic, corrosive and radioactive. Great care should be taken in their use, storage and disposal.

- Flammable substances are capable of burning or causing a fire.
- Corrosive substances destroy living tissue or eat away at other materials with which they come into contact.
- Toxic substances will cause poisoning either immediately (acutely toxic) or over a period of time (chronically toxic).
- Explosive substances can cause an explosion or release poisonous gases on exposure to air, water or other chemicals.
- Radioactive substances are known to cause changes to cells and chromosomal material leading to cancer, mutations and harm to unborn babies.

poisonous

corrosive

flammable

radioactive

▲ **Fig B7.18** Symbols warn of dangerous chemicals in household products

To do – Identifying dangerous substances

1 Examine the labels of as many household chemicals as you can.

2 Make a table to show those that are flammable, corrosive, explosive and radioactive.

3 Share your list with your friends at school and expand your list to include all the chemicals found by your classmates.

4 Talk about the ways these chemicals are stored or used in your household.

5 Make a list of the steps that you can take to improve the way these chemicals are stored, handled or disposed of in your household.

Using household chemicals safely

Become an informed consumer

When handling any chemical we should first read the label to get information on how it should be used. The label will indicate the following:

- any active chemicals present
- descriptions of any harm that could take place with improper use or in case of accidents
- precautions to be taken in case of harm
- storage recommendations
- possible First Aid instructions where appropriate

A list of some of the hazards associated with common household products and suggestions for their safe use are given in Table B7.9.

hazard	examples	suggestions for personal safety
flammable	furniture polish oil-based paint varnish paint thinner	keep away from heat and flame
corrosive	drain cleaner household batteries oven cleaner	keep out of reach of children avoid contact with skin store upright in air tight container
poisonous	ammonia rat poison bleach insecticide herbicide	avoid inhalation and skin contact do not swallow beware of leakage store upright in air tight container

Table B7.9 Common household products and suggestions for personal safety

Follow the manufacturers' guidelines for using the product

Some substances should not be used around fires. For example, it is unwise to be smoking a cigarette when handling petrol or paint.

Protective materials such as goggles and gloves should be worn when handling some insecticides used in agriculture.

You should avoid mixing chemicals unless the label permits this practice. For example, bleach should not be mixed with ammonia cleaners as the active ingredient in bleach reacts with the ammonia to produce toxic chlorine gas.

Food should not be eaten when handling chemicals and if spills occur, these should be effectively cleaned up. Substances with highly toxic fumes should be opened and used in a well ventilated area, then should be carefully sealed to prevent leakage after use.

Sometimes accidents do occur and chemicals may be inhaled, swallowed or come in contact with the skin. Immediate steps should be taken to counteract any lasting harm. Bleach, for example, is corrosive, causing irritation of the skin, eyes and respiratory tract. If ingested, it may cause vomiting. If bleach is swallowed, one could drink a mild alkali such as milk, milk of magnesia or the white of an egg to reduce the likelihood of any permanent harm.

Ammonia is widely found in disinfectants, floor waxes and window cleaners. Ammonia acts as an irritant when it comes in contact with the skin or the eye. Where this occurs, water should be used to wash the area for as long as it takes to remove the sensation. If swallowed, large doses of lime squash or lemon juices could be drunk.

Oven cleaners often contain sodium or potassium hydroxide, which are caustic and could result in burns to skin and eyes. They are also poisonous and if swallowed could lead to severe tissue damage. Many oven cleaners advise on the use of large quantities of water if accidentally swallowed.

Q7 Why is it unwise to be use a naked flame around petrol and paints?

Q8 Given what happens when acids react with alkalis, why is not advisable to drink a strong alkali if you have accidentally swallowed some bleach?

Safety tips for storing hazardous chemicals

- Follow the label's directions for proper storage conditions.
- As far as possible, leave the product in its original container with the original label attached.
- Never store hazardous products in food or beverage containers in case someone mistakes them for food or drink. Where it is necessary to change the original container, ensure that the new storage container is properly labelled.
- Store upright with tightly sealed lids and caps.
- Keep flammables away from corrosive substances and away from heat, sparks and sources that could ignite.
- Store products that emit vapours or fumes in a well-ventilated area, out of reach of children and pets.
- Keep metal containers dry to prevent corrosion, which could cause leaks.
- Store rags used with flammable products (furniture stripper, paint remover) in a sealed, marked container.

Disposal of household chemicals

Serious environmental damage can also occur if chemicals and their containers are not disposed of properly. Some insecticides contain heavy metals such as lead, mercury and arsenic. If these leak into the ground water supply they could cause serious harm to plants and animals.

The best advice that can be given for disposal of household chemicals is to use the chemical up or give it to someone who can. This will prevent harmful substances from contaminating the environment. Many household cleaners and disinfectants, such as bleach, can be flushed down the drain in small quantities. Some wastes, such as used motor oil, should be taken to an authorised collection site. Unused paint and varnish should be allowed to solidify and then be double wrapped before disposal. Under no circumstances should containers be thrown on dumps, buried in the back yard or burnt.

End-of-unit questions

1 Use substances A, B, C, D, E and F to answer questions **a–e**

A – vinegar, B – milk of magnesia,
C – grapefruit, D – Epsom salts,
E – rain water, F – ammonia cleaner

a Which substance(s) will have a pH above 7?

b Which substance(s) will have pH about 7?

c Which substance(s) will likely turn blue litmus paper red?

d Which substance(s) will react with baking soda?

e Identify the gaseous product that will be formed if the reaction in part **d** takes place.

2 Copy and complete Table B7.10 by writing the name of the salt formed when the acids combine with the bases present. The first one has been done for you.

acid bases	sulphuric acid	nitric acid	hydrochloric acid
sodium hydroxide	sodium sulphate		
calcium hydroxide			

Table B7.10

3 Write a word equation to show the products formed in the reaction between nitric acid and calcium carbonate.

4 Kerry accidentally added baking soda to a glass of tamarind juice. She noticed that there was much effervescence (bubbles were formed). Provide a suitable explanation for this observation.

5 Your classmate suggested that the following pairs of substances could be distinguished by using universal indicator paper. Comment on the feasibility of those suggestions.

a lemon juice and vinegar

b muriatic acid and vinegar

c ammonia cleaner and vinegar

By the end of this unit you will be able to:

- discuss the use of good and poor conductors of electricity.
- explain the relationship between current, voltage and resistance in circuits.
- describe the correct wiring of a plug.
- draw circuit diagrams, including components in series and in parallel.
- describe the magnetic effect of a current and give some of its uses.

Electrical hazards

We use electricity every day. Electricity brings us energy to make appliances such as lights, fridges and computers work. While electricity is very useful to us, it can also be dangerous.

The 1.5-volt cell that you put in a flashlight cannot do you any harm. The 9-volt battery used in a radio is also safe to handle, and you won't get a shock from a 12-volt car battery. But you need to be extremely careful with the mains electricity that comes from sockets in the wall. Both the 110- and 240-mains voltages have been responsible for many deaths.

voltage – the 'push' provided by a source of electricity to make a current flow

electric current – a flow of electric charge

volt (V) – the unit of voltage

Why is mains electricity more dangerous than a 12-volt car battery? The clue is in the voltage. When you touch an electricity supply, the **voltage** pushes an **electric current** through you. It is the current that harms you, but it is the voltage that makes the current flow. The greater the voltage, the greater the current. The unit of voltage is **volts** (symbol V).

Getting a shock

Children are often warned not to fly kites near electric power lines as these wires sometimes carry voltages as high as 500,000 volts, which can produce fatal shocks. Shock is the best known danger associated with electricity.

What would happen if you were to touch the mains electricity supply in your home or school? The high voltage would make an electric current flow through you. This has two effects:

▲ **Fig C1.1** This engineer is repairing electrical power lines in Trinidad. He knows how to handle high voltages safely

- The current makes your muscles contract violently, so that you are thrown across the room. The muscles of your hand may cause your hand to grip tightly to the wire, making it difficult for anyone to pull you to safety.

- The current heats your skin where it enters and leaves your body. It may also cause the watery fluid in your cells to boil. The result is a burn.

Either of these effects can prove fatal. That is why it is important to understand how electricity works and how to use it safely. This unit should help you to do so.

To do – Staying safe with electricity

Discuss the following questions:

1 Where have you seen signs warning of the dangers of electricity?

2 The mains supply can give a fatal shock. What advice have you been given about how to use it safely?

DANGER OF DEATH

KEEP OUT

▲ Fig C1.2

Q1 Explain why a 240-volt mains supply is more dangerous than a 110-volt mains supply.

Q2 Which is less dangerous to be shocked by, a wire touching the palm of the hand or one touching the back of the hand? Give a reason for your answer.

Electrical measurements

A voltage is needed to make a current flow. Voltage is measured in volts (V). A voltage can be supplied in several ways:

cell – a single 'battery' used to provide a voltage in a circuit

A **cell** is a single 'battery' and usually provides about 1.5 V.

battery – two or more cells connected end-to-end

A **battery** is made of two or more cells connected end-to-end; two cells provide 3.0 V, three provide 4.5 V, and so on.

A power supply is connected to the mains supply and gives a low, safe voltage. A mobile phone charger is a type of power supply.

The mains supply may be 110 V or 240 V, depending on where you live.

Measuring current and voltage

When we think of an electric current, we imagine a flow of electric charge. The bigger the current in a wire, the greater the flow of charge in the wire. As the charge flows, it carries energy. Electric current is measured in **amperes**, often called **amps** (symbol A).

ampere or amps (A) – the unit of electric current

ammeter – the instrument used to measure electric current

An **ammeter** is used to measure electric current. It has to be connected correctly in the circuit so that the current flows through the ammeter.

Fig C1.3 shows how to measure the current flowing through a lamp. We say that the ammeter is connected in series (end-to-end) with the lamp. It does not matter whether the ammeter is before or after the lamp; it will show the same reading.

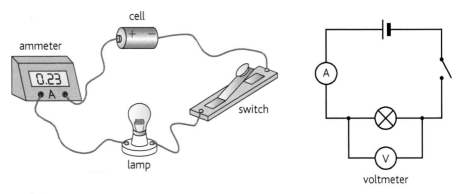

▲ **Fig C1.3**

voltmeter – the instrument used to measure voltage

The circuit diagram shows the symbols used for each component in the circuit. It also shows how a **voltmeter** must be connected to measure the voltage in volts provided by the cell. One end of the voltmeter is connected to each end of the cell.

Q3 Draw the circuit symbols for the following components:

  a cell

  b lamp

  c switch

Q4 Copy and complete the following table:

quantity	unit	name of meter	symbol for meter
current	( )		
	volt (V)		

conductor – a material that allows electric current to flow freely

insulator – a material that does not allow electric current to flow freely

semiconductor – a material that allows electric current to flow only weakly

Materials that conduct electricity

Some materials allow electric current to flow easily – these are called **conductors**. Most metals are good conductors of electricity. Materials that are poor conductors of electricity are called **insulators**. **Semiconductors** are materials that conduct electricity only weakly; they are important in making transistors and computer chips.

You can make a simple conductivity tester by connecting a dry cell in series with a low voltage lamp as shown in Fig C1.4 (the apparatus can be mounted in a hollow tube or box to make it easy to handle). The material to be tested is placed between points A and B.

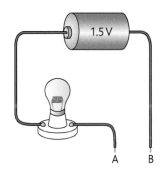

▲ **Fig C1.4** A simple conductivity tester

If the lamp lights when a material is placed across ends A and B of the wires, it means that current is flowing through the material. In this case the material is a good conductor of electricity. If the lamp does not light we cannot say for sure that no current is flowing through the material. This is because a certain amount of current must flow before the lamp can light. All we can say this time is that the material is a poor conductor of electricity.

SBA Skills

ORR	D	MM	PD	AI
✓		✓		✓

Experiment – How well do solids conduct electricity?

Start by making a conductivity tester:

1 Collect a cell (a 1.5 V 'battery'), a lamp in a holder, and three connecting wires.

2 Connect the cell and lamp together as shown in Fig C1.4.

3 Connect the two test wires and label their free ends A and B.

4 Use the conductivity tester to investigate some different materials. Collect a set of objects made from different materials. Test the materials to find out which types are good conductors, which are poor conductors, and which are in between.

5 Find out if a thick wire conducts electricity better than a thin wire:

    a Take a 30 cm length of thick nichrome wire and connect it to your conductivity tester. Note the brightness of the lamp.

    b Repeat with a 30 cm length of thin nichrome wire. Which wire is a better conductor?

6 Explain why both nichrome wires must be the same length if the experiment is to be a fair test.

7 Find out if a long wire conducts electricity better than a short wire. Again, use nichrome wires. How can you make this a fair test?

By looking at the brightness of the lamp we can judge whether the material that is being tested is a good conductor, a poor conductor, or somewhere in between.

SBA Skills

ORR	D	MM	PD	AI
✓				✓

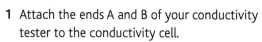

Experiment – Do liquids conduct electricity?

To test liquids to see if they are good or poor conductors, you will need a conductivity cell. This is a glass beaker (as shown in Fig C1.5) with two carbon rods that conduct the current into the liquid.

You are going to test various liquids to find out if they are good, poor, or in-between in their conduction properties (some suggestions for liquids include: rain water, distilled water, tap water, trench water, sea water, water to which salt is added, water to which dirt is added, vinegar, milk, coconut oil). You should not use your tester on highly flammable liquids because they may catch fire if there are any sparks.

1 Attach the ends A and B of your conductivity tester to the conductivity cell.

2 Place each liquid to be tested in the vessel so that it comes half-way up the carbon rods.

3 By judging the brightness of the lamp, put the liquids in order from best conductor to poorest conductor.

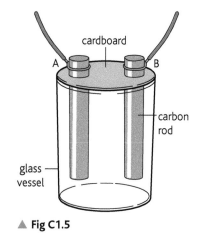

▲ **Fig C1.5**

Q5 A student uses a conductivity tester and finds that the lamp is very bright when testing a copper wire, dim when testing tap water, and does not light when testing paper. Put these three materials in order, from best conductor to worst conductor.

Using conductors and insulators

We need good conductors of electricity to transport electrical energy from the power stations to our homes and to our appliances. Metals such as copper and aluminium are widely used for this purpose. But we also need to be careful not to let these good conductors touch us when they are connected to a voltage supply. We must never touch bare live wires, and we should avoid doing electrical work in damp areas because water can be a good conductor, as your experiments should have shown. For this reason, electrical fires should never be put out with water.

Rubber, cotton, plastic and glass are examples of insulators. They are commonly used to insulate the conducting parts of appliances so that we cannot get shocked. However, if there is a very high voltage, some of these 'insulators' may conduct enough electricity to kill a person in contact with them.

Although metals are usually very good conductors, very long or very fine wires tend to behave like poor conductors. That is why fat wires are used when connecting heavy-duty equipment (e.g. an electric cooker) to the voltage supply. If the connecting wires are too fine, the appliance might not be able to get the current it needs, and will not work.

Q6 Name one material that is found in homes and that is:

    a a good conducting solid

    b a good conducting liquid

    c a poor conducting solid

Plugging in

Nowadays, electrical appliances are generally sold with a three-pin plug attached. This is because the manufacturers want to be sure that the plug is properly fitted; otherwise the appliance will not work correctly. However, it is still useful to know how to fit a plug to the flex (cord) of an appliance. You should also understand the choice of materials used for each part of the plug.

▲ **Fig C1.6** A correctly wired plug

Two of the wires, the 'live' and the 'neutral', carry the electric current. The third, the 'earth', is put there for safety. The colour-code used nowadays by most manufacturers is this: the brown wire is to be connected to the part of the plug marked L (live), the blue to the part marked N (neutral), and the yellow/green to the part marked E (earth).

- Each wire is made of metal, usually copper or steel because these are good conductors, with coloured plastic insulation around it.
- The wires are connected to the three brass pins. Brass is another good conductor. Also, it is stiff and does not corrode easily.
- The outside casing of the plug is made of plastic, an insulator, which protects the user from the high voltage inside.
- The earth wire connects the metal casing of an appliance directly into the ground. This means that if there is a fault and the case becomes connected to the mains voltage, the current will flow safely into the ground and the user will be safe.
- The plug also contains a fuse. You will find out the purpose of the fuse shortly.

To do – Wiring up

Practise wiring up a plug with a short length of flex. Some points to note:

- When cutting the insulation, take care not to cut any strands of wire as well.
- Wrap the bare wires clockwise round the 'posts' so that they are held firmly when screwed down.
- Ensure that the flex is held tightly by the cord grip.

 SAFETY: Ask your teacher to check your work. Do not plug into a socket – you could get a shock. It is best if the lab supply is turned off during this activity.

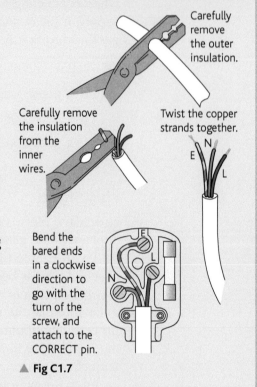

Carefully remove the outer insulation.

Carefully remove the insulation from the inner wires.

Twist the copper strands together.

Bend the bared ends in a clockwise direction to go with the turn of the screw, and attach to the CORRECT pin.

▲ **Fig C1.7**

Q7 Copy and complete the table below to show how the wires are connected in a plug.

name of wire	label on pin	colour of wire	position of pin
live	L		right
neutral			
			top

The heating effect of a current

If you look closely at a filament lamp (such as the bulb of a flashlight) you will see that the filament is a thin wire. When the lamp is switched on, the current flows through the filament and makes it so hot that it glows brightly.

▶ **Fig C1.8** The filament of a lamp gets hot when a current flows through it

This shows that an electric current has a heating effect when it flows through a conductor. Two other places where we make use of the heating effect of a current are:

- In an electric heater, current flows through wires that get hot and may glow.
- The fuse in a plug contains a thin wire. When the current gets too big, the fuse wire gets so hot that it melts ('blows') and the circuit is broken, stopping the excessive current from flowing.

SBA Skills

ORR	D	MM	PD	AI
✓		✓		

Experiment – Electric heat and light

1. Collect some different thicknesses of fuse wire. Cut a 10 cm length of each type.

2. Lay the thinnest wire on a piece of wood (to protect the bench). Connect the wire into a series circuit with a power supply (or battery) and a variable resistor.

 SAFETY: Wear eye protection before you switch on.

3. Switch on the power supply to give about 2 V. Adjust the variable resistor so that the current through the fuse wire increases. You should see the wire start to glow and eventually melt.

4. Repeat with the other thicknesses of wire. You can also use a piece of wire wool in place of the fuse wire.

5. If you have an ammeter, measure the current that makes the fuse wire melt.

Understanding fuses

fuse – a component in a circuit; the fuse wire melts when the current is too big, breaking the circuit

A **fuse** is included in an electrical circuit where there is a danger that the current may become too big. Imagine that you are using an electric iron or drill. If the current is too big, it will burn out the iron's element or the drill's motor. Then the appliance will need to be repaired.

A fuse is a weak link in the circuit. If the current gets too big, the fuse wire will melt and stop the current flowing – it's easier to replace a fuse than to repair an appliance.

Electrical resistance

Each fuse wire in the photograph is labelled with the current needed to melt it. It takes a smaller current to melt a thin wire than a thick one. Why is this?

resistance – the ratio of the voltage across a component to the current through it: $R = \frac{V}{I}$

Every component in a circuit has electrical **resistance**. As a current flows through a component, it must push past the component's resistance. This takes energy, and the energy lost by the current causes the component to get hot.

It is easier for a current to flow through a thick wire than through a thin wire.

▲ **Fig C1.9**

This is because a thick wire has less resistance than a thin wire, and that is why a thin wire gets hotter for a smaller current than a thick wire.

Q8 Which is thinner, a 10 cm length of 5 A fuse wire or a 10 cm length of 15 A fuse wire? Which has greater resistance?

Q9 Explain how the heating effect of an electric current is made use of in a clothes iron.

Calculating resistance

ohm (Ω) – the unit of electrical resistance

The resistance R of an electrical component is measured in **ohms (Ω)**. To measure the resistance, you need to connect a voltage V across it and measure the current I that flows through it. Fig C1.10 shows a suitable circuit.

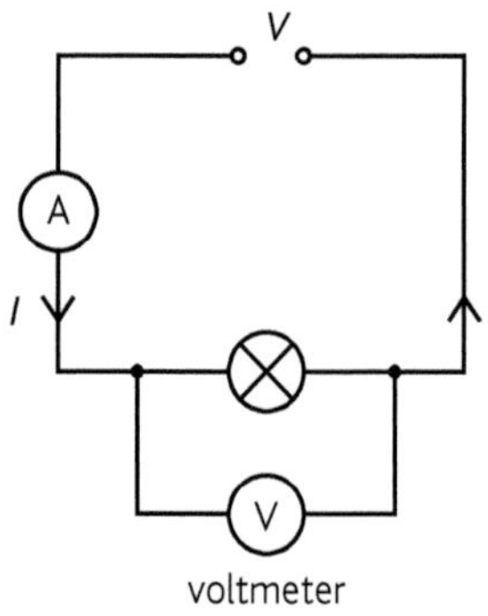

Here is the formula used to calculate resistance:

$$\text{resistance} = \frac{\text{voltage}}{\text{current}} \qquad R = \frac{V}{I}$$

This formula can be rearranged in two ways so that we can calculate voltage or current:

$$\text{voltage} = \text{current} \times \text{resistance} \qquad V = IR$$

$$\text{current} = \frac{\text{voltage}}{\text{resistance}} \qquad I = \frac{V}{R}$$

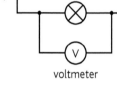
voltmeter

▲ **Fig C1.10**

Example 1: What is the resistance of a lamp if a current of 3 A flows when it is connected to the 240 V mains?

We are trying to find the resistance, so we use the first version of the formula:

$$R = \frac{V}{I}$$

Substituting in values for V and I gives:

$$R = \frac{240}{3} = 80\,\Omega$$

The lamp has a resistance of 80 Ω.

Example 2: What voltage is needed to make a current of 2 A flow through a 5 Ω heater?

Substituting in $V = R$ gives $V = 2 \times 5 = 10\,\text{V}$

So 10 V is needed to make 2 A flow.

Q10 What is the resistance of a lamp if a current of 0.1 A flows through it when it is connected to a 12 V car battery?

Q11 What current will flow through a 200 Ω heater when it is connected to a 10 V supply?

Resistors

A **resistor** is a component with a known value of resistance. It is included in the circuit to ensure that the current flowing has the correct value.

Fig C1.11 shows a selection of resistors together with the circuit symbol for a resistor.

▲ **Fig C1.11** These resistors have colour-coded stripes that indicate their values

SBA Skills

ORR	D	MM	PD	AI
✓		✓		

Experiment – Measuring resistance

1 Collect a selection of lamps and resistors.

2 Connect one of the resistors in a circuit that includes a battery, an ammeter and a voltmeter. Before you complete the circuit, ask your teacher to check that it is correct.

3 Measure the current through the resistor and the voltage across it. Copy Table C1.1 below and record your results.

4 Complete the final column by calculating the resistance.

component	current /A	voltage /V	resistance /Ω
resistor 1	0.25	6.0	$\frac{6.0}{0.25} = 24$
resistor 2			

Table C1.1

If you have a multimeter, you can use this to measure the resistance of a component directly. You will need to switch it to the 'ohms' setting.

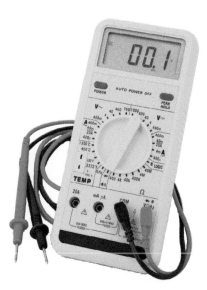

▶ **Fig C1.12** A multimeter can be switched to measure current, voltage or resistance

Series and parallel

in series – describes components connected end-to-end in a circuit

in parallel – describes components connected side-by-side in a circuit

There are two ways to connect components in a circuit, **in series** or **in parallel**.

Fig C1.13 shows two resistors connected in series (end-to-end). The current must pass through one resistor and then the other. This means that the current is smaller than if there was just one of the resistors – the circuit resistance is greater.

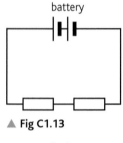

▲ Fig C1.13

Fig C1.14 shows two lamps connected in parallel (side-by-side). Each lamp feels the full push of the battery. The current flowing is greater than if there was just one lamp because now there are two paths for it to travel round the circuit – some flows through one lamp, some through the other lamp. This means that you get more current from the same voltage, and so the circuit resistance is smaller.

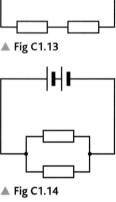

▲ Fig C1.14

SBA Skills

ORR	D	MM	PD	AI
✓		✓		

Experiment – Series and parallel circuits

1 Set up a circuit with a cell and a single lamp. Add a second lamp in series with the first. (You will have to disconnect the circuit to do this.) How can you tell that the current in the circuit has decreased?

2 Repeat this experiment but add the second lamp in parallel with the first. How can you tell that the current in the circuit has increased?

3 If you have an ammeter, repeat both experiments with the ammeter added to the circuits so that it measures the current from the cell.

Q12 Draw two circuits, each showing a lamp and a resistor connected to a cell. In the first, show the lamp and resistor in series with each other. In the second, show them in parallel.

Q13 a Calculate the current in a circuit when a $60\,\Omega$ resistor is connected to a $12\,V$ supply.

b A second resistor is connected in series with the first. The current decreases to 0.10 A. What is the total resistance in the circuit?

c What is the resistance of the second resistor? Explain your answer.

magnetic field – a region where a magnetic material feels a force

compass – a device with a magnetised needle that turns to show the direction of the magnetic field

The magnetic effect of a current

An electric current heats things up, but it also has another effect – it produces a magnetic field. Wherever an electric current flows, there is a **magnetic field** around it. Just like the magnetic field of a bar magnet, you can show up the field using iron filings or a **compass**.

In Fig C1.15, a wire carries a current upwards through a sheet of card. The compasses turn so that they all line up round the current.

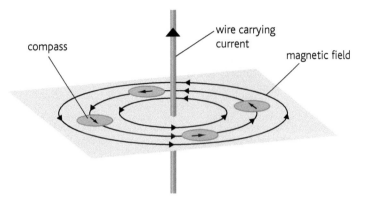

▲ **Fig C1.15** The magnetic field around a current; the black lines show the magnetic field

SBA Skills

ORR	D	MM	PD	AI
✓				✓

Experiment – Observing the field around a current

1 Place a compass on the table. Its needle will point North–South.

2 Lay a length of wire over the compass in the same direction as the needle.

3 Connect the ends of the wire to a cell. What happens?

4 Try reversing the connections to the cell. Can you explain what you observe?

A stronger field

You can make the magnetic field stronger by winding the wire into a coil. This concentrates the field into a smaller space. A coil of wire like this is called a **electromagnet** or **solenoid**.

The magnetic field around an electromagnet is like that of a bar magnet.

* There is a North pole at one end of the coil and a South pole at the other.
* Field lines come out of the North pole and go round to the South pole.

electromagnet – a coil of wire, often with an iron core, that becomes magnetised when a current flows through it

solenoid – another name for the coil of an electromagnet

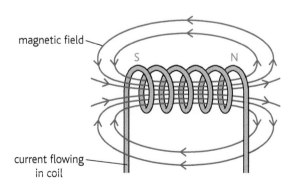

▲ **Fig C1.16** The magnetic field around an electromagnet

Here are three ways to increase the strength of an electromagnet:

- Increase the current in the coil.
- Increase the number of turns of wire used.
- Add an iron core inside the coil. This makes it much stronger.

▶ **Fig C1.17** This electromagnet has an iron nail as its core

SBA Skills

ORR	D	MM	PD	AI
✓		✓		✓

Experiment – Making and testing an electromagnet

Fig C1.18 shows how to make an electromagnet and test it:

1 Wind a coil of insulated wire around an iron nail.

2 Connect the bare ends to a battery or low-voltage power supply. Include a switch if possible.

3 Hold a compass near the electromagnet. Can you detect its field?

4 Hang paperclips end-to-end from the nail. How many will it hold?

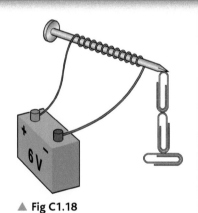

▲ **Fig C1.18**

5 Add a variable resistor to the circuit so that you can vary the strength of the electromagnet. Test its strength as you vary the current.

Q14 Which of the following will reduce the strength of the magnetic field of an electromagnet?

    a decreasing the current

    b adding more turns of wire

    c removing its iron core

Q15 If the direction of the current flowing through an electromagnet is reversed, the magnetic field is also reversed. How could you show this using a compass?

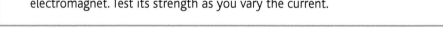

Using electromagnets

permanent magnet – a magnet such as a bar magnet or horseshoe magnet which retains its magnetism

temporary magnet – a magnet such as an electromagnet which is only magnetised when an electric current flows

An electromagnet can be more useful than a **permanent magnet**. You can adjust its strength by changing the current and you can switch it on and off – it is also known as a **temporary magnet**.

The giant electromagnet in the photograph is used at a scrap yard to move heavy loads of metal. When the load is in the right place, the operator switches off the current and the metal falls onto the heap.

Electromagnets are also used in relays (electromagnetically operated switches), in doorbells and in the recording heads of a tape recorder.

▲ **Fig C1.19** An electromagnetic crane moving metal at a scrap yard

Electric doorbell

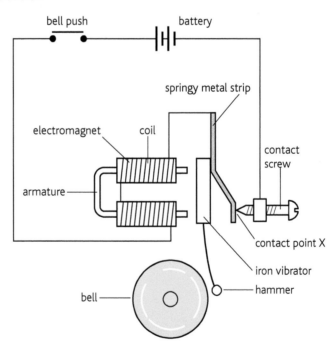

▲ **Fig C1.20** An electric bell has two coils mounted on an iron core (the armature)

The hammer of the doorbell vibrates back and forth automatically when someone presses the switch. This happens because:

- When the switch is closed, the circuit is complete and current flows through the electromagnet coils.
- This attracts the iron vibrator, causing the hammer to strike the gong.
- At the same time, the circuit is broken at point X. The current stops and the electromagnet stops working. The springy metal strip pulls the hammer back.
- Now contact is made again at X and the circuit is complete. Current flows again, and so on.

SBA Skills

ORR	D	MM	PD	AI
✓				✓

Experiment – Ring that bell!

Examine an electric bell and observe how it works. Try to identify each of the steps in the explanation above. You may even see sparks jump through the air at point X.

SBA Skills

ORR	D	MM	PD	AI
		✓		

Experiment – Make a buzzer

1 Use the electromagnet you made earlier. Clamp it horizontally, as shown in Fig C1.21.

2 Connect the ends to a low-voltage a.c. (alternating current) supply. (It *must* be a.c.)

3 Clamp a flexible steel hacksaw blade as shown, so that it will bend back and forth horizontally. Position it so that the blade doesn't quite touch the iron nail at point X.

4 Switch on. Because you are using a.c., the magnetic field of the electromagnet keeps increasing and decreasing. This makes the blade vibrate and bang on the nail.

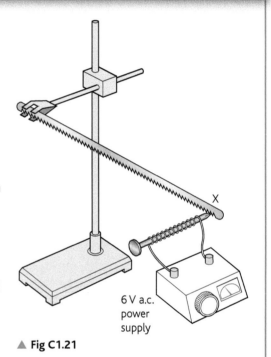

6 V a.c. power supply

▲ **Fig C1.21**

End-of-unit questions

1 Which row in Table C1.2 correctly shows the colours of the wires connected to a three-pin plug?

	live	neutral	earth
A	red	black	green
B	blue	brown	green/yellow
C	red	black	green/yellow
D	brown	blue	green/yellow
E	blue	brown	green

Table C1.2

2 What voltage is needed to make a current of 0.5 A flow through a 20 Ω resistor?

A 0.025 V

B 1.5 V

C 10 V

D 20.5 V

3 Which row in Table C1.3 correctly shows examples of materials that are electrical conductors and insulators?

	conductors	insulators
A	steel, glass	paper, plastic
B	copper, water	plastic, glass
C	copper, plastic	water, steel
D	water, plastic	glass, paper
E	aluminium, copper	water, plastic

Table C1.3

4 A lamp is connected to a 6.0 V battery. When the lamp lights, a current of 0.25 A flows through it.

a Calculate the resistance of the lamp.

b Draw a circuit diagram to show how the resistance of the lamp could be measured. Label the meters used with their names.

c Draw another circuit diagram to show how two resistors could be connected to a cell so that the resistors are in parallel with each other.

d If one of the resistors is disconnected from the circuit, will the current flowing from the cell increase, decrease or stay the same?

5 Some food and drink cans are made of steel (magnetic). Others are made from aluminium (non-magnetic). How could an electromagnet be used to separate the two types at a recycling centre? Illustrate your answer with a diagram.

6 Mrs Elias is using an electric iron to iron a new skirt.

a The handle of the iron is made of plastic. Explain why this is a suitable material for the handle.
Unfortunately, the iron does not get hot. Mrs Elias pulls the plug from the socket and unscrews the cover of the plug. Inside the plug, Mrs Elias can see three coloured wires. The blue wire has become disconnected from its pin.

b Which wire in a plug is coloured blue?

c What colour is the earth wire?
Mrs Elias screws the wire back into place. She decides that she should also test the fuse.

d Describe how Mrs Elias could test the fuse using a battery, a bulb and some wires. Draw a circuit diagram to illustrate your answer and say how she would know if the fuse was still in working order.

Electrical power

energy – what we need in order to make anything happen

We use electricity because it is a convenient way to move **energy** to where we want it. When you switch on a lamp at night, you get energy from the lamp in the form of light and heat. The energy comes from fuel burning at the power station (or from a wind turbine, etc.). It is carried by electric current along the wires all the way from the power station to your lamp.

power – the rate at which energy is used

watt, W – the unit of power

Electrical appliances are often labelled with the **power** they use when working normally. This is shown as W, which stands for **watt**, the unit of power. For example, the label on a lamp may say 40 W, 60 W or 100 W.

Power is the rate at which energy is supplied. Power is sometimes called 'wattage'. The higher the wattage of an appliance, the greater the rate at which it uses electricity.

Fig C1.22 shows the label on an electric heater. Its power rating is shown as 2000 W when it is connected to the 240 V mains supply. This could also be shown as 2 kW (kilowatts).

$$1 \, kW = 1000 \, W$$

▲ **Fig C1.22**

To do – Power ratings

Inspect some electrical appliances and find out their power ratings. Make a note of any other information on the labels, such as the voltage or current. Put your findings in a table; arrange them from lowest to highest power.

SBA Skills

ORR	D	MM	PD	AI
✓				

Typical power ratings

Table C1.4 shows the typical power ratings of some electrical appliances. You should notice that the appliances that produce heat are often the ones with the highest power ratings.

appliance	power rating / W
calculator	0.01 W
car sidelight	12 W
car headlight	36 W
electric drill	50 W
filament lamp	40 W, 60 W, 100 W, 150 W
television set	200 W
electric heater	1000 W, 2000 W
clothes iron	1500 W
electric stove	8000 W

Table C1.4

Q1 Look at Table C1.4. Explain why the car sidelight is less bright than the car headlight.

Calculating power

The power P of an appliance is related to the voltage V it is connected to and the current I that flows through it.

power = voltage × current $\qquad P = VI \qquad$ watts = volts × amps

Often we want to use this formula to calculate the current in an appliance, so we need to rearrange it like this:

$$\text{current} = \frac{\text{power}}{\text{voltage}} \qquad I = \frac{P}{V} \qquad \text{amps} = \frac{\text{watts}}{\text{volts}}$$

Example: What current flows in a 36 W car headlight when it is connected to a 12 V car battery?

Substituting in amps = $\frac{\text{watts}}{\text{volts}}$, we have:

$$\text{current} = \frac{36}{12} = 3\,\text{A}$$

So there is a current of 3 A in the headlight.

It is useful to be able to calculate the current that flows because this can help us to choose suitable wires. You may have noticed that a radio connected to the mains supply has a much thinner flex (cord) than a clothes iron. This is because the radio is a low-power appliance (perhaps 20 W), so only a small current flows. An iron is a high-power appliance (perhaps 1500 W), so a much bigger current flows.

Electricians know this, so they ensure that the correct cables are fitted in a house.

- The cables that bring electricity to the lights are thin, because lights are low-power and so the current is small.
- The cables for a stove are thick because the stove needs a big current when it is fully switched on.

It would be dangerous to connect a stove or a heater to a lighting point because the current would be too great for the wiring. The wires would get hot; the insulation might melt and catch fire. Many house fires start like this.

Fortunately, each wiring circuit in the house is fitted with a fuse, which should blow if the current is too great. If there is a fault, the fuse may keep blowing. To get round this problem, some people put something metal in place of the fuse – a very dangerous thing to do!

Q2 Calculate the power rating of a lamp if a current of 0.25 A flows in it when it is connected to the 240 V mains.

Q3 What current flows through a 60 W lamp when connected to the 110 V mains?

Q4 It is dangerous to attach many appliances to a single power point, as shown in Fig C1.23. This may blow the fuse in the circuit that brings current to the power point. Explain why.

▲ **Fig C1.23** It is dangerous to plug too many appliances into a socket like this – the current flowing may be too high

Fuse and cable calculations

Because we can calculate the current that flows in an appliance, we can decide which cables are suitable and which fuse value to use in a circuit. Manufacturers make cables of different thicknesses. Fig C1.24 shows cables that can be used for different values of current. The thinnest is suitable for currents up to 1 A, and the thickest for currents up to 30 A.

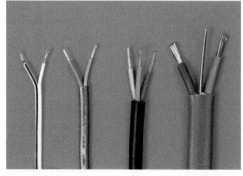

▲ **Fig C1.24** If the current in a circuit is likely to be high, a thick cable must be used

Example: A 1200 W heater is connected to a 240 V mains supply. What fuse value is suitable? Choose from 3 A, 5 A, 10 A, 30 A.

The current in the heater is given by:

$$\text{amps} = \frac{\text{watts}}{\text{volts}} = \frac{1200}{240} = 5\,\text{A}$$

The 3 A fuse is too small. The 5 A fuse is also too small – there would be a danger of it blowing if the current were just a little bigger. So the 10 A fuse is best. The 30 A fuse would be much too big – the current might rise to a dangerous level without blowing the fuse.

Q5 A 800 W freezer runs on a 110 V mains supply. What fuse value is suitable? Choose from 3 A, 5 A, 10 A, 30 A.

To do – Cable collection

Collect pictures or samples of electrical cables with different current ratings. For each, find out a suitable use.

Transformers

Some appliances are set to operate from the 110 V mains. If you were to plug it in to the 240 V mains the fuse would burn out. All that is needed to avoid this happening is a transformer to convert the 240 V mains electricity into 110 V.

But be careful! Transformers have also got current ratings. This is because transformers are made from coils of wire, each of which has an ampere rating. So transformers with coils made of thinner wire will only be able to cope with a lower current.

Sometimes, instead of stating the current rating of a transformer, the manufacturer gives the current × voltage rating. This is called the power rating of the transformer. A 1000 W polisher, for example, should be used with a 1000 W (or greater) transformer. A transformer of a lower power rating will overheat and burn out because the current flowing through the wires will be too great.

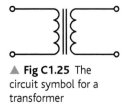

▲ **Fig C1.25** The circuit symbol for a transformer

One final point about transformers. You must know where to connect the input and output voltage wires on them. The manufacturers usually indicate this. A transformer connected one way may step down 220 V to 110 V. Connected the reverse way it will step up 220 V to 440 V.

Q6 A man once borrowed a 110 V projector to give a film show in a 240 V area. He used a 110/240 V transformer, as shown in Fig C1.26. When he switched on the projector its lamp blew and the projector went up in smoke. Can you explain how it happened?

240 V

110/240 V

110 V

▲ **Fig C1.26**

Paying for electrical energy

Why do we use electricity? It is a convenient way of transferring energy from one place to another, and it allows us to do things that we wouldn't be able to do otherwise. But we have to pay for that energy, and the more we use, the more we must pay.

energy consumption
– the amount of energy used when an electrical appliance is in operation

The amount of electricity used is called **energy consumption**. The rate of energy consumption depends on how fast you are using electricity (the power) and how long you use it for.

- Appliances with a higher power rating use energy more quickly.
- The longer the appliance runs for, the greater the amount of energy used.

So we can calculate the amount of energy used like this:

$$\text{energy} = \text{power (in kW)} \times \text{time (in hours)}$$

Example: A 5 kW oven is switched on for 4 hours. How much energy does it use?

$$\text{energy} = \text{power} \times \text{time} = 5 \times 4 = 20\,\text{kWh}$$

kilowatt-hour (kWh)
– a unit of energy consumption, equal to the energy used by a 1 kW appliance in 1 h

The unit of energy is the **kilowatt-hour (kWh)**, sometimes simply called a 'unit' of electricity. It is the amount of energy used by a 1 kW appliance in 1 h.

Table C1.5 shows how to calculate the energy used by electrical appliances running for 20 h. Note that the energy in Wh (watt-hours) must be divided by 1000 to give kWh.

appliance	power	energy used in 20h
3 lamps	3 × 60 W	180 W × 20 h = 3600 Wh = 3.6 kWh
1 refrigerator	300 W	300 W × 20 h = 6000 Wh = 6 kWh
1 iron	700 W	700 W × 20 h = 14,000 Wh 14000 Wh = 14 kWh
1 radio	2 W	2 W × 20 h = 40 Wh = 0.04 kWh

Table C1.5 Calculating the energy used by different appliances over 20 hours

Q7 How many kWh of energy are used by a 2.5 kW heater in 12 h?

Q8 How many kWh of energy are used by a 100 W lamp in 5 h?

Reading the meter

The electricity meter shows the amount of energy used, in kWh. It is useful to know how to read the meter and to keep a record. Then you will not be surprised when the bill comes!

There are two types of electricity meter, analogue (with dials) and digital (just numbers). Fig C1.27 shows both types of meter, each with the same reading.

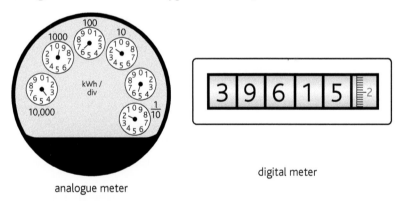

kWh / div

3 9 6 1 5

analogue meter

digital meter

▲ **Fig C1.27** Two types of electricty meter

Digital meters are very easy to read. To read an analogue meter, you have to look carefully at the little needles and dials. Half of the dials go clockwise, the others anticlockwise. If a needle is between two numbers, you must read off the number that the needle has just passed.

Q9 What is the reading on the electricity meter shown in Fig C1.28?

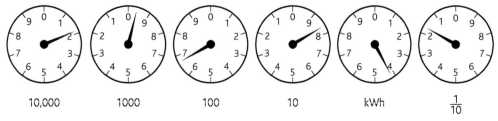

| 10,000 | 1000 | 100 | 10 | kWh | $\frac{1}{10}$ |

▲ **Fig C1.28**

To do – Meter reader

1 Ask an adult to show you where the electricity meter is at home. Take a reading.

2 You may be able to see a spinning disc. The more power that is being used, the faster the disc turns. Try turning off as many appliances as possible in the house – does the wheel stop spinning? Do the numbers on the display stop changing?

3 Take readings before and after a high-power appliance is used. Can you work out how many kilowatt-hours have been used?

Calculating the cost

After the meter has been read, you will have to pay the supplier who will send you a bill. You pay for each kWh of energy you use – that is the 'energy charge'. To discourage people from using too much electricity, there is an extra charge after the first 50 kWh – that is the 'fuel charge'. Here is what to look for in the bill shown in Fig C1.29.

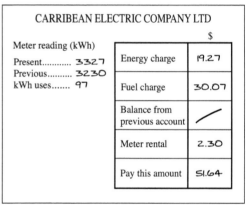

	$
Energy charge	19.27
Fuel charge	30.07
Balance from previous account	╱
Meter rental	2.30
Pay this amount	51.64

CARRIBEAN ELECTRIC COMPANY LTD

Meter reading (kWh)
Present............ 3327
Previous.......... 3230
kWh uses....... 97

▲ **Fig C1.29** A typical electricity bill

- The energy used is calculated from the difference between the current meter reading and the previous one. The difference in readings is 3327 − 3230 = 97 kWh.
- The fuel charge is calculated at 31c per kWh for every unit used: 97 × 31 = $30.07
- There is no energy charge for the first 50 kWh, but each unit after that costs 41c: 47 × 41 = $19.27
- There is also a meter rental charge of $2.30 per month.

So the total cost is $30.07 + $19.27 + $2.30 = $51.64.

To do – Billing clerk

1 Find out the different charges made by your electricity company. How much are the fuel and energy charges? Are there any standing (fixed) charges such as meter rental?

2 Now imagine that you are the clerk who makes up the bill. Draw up a bill for a customer whose meter reading has increased from 3407 in January to 3674 in April.

Q10 On 4 January and 4 February an electricity meter read 18431 and 18511 kWh respectively. If the energy charge is 30c per kWh and the fuel adjustment charge is 40c per kWh, and the monthly meter rental is $1.80, calculate the electricity bill for the month. (Assume all kWh units have to be paid for.)

Energy-saving habits

Electricity is expensive, so it is a good idea to practise energy-saving habits. It makes sense to switch off appliances when not in use, turn off the lights when you leave the room, and not leave appliances such as televisions and computers on standby for long periods of time – switch them off completely.

Look at Table C1.5 again.

- Using an iron for 20h uses the most energy.
- Using a radio for 20h uses very little energy.

You shouldn't be surprised to find that it is the things that produce heat that use most electricity.

A fridge, also, can waste a lot of energy. If you open it often or leave it open for long, more energy will have to be used up to keep the things inside cold. And if the rubber seal (the gasket) around the edge of the door does not fit tightly, heat from the surroundings will get into the fridge and so the fridge will have to use more energy to keep the things inside cold. Most high energy bills result from a badly fitting gasket on a fridge. It's like running an air conditioner with the windows and doors open.

Also, if an appliance becomes rusty, a circuit may be formed across the wires and some current will flow through this 'short circuit'. Some energy will be wasted in the short circuit. It will not be used by the appliance.

Q11 Name two household electrical appliances that operate at high power and two that operate at low power.

Q12 What are two ways of avoiding wasting electricity when using a refrigerator?

End-of-unit questions

1 Which formula shows correctly how to calculate power?

 A power = energy × time

 B power = $\dfrac{\text{time}}{\text{energy}}$

 C power = voltage × time

 D power = voltage × current

2 An electric iron has a power rating of 1500W. It runs from a 110 V mains supply. What would be a suitable fuse to fit in the plug?

 A 3 A

 B 10 A

 C 13.6 A

 D 15 A

3 An electric fan with a power rating of 100 W and an electric lamp rated at 120 W are run for 10 h. How many units (kWh) of electricity are used?

A 0.22

B 2.2

C 22

D 220

E 2200

4 A domestic electricity supply must have fuses, and the cables which carry electricity around the house must be of the correct thicknesses.

a Explain why the cables of the lighting circuits can be thinner than those that go to an electric cooker.

b Describe and explain what can go wrong if an excessive current flows in the cables that carry electricity around the house.

c Describe and explain what happens inside a fuse when an excessive current flows. How does this protect the user?

5 Imagine that you are an energy-saving adviser working for your local council. Your task is to visit local people and advise them on how to reduce their electricity bills. What advice would you give them?

6 John works in a factory which makes electrical heaters. He tests the heaters by using an ammeter to measure the current flowing through each heater when it is connected to a 12 V supply.

The table shows his results for three different heaters.

Heater	voltage / V	current / A
A	12.0	0.40
B	12.0	0.60
C	12.0	0.20

a Which heater has the greatest electrical resistance? Explain your answer.

b Calculate the resistance of Heater B.

c When in use, the heaters are connected to a 240 V supply. Will the current flowing through each heater be greater or less than when connected to the 12 V supply? Explain your answer.

John carries out another test on the heaters. He places each one in turn in a beaker of cold water and connects it to the 12 V supply.

d Which heater will cause the water to heat up most quickly? Explain your answer.

e Give two ways in which John can ensure that this test is a fair comparison of the heaters.

C1c Light and Lighting

Electric lighting

Electricity can be made to produce light in a variety of ways, e.g. filament lighting, fluorescent lighting, the discharge tube, the electric arc, light-emitting diodes (LEDs) and lasers. In this unit the principles involved in the production of filament and fluorescent lighting will be described.

▲ **Fig C1.30** Some different light bulbs; the spiral one on the right is an energy-saving fluorescent light bulb

Filament lamps

The filament lamp is in common use in most households. We have already seen on page 288 that it makes use of the heating effect of an electric current which makes it hot so that it glows. This is called **incandescent** lighting.

incandescent – describes something which glows because it is hot

SBA Skills

ORR	D	MM	PD	AI
✓		✓		

Experiment – The glowing wire

1 Connect a piece of nichrome wire with a rheostat (variable resistor) and a power pack.

2 Adjust the rheostat to control the electric current in the wire to see how it produces a lighting effect.

⚠ **SAFETY:** Be sure to place the nichrome wire on a wooden board to protect the bench from burning. Do not touch the nichrome wire during the experiment as it will become very hot (even before it starts to glow).

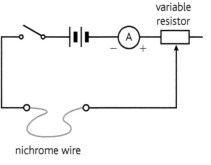

variable resistor

nichrome wire

▲ **Fig C1.31**

The tungsten filament

Tungsten wire is usually used as the filament in an incandescent lamp because it can reach very high temperatures without melting. Tungsten lamps are usually filled with a mixture of nitrogen and argon gases to help preserve the filament. If air were used the hot metal would burn up rapidly. If there were a vacuum inside the lamp the tungsten wire would evaporate quickly when hot.

Filament lamps are usually specified by their power (wattage) and operating voltage. The higher the wattage, the brighter the lamp. Thus a 110 V 100 W lamp is brighter than a 110 V 40 W lamp.

Unfortunately, only a small percentage of the electrical energy is converted to light in an incandescent lamp. Most of it is radiated from the filament and lost as heat. In the fluorescent lamps there is much less energy loss.

SBA Skills

ORR	D	MM	PD	AI
✓		✓		

To do – Comparing light bulbs

1 Observe some different light bulbs that have different power ratings (wattages).

2 If you have a light meter, measure the brightness of each bulb. Make sure that you hold the meter at the same distance from each bulb.

Fluorescent lamps

Inside a fluorescent tube there is a gas, usually mercury vapour, at low pressure (mercury is very poisonous, so if a fluorescent tube breaks, you should provide plenty of ventilation and try not to inhale the vapour). Fig C1.32 shows how the fluorescent lamp works.

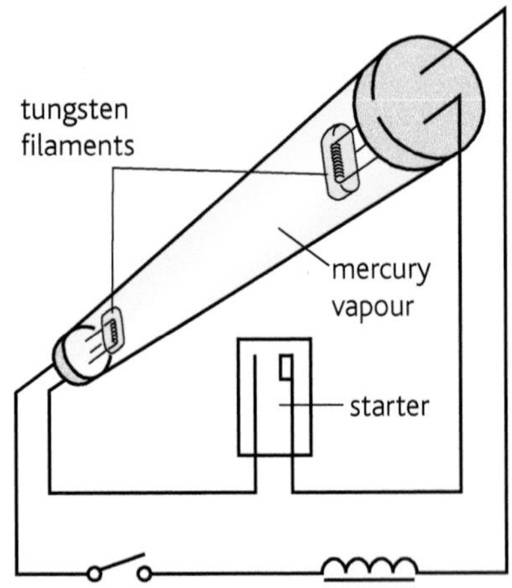

The light given off by a mercury lamp is an eerie, bright, greenish white. A large amount of invisible, dangerous ultraviolet (UV) light is also given out, which can not only damage the retina of the eye, but also represents a wastage of energy. The inside of the tube is therefore coated with a powder that absorbs the UV, rendering it safe. The powder glows white when it is hit by UV light; it is said to **fluoresce**, and this is how the fluorescent lamp gets its name. The powder used in some types of fluorescent lamps is highly poisonous due to the presence of beryllium compounds.

▲ **Fig C1.32** A fluorescent lamp has a starter. This causes the coil to produce a momentarily high voltage that starts the current flowing through the mercury vapour in the tube

fluoresce – to give out light after absorbing ultraviolet radiation

Many modern 'low-energy' light bulbs are small fluorescent tubes, curled up into a small volume, as Fig C1.33 shows. These convert about 15% of the electrical energy supplied into light, about five times as much as a filament lamp. In future, we will use more lamps based on energy-efficient light-emitting diodes (LEDs).

Quality of light

We are used to seeing things in **natural light** (daylight). For example, if you buy a blue shirt, it will look blue out of doors. However, indoors, under **artificial light**, its colour may look rather different. Why is this?

natural light – the light from the Sun and sky; daylight

artificial light – the light produced by an electric lamp

Household fluorescent light looks white, but if you ever take colour pictures under fluorescent lights you will find that the pictures come out greenish. This is because fluorescent white light has a lot of green light in it, more than natural light has. Household incandescent white light has a lot of yellow in it, so the pictures taken in this light come out with an orange tinge.

▲ **Fig C1.33** A low-energy fluorescent light bulb

Colour temperature (the temperature of a hot filament) is sometimes used to describe the quality of a light. To produce light that resembles daylight in quality, an incandescent lamp filament has to be run at the very high temperature of 6000°C. Such a high temperature destroys the tungsten filament very quickly, so the incandescent photoflood lights used by TV cameramen and in photo studios do not last long.

Because the quality of light depends on the colour temperature, it is not surprising that a garment bought in a store lit by fluorescent lamps looks a slightly different shade when taken home and viewed in natural daylight. Fluorescent lights used in the store display make the garment look a tinge greener. The store walls also have an effect. They reflect light of their colour onto the garment.

Digital cameras often have a control that the user can set to indicate whether they are taking pictures in natural or artificial light. The camera then automatically adjusts the balance of colours to look more natural.

Table C1.6 compares incandescent (filament) and fluorescent lamps.

incandescent (filament)	fluorescent
waste most of the energy supplied as heat, hence expensive to run	wastes less energy; cheaper to run
filament usually small, hence sharp shadows	often long, hence very soft shadows
brightness easily controlled with a variable resistor or variable inductor	brightness not easily controlled
very yellowish light unless driven at high temperatures, in which case does not last long	light resembles daylight in quality, though slightly greener; will maintain this quality for a long time.

Table C1.6

Q1 What type of lighting (filament or fluorescent) would you use in the following situations and why?

    a To light up the subjects being photographed in a photo studio.

    b To light up a classroom being used for evening classes.

    c To light up a theatre stage.

Q2 Why do colours of objects look different in natural light and in artificial light?

Q3 Which is brighter, a 40 W fluorescent light or a 40 W incandescent light? Give a reason for your answer.

Colour

White light, whether artificially produced or from the Sun, is actually made up of a wide range of colours. If this were not so, our world would appear far less colourful than it does.

The colours of white light

A ray box, glass prism and white screen can be used in a darkened room to show that white light consists of a whole range of colours, as shown in Fig C1.34. This is called **dispersion**.

dispersion – the splitting of white light into the different colours of which it is composed

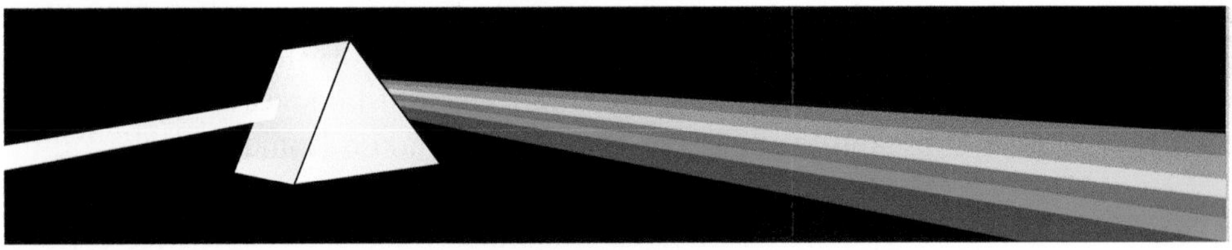

▲ **Fig C1.34** Forming a spectrum of white light using a prism

spectrum – all the colours of which white light is composed, spread out side-by-side

The spread of colours produced by dispersion is known as the **spectrum**. We can identify seven colours: red, orange, yellow, green, blue, indigo and violet, but there are in fact an infinite variety of shades between the red and violet ends.

SBA Skills

ORR	D	MM	PD	AI
✓		✓		

Experiment – Splitting light

You are going to use a prism (glass, plastic or water) to split white light into a spectrum.

1 Shine a narrow beam of light from a ray box or flashlight on to the prism.

2 Adjust the angle of the beam until you can see a clear spectrum.

3 Position a piece of white card so that the spectrum shows clearly on the card.

4 Now place a coloured filter in the path of the white beam. (You can use coloured cellophane wrappers from sweets as filters.) Do some colours disappear from the spectrum? Try different colours of filter. What colours do they absorb?

To do – Make a rainbow

Like the glass prism, rain drops can also split white light into its component colours. That is how a rainbow forms. You can make your own rainbow like this:

1 Set up a garden hose so that it produces a fine spray of water droplets in the air.

2 Stand with the Sun behind you. You should see a rainbow in the spray.

Q4 Which colours are at opposite ends of the spectrum of white light?

Q5 What colour comes between orange and green in the spectrum?

Primary and secondary colours

We have seen that white light is a mixture of coloured lights. Now we can look at what happens when we mix two or more colours together.

SBA Skills

ORR	D	MM	PD	AI
✓				

Experiment – Mixing coloured lights

Three ray boxes (or flashlights) with colour filters in front of them can be used to investigate the effects of mixing coloured lights.

1 Set up two ray boxes so that their light shines on a screen.

2 Put a red filter in front of one ray box and a green filter on the other. What colour do you see where their beams overlap on the screen?

3 Try different combinations of red, green and blue light. Record the colours you see.

4 Add a third ray box. Use red, green and blue filters. What do you see?

primary colours – colours that cannot be produced by mixing other colours

secondary colours – colours produced by mixing two primary colours

Red, green and blue lights are three colours that cannot be produced by mixing coloured lights. They are called **primary colours**. Magenta, yellow, and cyan are the **secondary colours** and are produced by mixing primary colours in pairs. Notice that to produce light that appears white to the eye only the primary colours red, green and blue are needed. The entire spectrum is not necessary.

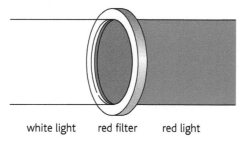

white light   red filter   red light

▲ **Fig C1.35** A red filter lets through red light. The other colours in white light are absorbed

red + green = yellow

red + blue = magenta

green + blue = cyan

Colour mixing is made use of in television. The screen of a colour TV is made up of many tiny dots (or lines) of powders that glow red, green or blue to make the picture. When these glow, the varying colours in a scene are reconstructed.

▲ **Fig C1.36** Mixing coloured lights

If a printed colour photograph is examined under a magnifying glass, dots of various colours can be seen. First the magenta dots are printed, then the cyan, then the yellow, then the black. This produces the complete range of colours seen.

SBA Skills

ORR	D	MM	PD	AI
✓		✓		

To do – Colours making pictures

1 Use a hand lens (magnifying glass) to examine the screen of a TV set or computer monitor. Can you see the coloured dots that make the picture? Try to estimate their size, and how many there are.

2 Examine a coloured photograph in a newspaper, magazine or book. How is this made up?

Q6 Name the three primary colours of light.

Q7 Which two primary colours of light combine to give yellow light?

Q8 Can you predict what colour you will get when secondary yellow light is mixed with blue light? (The colour mixing diagram may help you.)

How we see colour

A red dress seen in white light appears red. This is because red light is reflected by the dye in the dress. All other colours are absorbed by the dye. In red light the dress also looks red. But in blue light the dress looks black. This is because the red dye absorbs blue light.

And a white dress looks red in red light. White reflects all colours of light, but if there is only red light falling on it, it can only reflect red.

Q9 What colour will a red dress appear under green light?

Q10 What colour will a white dress appear under blue light?

Q11 How will the following look in secondary yellow light?

    a a white blouse with red polka dots

    b a white blouse with green polka dots

    c a blue blouse with black polka dots

SBA Skills

ORR	D	MM	PD	AI
✓				

To do – Using coloured filters

1 Using coloured pens, draw lines of different colours (including black) on a sheet of white paper.

2 Observe the lines through a coloured filter (glass, plastic or cellophane).

3 Record how the appearance of the lines changes and try to explain what you see.

Now use what you have learned to devise a method of secret writing.

4 Write a message in colour on white paper.

5 Add blue lines to confuse the appearance.

6 Ask a class member to look at your message through different filters. Which filter allows them to see the message?

Pigments

pigments – substances that give colour to materials

Dyes and paints are examples of **pigments**, substances we use to give things colour. The primary colours of paint are red, yellow and blue.

Most of the pigments in common use are 'impure' in the sense that they reflect more than one colour of light.

- Yellow pigments, for example, reflect red and green light as well as orange and primary yellow. All other colours in the spectrum are absorbed.
- Blue pigments are also impure. They reflect mainly blue and green light, and absorb the other colours.

From this, we can work out what happens when we mix yellow and blue pigments. Since green is reflected by both pigments, a yellow and blue paint mixture looks green. Between them, the yellow and blue absorb all the colours of the spectrum except green. Mixing red, green and blue paint gives nearly (but not exactly) black paint.

You can see that pigments work differently to light. When they are mixed, the colour that results is produced by subtraction. The colour seen is the colour of light *not* absorbed by any of the pigments.

So the mixing of primary pigments works like this:

- Mixing blue and yellow gives green.
- Mixing red and yellow gives orange.
- Mixing red and blue gives purple.
- Mixing all red, yellow and blue gives nearly (but not exactly) black, because the mixture contains pigments that will absorb all the colours of light.

Fig C1.37 shows each of these combinations.

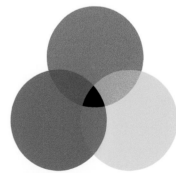

▲ **Fig C1.37** A pigment-mixing chart. The big circles show the three primary colours; where they overlap shows the result of mixing these colours

Chromatography

chromatography – a
technique used to
separate mixtures of
pigments

The process of **chromatography** can be used to separate the pigments that make up a particular colour. For example, black ink from a felt-tip pen can be shown to be made up of a variety of coloured dyes. To do this, a small amount of ink is placed on filter paper and a solvent added (Fig C1.38). As the solvent spreads out, the different colours are separated because they are carried at different speeds by the solvent.

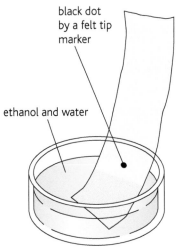

black dot
by a felt tip
marker

ethanol and water

▲ **Fig C1.38** Using chromatography to separate the different pigments in black ink

SBA Skills

ORR	D	MM	PD	AI
✓				

Experiment – Chromatography

1 Pour a mixture of ethanol and water into a shallow dish.

2 On a strip of blotting paper or filter paper, mark a small black dot using a felt-tip pen.

3 Fix the strip of paper so that its lower end dips into the liquid and the black dot is half a centimetre above the liquid.

4 Watch how the colours are separated as the solvent moves up the paper.

5 Try to analyse some other pigments. For example, flowers can be crushed and their dyes extracted using a suitable solvent, e.g. methylated spirits with a few drops of hydrochloric acid, or white rum. The dye can then be analysed by chromatography using the same solvent.

Q12 Which secondary colour is produced when red and yellow pigments are mixed?

End-of-unit questions

1 Which statement about electric lights is *correct*?

   A Fluorescent lamps are an example of incandescent lighting.

   B Fluorescent lamps make more efficient use of the energy supplied to them than filament lamps.

   C Filament lamps are cooler than fluorescent lamps.

   D It is easier to control the brightness of fluorescent lamps than filament lamps.

2 Which statement about the spectrum of white light is *incorrect*?

   A Indigo is next to blue.

   B Orange is between yellow and red.

   C Red and violet are at opposite ends.

   D Green and blue are on opposite sides of blue.

   E Green and blue are on opposite sides of yellow.

3 Which row in the table correctly shows the colour that results from mixing red and blue light, and red and blue pigments?

	red and blue light	red and blue pigments
A	magenta	purple
B	yellow	magenta
C	purple	magenta
D	green	purple
E	magenta	black

4 A mixture of red and blue light is passed through a red filter.

   a What colour of light will pass through the filter?

   b What happens to the rest of the light?

   c The filtered light now falls on a piece of white paper with blue writing on it. How will the paper and writing appear?

   d If the filtered light is mixed with green light, what colour will be seen?

5 How could you show that the colours of flowers are due to mixtures of pigments? Describe a suitable experiment. Include a diagram and state what you would expect to observe.

6 Joanna is buying a new dress to wear to a party. It is red with yellow flowers on it. She takes the dress out into the street so that she can see the colours more clearly.

   a Explain why the colours look different when she takes the dress outside the shop. Joanna wears the dress to the party. She stands near to a red light.

   b What colour will the red dress appear? Explain your answer.

   c Explain why the yellow flowers appear black under red light.

   d Joanna's friend Bob is wearing a white shirt and black trousers. How will his clothes appear when he stands next to the red light?

   e Explain why Bob's trousers look black whatever colour of light is shining on them.

By the end of this unit you will be able to:

- identify the various types of fossil fuels.
- discuss the problems associated with using fossil fuels.
- identify alternative sources of energy.
- describe how solar energy can be used.
- appraise the extent to which solar energy can be used.

The fuels we use

fuel – a substance that is burned to release its store of energy

A **fuel** is a substance that is burned to release its store of energy. Much of the energy we use comes from burning fuels. The following pictures show some examples of fuels in use. Which of these fuels are important for you?

▲ **Fig C2.1** Wood for cooking on an open fire

▲ **Fig C2.2** Natural gas for a stove in the kitchen

▲ **Fig C2.3** Coal for a power station to generate electricity

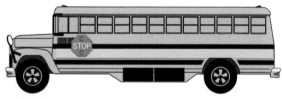

▲ **Fig C2.4** Diesel for a bus to take children to school

▲ **Fig C2.5** Petrol for a health visitor's car

stored energy – energy that is stored in an object or substance

chemical energy – stored energy that can be released by a chemical reaction such as burning

We use fuels because they have **stored energy**. They store energy in the form of **chemical energy**. To release the energy stored in a fuel, a chemical reaction must happen. The fuel must burn and, to do this, it needs oxygen from the atmosphere. If a fire doesn't get enough oxygen, it will go out.

When a fuel burns, new substances are formed. These usually include carbon dioxide (CO_2) and water.

Sometimes we are not aware that a fuel is burning. For example, in the engine of a car, petrol and air are mixed together. When you turn the key in the ignition, a spark causes an explosion – the petrol burns and the hot gases produced push on a piston to make the car go.

SBA Skills

ORR	D	MM	PD	AI
✓				

Experiment – Burning fuels

Try burning small samples of fuels such as charcoal, motor oil and ethanol (methylated spirits).

 SAFETY: Wear eye protection when burning fuels.

1 Place a small quantity (about 5 g) of fuel in a metal dish – an old tin can will do.

2 Ignite the fuel using a match or gas burner.

3 Record what you observe as the fuel burns.

4 Fill a test-tube with ice-cold water and hold it a safe distance above the flame. Can you see water condensing on the outside of the cold tube?

Burning equations

When a fuel is burned, the chemical energy stored by the fuel is changed to heat energy, and often some light energy as well. Here are two equations to show how the substances change in the chemical reaction and how the energy changes:

$$\text{fuel} + \text{oxygen} \longrightarrow CO_2 + \text{water} + \text{other substances}$$

$$\text{chemical energy} \longrightarrow \text{heat energy} + \text{light energy}$$

Q1 What two forms of energy may be released when a fuel burns? Where does this energy come from?

Fossil fuels

fossil fuel – a fuel obtained from materials that have been buried underground for millions of years

Coal, oil and gas are called **fossil fuels** because they are made up of the remains (fossils) of ancient plant and animal matter. Reserves of fossil fuels have been in the ground for millions of years. Most of the energy used in countries with developed economies comes from burning fossil fuels.

There are places on the Earth where oil bubbles up from

▲ **Fig C2.6** Natural gas is a fossil fuel. This is the liquefied natural gas plant at Port Fortin, Trinidad

317

the ground, and where coal is easily dug up. People have used these fossil fuels for centuries. It's much easier than finding firewood, and fossil fuels are concentrated stores of energy that burn with a hot flame.

Nowadays, most coal, oil and gas are extracted in large-scale mines and wells from deep underground. When oil has been extracted from underground it must be refined to produce different sorts of fuel – petrol for cars, diesel for trucks, kerosene for aircraft, for example.

▲ **Fig C2.7** Many people use gas for cooking and heating. This woman in Tobago supplies gas in cylinders that can be refilled when they run out

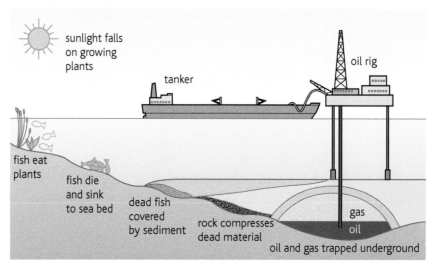

▲ **Fig C2.8** How fossil fuels are formed

Fossil fuels come from materials that were living a long time ago – millions of years ago. Dead plants and animals fell into boggy ground where they rotted. Or they sank to the seabed and rotted there. Gradually these rotting remains became trapped under layers of rock. They turned into coal, oil and gas.

The original plants got their energy from sunlight. The animals ate plants, so their energy also came from sunlight. So, when we burn fossil fuels, we are releasing the energy of sunlight that fell on the Earth millions of years ago.

Biofuels

For many people in the world, wood is a very important fuel. They use it to heat their homes and to cook their food. Wood can also be converted into charcoal. When it burns, charcoal produces less smoke and ash than wood, and it burns with a hotter flame.

biofuel – a fuel that comes from living materials

Wood and charcoal are two types of **biofuel**. Biofuels come from materials such as plants that were living recently. ('Bio' means 'living'.) Some farmers grow trees such as willows especially for fuel.

Other types of biofuel include:

- straw
- various quick-growing crops such as sugar cane
- animal dung and human sewage
- rubbish.

How can sugar cane be a fuel? The answer is that the sugar must be extracted and then fermented to produce alcohol. This can be burned, or mixed with diesel to produce biodiesel, used in trucks, buses and other vehicles.

biogas – a gas fuel made by digesting plant and animal materials

How can dung and sewage be fuels? They, too, must be processed by putting them in a digester. This is a big tank where bacteria attack the waste materials so that they rot to produce **biogas** (methane). The solid residue can also be burned as a fuel.

▲ **Fig C2.9** This farmer grows canola. The seeds will be processed to make biofuel

To do – Finding fuels

SBA Skills

ORR	D	MM	PD	AI
✓				

1 Think about your everyday life. What fuels do you use regularly?

2 What other fuels can you identify? What are their uses?

3 Summarise your ideas in a table. List the fuels and their uses. Which ones are fossil fuels and which are biofuels?

Q2 Put the following fuels into two lists, *Fossil fuels* and *Biofuels*.

coal   wood   straw   natural gas   oil

Q3 Explain how the energy released when dung is burned came originally from the Sun.

Renewable and non-renewable energy sources

If we cut down a tree to use its wood as fuel, we can plant another tree. Eventually this, too, can become fuel. Because we can always go on replacing the crops we burn, biofuels are described as **renewable** energy sources. We need never run out of them provided we keep replanting them.

renewable – describes an energy source that can be replaced after it has been used

non-renewable – describes an energy source that cannot be replaced after it has been used up

After a fossil fuel has been burned, there is no way to get its energy back again. It has been used up. For this reason, fossil fuels should be used carefully and not wasted. One day, they could be completely used up. They are **non-renewable** energy sources.

Table C2.1 shows how much remains in the world's known reserves of fossil fuels.

fossil fuel	how much remains in the ground?
gas	enough for about 30 years
oil	enough for about 40 years
coal	enough for about 200 years

Table C2.1 The world's known stores of fossil fuels

To do – Charting energy

Make an illustrated chart to show how the fuels we use get their energy from the Sun. Your chart should show how light coming from the Sun ends up being used for cooking, heating, transport, etc.

Make sure that you include both biofuels and fossil fuels.

SBA Skills

ORR	D	MM	PD	AI
✓				

Q4 Copy and complete the sentences with the word 'renewable' or 'non-renewable'.

    a When the last lump of coal has been burned, you can never have any more. Coal is a _____ fuel.

    b When wood is burned, you can always grow some more. Wood is a _____ fuel.

Q5 Explain why dung can be described as a renewable energy resource.

Problems with fossil fuels

Although fossil fuels are used throughout the world every day, there are some serious problems associated with using them.

Global warming

Most scientists who have studied the Earth's climate agree that our use of fossil fuels has caused changes in the Earth's atmosphere.

The air is a mixture of gases, mainly oxygen and nitrogen. There is a small amount of carbon dioxide, which is present naturally and helps to keep the Earth warm. It acts like a blanket, stopping heat from radiating away (escaping) into space. This is called the **greenhouse effect**.

greenhouse effect – the raising of the Earth's temperature because gases in the atmosphere trap heat

The greenhouse effect is a good thing. Without it, the Earth would be about 15°C cooler, as cold as the Moon. The atmosphere helps to make the Earth more comfortable for life.

▲ **Fig C2.10** This power plant burns coal to generate electricity. Its chimneys pour harmful gases into the atmosphere

However, over the last 200 years a lot of the Earth's fossil fuels have been burned, putting more and more carbon dioxide into the atmosphere. This is making the 'blanket' thicker so that the Earth is getting warmer – its average temperature is increasing. This is called **global warming**.

global warming – the slight increase in the Earth's average temperature caused by an increase in the greenhouse effect

climate change – changes in the Earth's climate resulting from global warming

This could result in **climate change**, with changing weather patterns, including more violent hurricanes and droughts in some tropical areas. Sea levels may rise because of melting ice, flooding low-lying countries. There is already some evidence that this is beginning to happen.

Acid rain

Burning fossil fuels has another harmful effect on the environment. Fossil fuels are not pure substances, and when they burn, they produce a mixture of substances. Two harmful substances they release are oxides of nitrogen and sulphur.

When these oxides get into the air, they dissolve in rain water to make acids. Then, when the rain falls, the acid harms the living organisms it falls on. For example, if acid rain falls on a forest, the trees may lose their leaves and die. If acid rain gets into rivers and lakes, the fish and invertebrate creatures living there may be killed.

▲ **Fig C2.11** These trees were killed by acid rain

Power stations that burn coal are particularly damaging because they cause a lot of acid rain. They can be adapted to remove the harmful oxides, but this is expensive and wastes some of the energy of the fuel.

Q6 Steve says, "I like warm weather. We should burn lots more fossil fuels to increase global warming." Give at least three reasons why someone might disagree with Steve.

Alternative energy sources

We have seen that there are problems with using fossil fuels. They may run out, and they harm the environment. They are likely to become increasingly expensive.

alternative energy sources – renewable energy sources that can replace fossil fuels

So it makes sense to look for **alternative energy sources** to replace fossil fuels. Biofuels are an example of an alternative energy source, because they are renewable.

Energy from water

Several Caribbean countries including Jamaica and St Vincent already make use of **hydroelectricity** ('hydro' means 'water'). Water in a river or stream is trapped behind a dam as it runs down a hillside. The water flows through pipes where it turns a **turbine**. (A turbine has blades attached to an axle, rather like the blades of a windmill.) The turbine turns a **generator**, which produces electricity.

hydroelectricity – electricity when water is released from behind a dam

turbine – a device that is made to spin round when water, steam or air pushes past its blades

generator – a device that produces electricity when it is forced to spin

gravitational potential energy – the energy stored by an object that has been lifted upwards

electrical energy – energy carried by an electric current

The water behind a dam has **gravitational potential energy**. This energy is released when the water flows down past the turbine. Some of the energy is converted to **electrical energy**.

It can be expensive to build a hydroelectric scheme but, once it is built, it costs very little to run. So long as it continues to rain in the mountains, water will flow down to fill the reservoir behind the dam. Rain water comes free!

▲ **Fig C2.12** A small hydroelectric dam like this can provide plenty of electricity for a small community

Other kinds of water power

Think of the huge amount of energy of all the water moving about in the sea. Anything that is moving has **kinetic energy**. In recent years, engineers have come up with some clever ways of using some of this energy.

kinetic energy – the energy of something that is moving

For example, the tides can produce large currents. An underwater turbine can be positioned where the current will make it spin round, causing a generator to turn. Alternatively, water can be trapped at high tide behind a

▲ **Fig C2.13** You can sense the energy of the seawater as it pushes its way through this tidal barrage, generating electricity as it does so

tidal barrage (rather like the dam of a hydroelectric scheme). When the tide goes out, the water is released to generate electricity.

The energy of waves can also be harnessed to generate electricity. A wave power generator, floating in the sea, moves up and down as the waves pass it. This up and down motion is used to generate electricity, which is carried along cables to the shore.

Q7 What form of energy is stored by water when it is trapped behind a tidal barrage?

Q8 Waves on the sea have energy. The surface of the sea moves up and down. What form of energy does moving water have?

Wind power

Wind turbines make use of another renewable energy source – the wind. Although the wind doesn't blow all the time, when it does blow, it can be used to turn the blades of a turbine. This makes a generator spin to produce electricity.

A small wind turbine can produce enough electricity for a small household. Bigger ones can produce enough electricity for hundreds of homes, or for a factory. Several turbines together make a wind farm.

Q9 What form of energy does wind (moving air) have?

▲ **Fig C2.14** A wind farm in Curaçao

SBA Skills

ORR	D	MM	PD	AI
✓		✓		

Experiment – Turning power

Many different sorts of power stations make use of a turbine and generator. An energy source (wind, water or steam) turns the turbine, which turns the generator to produce electricity.

You can recreate this process using a small electric motor as a generator, and making your own turbine.

1 Fix a small motor in a clamp so that its axle is horizontal.

2 Connect wires to the motor, and connect it to a voltmeter.

3 Design and make a turbine that will fit on the axle of the motor. It will look something like the turbine in Fig C2.15, but the number of blades is up to you. Try to design your turbine so that you can alter the angle of the blades.

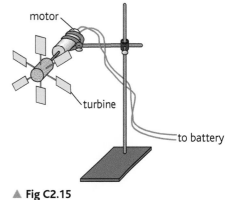

▲ **Fig C2.15**

4 Attach the turbine to the axle of the motor and use a fan to make it spin. Can you measure the voltage generated?

5 Replace the voltmeter with an ammeter to measure the current.

Geothermal power

In some parts of the world, there are hot rocks below the ground. Sometimes hot water and steam push up from underground. We can use this energy in a **geothermal power station**. In this case, steam turns a turbine that turns a generator.

Only a few places in the world have the right conditions to be able to make use of geothermal energy.

Geothermal energy is nothing to do with the Sun. The underground rocks are hot because they contain radioactive substances such as uranium that give out heat as they decay. Nuclear power stations also make use of the energy stored in uranium.

geothermal power station – a power station that generates electricity from the energy of hot rocks underground

▲ **Fig C2.16** This geothermal power station is in Iceland, a country with many volcanoes. Waste steam is pouring into the air

SBA Skills

ORR	D	MM	PD	AI
✓				

To do – From the Sun

1 In Unit B1 you studied the ways in which heat energy travels around. Use what you know about convection currents to explain how the energy produced by a wind farm comes originally from the Sun.

2 If you have studied the water cycle in Science or Geography, you should also be able to explain how the energy of hydroelectric power comes originally from the Sun.

3 Where does the energy of wave power come from?

Make a presentation to the class about one of these ideas. You should explain what makes these energy sources renewable.

Solar energy

As you have seen, most of the energy we use comes originally from the Sun. The Sun is a giant ball of hot gas, shining brightly in the sky. We can feel the energy of its rays when they warm our skin. This is called radiation.

We get three types of radiation from the Sun:

- **visible light** – this is the light we can see with our eyes
- **infrared radiation** – this is the heat we feel when our skin absorbs it
- **ultraviolet radiation** – this is the radiation that tans (and burns) our skin as it is the most energetic type

visible light – light that we can see

infrared radiation – heat radiation that can travel through empty space

ultraviolet radiation – invisible, energetic radiation that can travel through empty space

Ultraviolet radiation is hazardous and can cause burning of the skin. Fortunately for us, most of the ultraviolet radiation coming from the Sun is absorbed high up in the atmosphere by a layer of gas called the ozone layer.

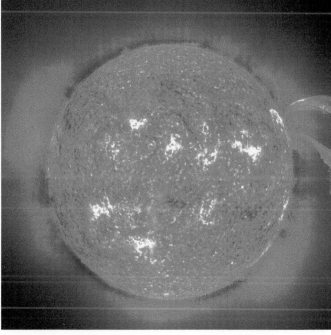

▲ **Fig C2.17** The Sun, photographed by a satellite in space

Making use of solar energy

Since most of the energy we use comes originally from the Sun, it's a good idea to try to use it directly. There are two ways of doing this:

- using the energy of sunlight to heat things up
- using the energy of sunlight to generate electricity.

Heating up

A solar water heater uses sunlight to heat water. The heater contains pipes, painted black so that they will absorb the Sun's rays. Water passes through the pipes and gets hot. The hot water is stored in a tank and can be used for heating or washing.

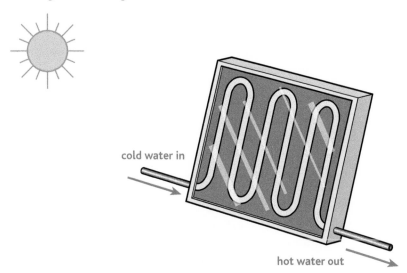

cold water in

hot water out

▲ **Fig C2.18** Capturing the Sun's rays – a solar water heating panel

Fig C2.19 shows people installing a solar water heater on the roof of their house. Fig C2.20 shows another way to collect the Sun's energy, using reflectors. These are mirrors that reflect the Sun's rays onto a kettle, which gets hot enough to boil.

▲ **Fig C2.19** Installing a solar water heater

▲ **Fig C2.20** A clever way to boil a kettle

Q10 Why is a solar water heater coloured black?

SBA Skills

ORR	D	MM	PD	AI
✓		✓	✓	✓

Experiment – Solar heat

1 Find two plastic bags, one black and one transparent. They should be about the same size.

2 Fill each bag with water, putting the same amount in each.

3 Put a thermometer in each bag and close it so that the thermometer sticks out of the top. (You may need to hang the bags from clamps.)

4 Place the bags side-by-side in bright sunlight, as in Fig C2.21. Record the temperature of each bag every 5 minutes. Which one heats up more quickly?

5 Think about this experiment. Why should the bags be the same size? Why should they have the same amount of water? What else must be controlled to make this a fair test?

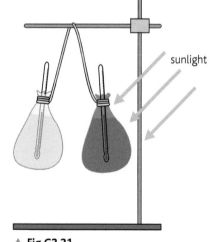

sunlight

▲ **Fig C2.21**

Solar electricity

Electricity can be generated directly from sunlight using a solar panel, better called a **photovoltaic cell, solar cell** or photocell. (Sometimes, when people talk about solar panels they mean solar water heaters.) A photovoltaic cell absorbs light and converts a fraction of the energy to electrical energy. The rest becomes heat.

photovoltaic cell or solar cell – a device that converts sunlight into electrical energy

Fig C2.22 shows a good use for photocells – generating electricity to light road signs at night.

- During the day, when the Sun is shining, the photocells generate electricity that charges up a battery.
- At night, the battery provides the electricity to light up the signs.

From Fig C2.22, you can understand one of the problems of photocells; quite a big panel is needed to power just one light. This is because most of the energy of the sunlight is wasted as heat – it doesn't become electricity.

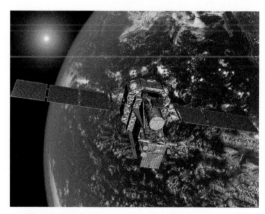

▲ **Fig C2.22**

Photocells are an excellent idea but they are not very efficient and they are expensive to install. But once they have been paid for, the electricity they provide is free.

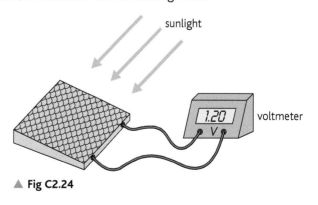

▲ **Fig C2.23** The International Space Station makes use of photocells – there is plenty of sunlight in space (and no petrol stations)

SBA Skills

ORR	D	MM	PD	AI
✓		✓		✓

Experiment – Electricity from the Sun

You will need: a small photocell, a voltmeter and connecting wires.

1 Connect your photocell to the voltmeter. Cover the photocell – what does the meter read?

2 Now experiment to find the greatest voltage you can produce. Put the photocell in full sunlight. Turn it round, tilt it – how should the photocell be positioned to make best use of the sunlight?

sunlight

1.20 V

voltmeter

▲ **Fig C2.24**

A solar future?

The map shows that the Caribbean is one of the sunniest areas of the world. In future, because of the problems that come from using fossil fuels, it will be important to make greater use of solar energy.

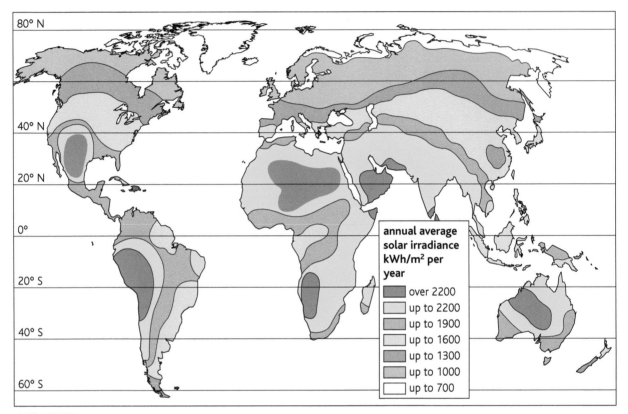

▲ **Fig C2.25**

The main problem is that the energy of sunlight is not very concentrated. A household of four people will need a large solar water heater, and they will need to cover a large part of the roof of their house with photocells. In periods of bad weather, when the Sun doesn't shine brightly for days on end, they may run out of hot water and electricity.

It will make sense for most countries to have a mixture of ways of generating electricity, including solar, wind, wave and hydro, with fossil fuels available to add in an emergency.

Q11 Explain why, at present, we make use of a lot of fossil fuel energy and very little solar energy.

End-of-unit questions

1 Which one of the following is *not* a renewable energy source?

   A natural gas

   B waves

   C sunlight

   D biogas

2 What form of energy is stored by petroleum?

   A heat

   B geothermal

   C chemical

   D fossil

3 Energy from the Sun travels through empty space to reach the Earth. Which method of heat transfer is this?

   A conduction

   B radiation

   C convection

   D reflection

4 Hydroelectricity is another renewable energy resource. The sentences below explain how electrical energy from a hydroelectric power station originally came from the Sun. Copy the sentences and write them in the correct order to show the whole process.

   • Rain falls in the hills and fills the rivers.
   • Moving water makes the turbine spin round.
   • The Sun's rays make water evaporate from the sea.
   • River water is trapped behind the dam.
   • The turbine makes the generator spin round.
   • Water vapour forms clouds.
   • The generator produces electricity.

5 The Caribbean is one of the sunniest areas on Earth. Sunlight could provide much of the energy we need.

   a Outline the principal ways in which the energy of sunlight can be used directly.

   b Suggest reasons why we do not make much use of this energy at present.

   c Why will it be important to make more use of solar energy in the future?

6 The government of a Caribbean island nation is thinking of generating electricity using the energy of water in a river flowing down from the hills. They propose building a dam across the valley.

   a What form of energy does the river water have when it is high up in the hills?

   b When the dam is built, water will be stored behind the dam. How can its energy be converted to electricity?

   c The island has a dry season of 6 months each year. How will this affect the availability of electricity from the scheme?

   d Explain whether the scheme can be described as renewable or non-renewable.

C3 Machines and Movement

Easing our daily chores with machines

Many of our daily tasks are made easier by machines. A machine is any device that enables a small force to overcome a large force, or that enables a force to be applied from a convenient position.

In this unit, we will look at some simple machines, such as levers and pulleys. We will see that many complex machines, such as bicycles, motor cars and lawnmowers, consist of many simple machines working together.

Levers

lever – a rod that turns about a pivot when forces act on it

pivot or fulcrum – the point about which a lever turns

effort – the force that the machine produces

load – the force applied to a machine

The simplest machine is the **lever**, which is a rod that turns about a **pivot** (or **fulcrum**).

In Fig C3.1 the piece of wood the carpenter is using to get out the stuck post is an example of a lever. The force the man is applying is called the **effort**. The total force he is trying to overcome is called the **load**. The carpenter cannot pull the post out directly because the load is too great. But, by using this machine, the little effort he can manage can overcome the large load. This machine arrangement is used as a 'force multiplier' because the lever has the effect of increasing the small effort force so that it equals the load force.

▲ **Fig C3.1**

The law of levers

The carpenter is using the fact that a small effort force acting at a large distance from the pivot can balance a small load force acting close to the pivot. We can write this as an equation:

effort × distance of effort from pivot = load × distance of load from pivot

$$W_1 x_1 = W_2 x_2$$

This is known as the law of levers. It is easily tested in the laboratory.

Experiment – Testing the law of levers 1

A half-metre rule, a round pencil and a set of weights can be used to check the law of levers, as shown in Fig C3.2.

1 Balance the rule on the pencil.

2 Place a weight W_1 at a distance x_1 from the pivot.

3 Place weight W_2 on the rule so that it balances again.

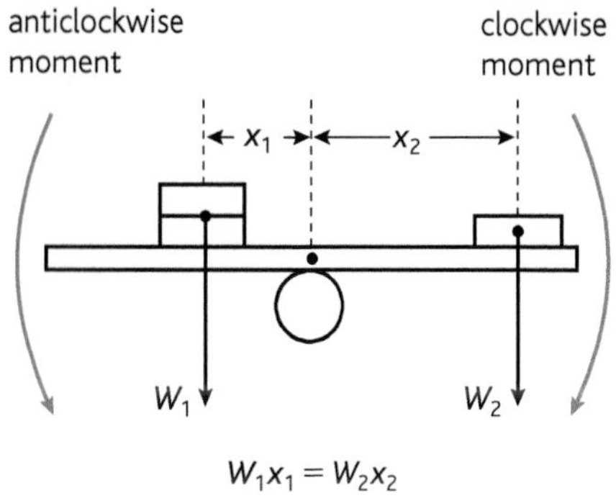

You should find that, within experimental limits, $W_1x_1 = W_2x_2$.

anticlockwise moment

clockwise moment

$$W_1x_1 = W_2x_2$$

▲ **Fig C3.2**

In other words, for the rule to balance, the weight on one side, multiplied by its distance from the pivot, must equal the weight on the other side, multiplied by its distance from the pivot.

Moment of a force

moment – the turning effect of a force

The products W_1x_1 and W_2x_2 are called **moments** or turning effects of the forces. The equation $W_1x_1 = W_2x_2$ is the law of levers, also known as the principle of moments. For a rule to balance, or for a load to be overcome, the clockwise moment of one force about the pivot must be equal to the anticlockwise moment of the other force about the pivot.

This law of levers must be satisfied when a small force balances a larger force. The equation tells us that by means of a large lever arm x_2, a small force W_2 can be used to balance a large force W_1.

If the rule is not pivoted at its midpoint, the weight of the rule W must be taken into account.

$$W_1x_1 = W_2x_2 + Wx$$

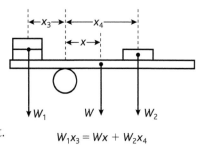

$$W_1x_3 = Wx + W_2x_4$$

▲ **Fig C3.3**

The moment of W must be calculated and then added to the moment of W_2 because both forces have clockwise moments – both are tending to turn the rule clockwise.

sum of clockwise moments = sum of anticlockwise moments

Q1 Explain why we say that the force W_1 has an anticlockwise moment.

Calculating moments

To calculate the moment of a force, we need to know two things:

• the size of the force, in newtons (N)
• the distance of the force from the pivot.

Then we calculate the moment of the force as below:

$$\text{moment of force} = \text{size of force} \times \text{distance from pivot}$$

Example: A diver of weight 800 N stands at the end of a diving board, 3.0 m from the fixed end. The moment of his weight is $800 \times 3 = 2400\,\text{N m}$. The unit of moment is the newton-metre (N m).

Q2   A child of weight 250 N sits at the end of a see-saw, 1.8 m from the central pivot. Calculate the moment of the child's weight about the pivot.

SBA Skills

ORR	D	MM	PD	AI
✓		✓	✓	

Experiment – Testing the law of levers 2

1   Take a long rule and balance it on a pencil or pen. Check that it balances at its midpoint.

2   Use 100 g masses as weights. (Each has a weight of 1.0 N.) Place a single weight at one end of the rule. Where must two weights be placed to balance it?

3   Try with different combinations of weights. Each time, test whether the law of levers is satisfied by checking that $W_1 x_1 = W_2 x_2$.

4   Using a single known mass, a metre rule and a pencil, devise a means of finding the mass of the metre rule.

Classes of levers

Lever arrangements are classed according to the arrangement of the load, pivot and effort.

- In Class 1 levers, the load and effort are on opposite sides of the pivot. Scissors are an example (they are like two levers joined at the pivot). You apply a force at one side of the pivot to produce a force that cuts at the other side.
- In Class 2 levers, the load and effort are on the same side of the pivot but the effort is further away than the load. The effort must therefore be smaller than the load. A wheelbarrow is an example.
- In the Class 3 lever system, the effort and load are again on the same side of the pivot, but this time the effort is nearer to the pivot and therefore is larger than the load. A fishing rod and tongs are examples.

Levers of all three classes can be found in the human body. The base of the skull, the foot and the forearm are examples of Class 1, Class 2 and Class 3 levers, respectively.

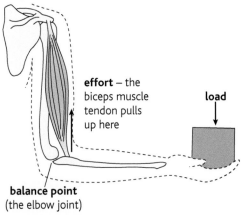

effort – the biceps muscle tendon pulls up here

load

balance point (the elbow joint)

▲ **Fig C3.4** Class 3 levers are common in the body. The bicep muscle provides a big force to lift the weight in the hand

Q3 Look at the pictures in Fig C3.5. Can you identify the load, effort and pivot in each case?

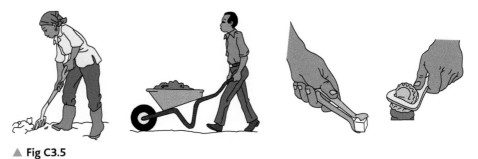

▲ **Fig C3.5**

Q4 Summarise what you have learned about classes of levers by copying and completing the table below. Your answer to **Q3** will help you with the last column.

lever system	description	examples
Class 1	pivot between load and effort	
Class 2	... between ... and ...	
Class 3	... between ... and ...	

Levers in vehicles

Levers are often used in vehicles. They may be parts of vehicles, or they may be tools used to make or mend them.

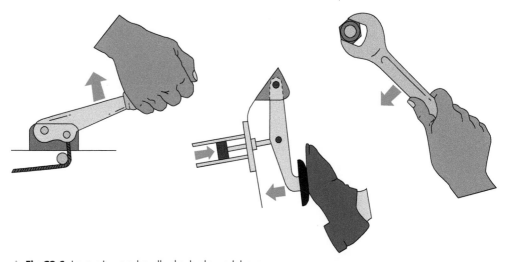

▲ **Fig C3.6** Levers in cars: handbrake, brake pedal, spanner

Look at the pictures in Fig C3.6 carefully. They all show levers in action. Can you identify the load, the effort, the load arm, the effort arm, and the pivot? Can you tell whether the levers are Class 1, Class 2 or Class 3? Are the levers being used mainly for their force multiplying effect or mainly to apply the effort from a convenient position?

More machines

Levers are just one type of simple machine. Remember: with any machine, a force is applied (the effort) and the machine produces a useful force (the load). Often, the load moved is greater than the effort (a force multiplier); sometimes the machine simply makes it easier to apply the force.

If a machine uses a small effort to overcome a large load we say that the machine has a large force multiplying effect. This force multiplying effect is called the **mechanical advantage (MA)**. Mechanical advantage is defined as the ratio of the load (L) to the effort (E). So, if a load of 80 N is raised by applying an effort of 20 N, the mechanical advantage of the machine is 4.

mechanical advantage –
the ratio of load to effort
for a machine:
$$MA = \frac{load}{effort}$$

$$MA = \frac{load}{effort}$$

Pulleys

A pulley is a wheel, usually grooved, that is pivoted at the middle so that it can turn. Fig C3.7 shows a simple pulley being used to hoist a load upwards. The rope is pulled downwards. The pulley is being used here mainly to allow the force to be applied from a convenient position. To lift a heavy load, the man can use his whole weight to pull downwards on the rope.

pulley

rope being
pulled from
below

▲ **Fig C3.7**

Look at Fig C3.8, which shows a block and tackle pulley system being used to hoist a very heavy load in a ship. The block is fixed and the tackle can move. The effort is applied along one rope that passes over the block. This rope passes three times between the block and tackle (this number depends on the number of pulleys in the block and tackle system), and so the load is held up by three portions of the rope. Each pulls upwards on the load.

Using a spring balance to measure load and effort, you can show that a simple fixed pulley has a mechanical advantage of approximately 1. That is, the load lifted and the effort applied are about equal. In a block and tackle system of pulleys, the mechanical advantage is greater than 1; it is approximately equal to the number of pulley wheels. That is why the block and tackle system is used when heavy objects are being hoisted.

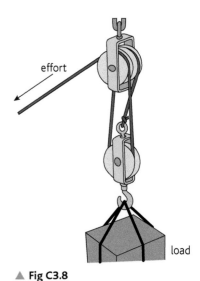

effort

load

▲ **Fig C3.8**

In Fig C3.8 the MA is equal to 3 because there are three pulleys. So, if you pull downwards with a force of 100 N, you can lift a 300 N load upwards. That makes the task a lot easier.

Q5 A block and tackle system is set up in which four strings raise the load. What is the mechanical advantage of this system?

Q6 A pulley system is used in which an effort of 150 N is enough to raise a load of 750 N. What is the mechanical advantage of this system?

SBA Skills

ORR	D	MM	PD	AI
✓		✓		

Experiment – Investigating pulleys

1 Set up a single pulley with a string passing over it. (If you don't have a pulley wheel, clamp a smooth wooden broom handle horizontally and pass the string over it.)

2 Use a spring balance to weigh a load such as a 1 kg mass. Hang the mass on one end of the string.

3 Attach the spring balance to the other end of the string and use it to pull the load upwards. It should rise at a slow, steady speed. What effort is needed?

4 Now investigate a block and tackle system. Remember to weigh the load each time, and to measure the effort needed to lift it.

The inclined plane

The inclined plane is yet another simple machine. It is easier to push or pull a heavy load up a slope than to lift it straight upwards.

If you pull a loaded trolley up the plane with a spring balance you can show that the mechanical advantage of this system is also greater than 1. That

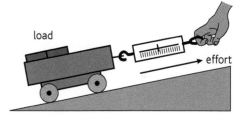

▲ **Fig C3.9** Measuring the effort used to pull a load up an inclined plane

is, the effort applied is less than the load being overcome. However, inclined planes are also used for the convenience they provide. It is easier to roll a heavy drum up a plank into a truck than to lift it straight up from the ground.

SBA Skills

ORR	D	MM	PD	AI
✓		✓		✓

Experiment – Up the slope

1 Set up a plank or board so that it slopes at an angle of about 30°.

2 Use a spring balance to weigh a load such as a wooden block.

3 Attach string to the load and use the spring balance to pull it slowly up the slope.

4 Record the effort needed. How does the effort compare with the load?

5 Try different loads, and try changing the angle of the board. Record your results and discuss what they tell you about the inclined plane.

Q7 A workman sets up a ramp so that he can more easily push a heavy load onto a platform. If he pushes with a force of 300 N to move a load of 840 N, what is the mechanical advantage of this arrangement?

Q8 The workman then decides it would be more convenient if he fixes a pulley wheel at the top of the ramp. He attaches a rope to the load and passes it over the pulley. He then pulls downwards on the free end. Draw a diagram to show this and explain why it makes it easier for the workman to move the load up the slope.

Machines transfer energy

If you use machines such as pulleys and levers for a long time, you may get tired. You are using up your stores of energy. Where does this energy go? Perhaps you have been lifting a heavy load. The load has gained gravitational potential energy (see Unit C2). Your energy has been transferred to the load.

The amount of energy transferred by a force depends on two factors:

- the size of the force
- the distance moved by the force.

The greater the force and the further it moves, the greater the amount of energy transferred. Hence we can write:

energy transferred = force × distance moved by the force

If the force is in newtons and the distance is in metres, the energy will be in joules (J).

Example: A donkey pulls a cart for 500 m. Its pulling force is 240 N. How much energy is transferred?

energy = force × distance = 240 × 500 = 120 000 J.

Q9 How much energy is transferred when a girl lifts a box weighing 15.0 N onto a table 0.8 m above the floor?

SBA Skills

ORR	D	MM	PD	AI
✓		✓		

Experiment – Energy on an inclined plane

Set up an inclined plane as in the previous experiment, *Up the slope*, on page 335. This time, you are going to make measurements to determine the energy transferred by the effort and the energy gained by the load.

1 Draw two lines across the plane, as shown in Fig C3.10. Pull the load up the slope, from the start line to the finish line. Record the pulling force (the effort). Measure the distance moved by the effort.

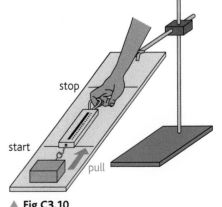

▲ **Fig C3.10**

2 Now you need to find the vertical distance moved by the load. Use a rule to find the height above the table of the start line and of the finish line. Subtract to find the vertical height risen by the load.

3 Calculate the energy transferred by the effort and the energy gained by the load. It will help if you do this as shown in Table C3.1.

effort /N	distance moved by effort/m	energy transferred by effort/J	load /N	distance moved by load/m	energy transferred by load/J
5.0	0.40	5.0 × 0.40 = 2.0 J	10.0	0.14	10.0 × 0.14 = 1.4 J

Table C3.1

4 Repeat the experiment using different loads.

5 Try changing the angle of the inclined plane.

Energy efficiency

We want our machines to make good use of the energy we put into them. We want them to have a high **efficiency***. If a machine wastes most of the energy we put in, its efficiency is low.

The efficiency of a machine is the fraction of the energy we supply that is usefully transferred to the load. Efficiency is usually given as a percentage, and is calculated like this:

$$\text{efficiency} = \frac{\text{useful energy converted}}{\text{energy supplied}} \times 100\%$$

or

$$\text{efficiency} = \frac{\text{load} \times \text{distance moved by load}}{\text{effort} \times \text{distance moved by effort}} \times 100\%$$

Example: A pulley system is used that transfers 40 J of energy to the load when the effort transfers 50 J of energy. Its efficiency $= \frac{40}{50} \times 100\% = 80\%$.

Q10 A worker uses a pulley system to lift some bricks. The worker's pulling force transfers 800 J of energy; the bricks gain 600 J of energy. What is the efficiency of the pulley system?

Q11 Look at the table of results for the experiment *Energy on an inclined plane* (Table C3.1 above). The example row shows typical results. Calculate the efficiency of the plane.

*****efficiency** – the fraction of the energy supplied by the effort that is usefully transferred to the load:
$$\text{efficiency} = \frac{\text{useful energy converted}}{\text{energy supplied}} \times 100\%$$

SBA Skills

ORR	D	MM	PD	AI
		✓		✓

To do – Efficiency of the inclined plane

1 Return to your results from the experiment *Energy on an inclined plane*. For each row in your table, calculate the efficiency.

2 Is the plane more efficient when it is steeper, or when it is less steep? You may need to make more measurements to answer this question.

Reasons for inefficiency

If a machine is less than 100% efficient, this means that some of the energy supplied by the effort is wasted, because it is not transferred to the load. Where has it gone?

With a block and tackle, it isn't just the load that is being lifted, the tackle is also being raised. This means that some of the energy is used in lifting the tackle and does not get transferred to the load.

friction – the force that acts between two surfaces as they slide over each other

An inclined plane may be rough, so that there is **friction** as the load slides upwards. Overcoming the friction wastes energy as heat; the slope may become slightly warm. Pulley systems and levers may also suffer from friction, particularly at the axles of the pulley wheels and at the pivot of a lever.

Any roughness causes friction, and friction wastes energy. Rusting and other forms of corrosion make surfaces rough; smooth surfaces have the least friction.

lubrication – adding a fluid to reduce the friction between surfaces

Lubrication can reduce friction. A drop of lubricating oil at the pivot of a lever is usually all that is needed to reduce friction. This will reduce the effort and so increase the efficiency of the machine. Oiling or greasing not only helps to reduce friction, but also helps to prevent the bearings at the pivots rusting. Replacing worn bearings with new ones also helps to increase efficiency, since worn bearings are rough and cause more friction than smooth new bearings.

Q12 How could you increase the efficiency of an inclined plane? How would you test your idea?

Q13 A simple pulley (a single pulley wheel) is used to lift a 20 N load through a distance of 1.0 m. The effort required is 40 N. After oiling the pulley, the effort required is reduced to 25 N. By how much has the efficiency improved?

Three more important machines

Here are some more important machines that are found in vehicles and elsewhere.

The wheel and axle

The steering wheel and column of a motor car form an example of another kind of simple machine – the wheel and axle. In Fig C3.11 the effort E is smaller than the load L because the effort arm R is large and the load arm r is small. If the machine is 100% efficient, we have:

$$E \times R = L \times r$$

Example: What load can be moved by an effort of 100 N if $R = 40$ cm and $r = 5$ cm? Substituting in the formula gives:

$100 \times 40 = L \times 5$, and so $L = \dfrac{4000}{5} = 800$ N.

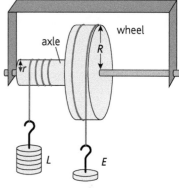

▲ **Fig C3.11** The principle of the wheel and axle

Screwdrivers also make use of the wheel and axle principle. The fatter the handle, the less effort has to be applied to overcome the opposing frictional load. (Notice that it's not the *length* of the screwdriver that is important – unless you are using it to lever something.)

The screw

The screw is yet another simple machine associated with vehicles. Some motor vehicle jacks are designed round the screw. When the effort E is moved through one complete revolution, the load L moves a tiny distance equal to the pitch p of the screw (the distance from one thread to the next). The energy supplied by the effort is equal to effort × distance moved by the effort. The energy gained by the load is equal to load × distance moved by the load. Assuming that the energy supplied is equal to energy gained, we have:

$$E \times 2\pi R = L \times p$$

So
$$E = \frac{Lp}{2\pi R}$$

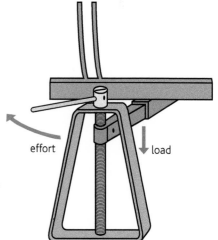

▲ **Fig C3.12** A motor vehicle jack makes use of a screw

So although the load may be very large, the effort can be much smaller than L as long as the pitch p is very small and the radius of the big wheel R is large.

Q14 What effort is needed to raise a load of 6000 N using a car jack whose screw has a pitch of 2 mm if the effort arm has a length of 150 mm?

The hydraulic press

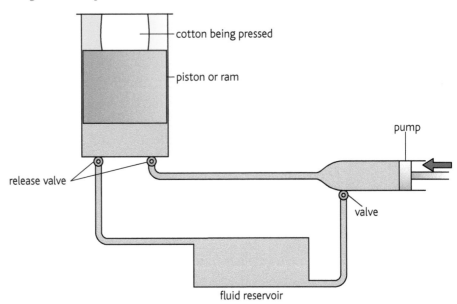

▲ **Fig C3.13** A hydraulic press, used to compress a bale of cotton

The hydraulic press is used in the hydraulic jack and in hydraulic braking. The hydraulic press acts as a force magnifier. A small force is applied on a piston in a tube of small diameter. This presses on hydraulic fluid (a special type of oil), which transmits the pressure to a piston of larger diameter. This piston pushes with a greater force.

Fig C3.14 shows a hydraulic brake in a car. When the driver presses the brake pedal, the piston presses on the fluid in the master cylinder. Hydraulic fluid transmits this pressure to the wheel cylinder, as shown. There is a piston at each end and these force out the brake shoes. These provide the force needed to slow the turning wheels. An advantage of this system is that equal force can be exerted on all four wheels at once.

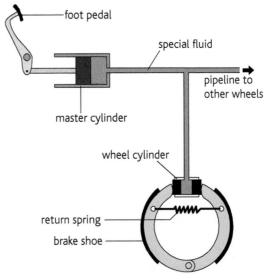

▲ **Fig C3.14** A hydraulic braking system

Experiment – A hydraulic lift

You can create your own hydraulic lift using everyday objects.

1 Collect a large polythene bag. Make sure that there are no holes in it.

2 Connect some rubber tubing to the bag, making sure that there is a tight seal between the bag and the tube.

3 Place a brick on top of the bag.

4 Blow into the bag. The brick is lifted.

▲ **Fig C3.15**

End-of-unit questions

1 A girl uses a wheelbarrow to lift a load of vegetables, as shown in Fig C3.16.

▲ **Fig C3.16**

The wheelbarrow acts as a lever. What class of lever is it?

A Class 1, because the load is between the pivot and the effort.

B Class 1, because the load and effort are the same side of the pivot.

C Class 2, because the pivot is between the load and the effort.

D Class 2, because the load is between the pivot and the effort.

E Class 3, because the load and effort are on opposite sides of the pivot.

2 A block and tackle is used to raise a 25 N load. The effort required is 10 N, and the load rises 2.0 m.
The mechanical advantage of this machine is:

A 0.2

B 0.4

C 2.0

D 2.5

E 15

3 A load is dragged 2.0 m up an inclined plane by a force of 200 N. The load gains 300 J of energy.
The efficiency of the inclined plane is:

A 33%

B 67%

C 75%

D 100%

E 150%

4 **a** What type of machine is each of the following?
- a door knob
- the steering wheel of a car
- the wheels on the top of a dragline
- the tray and the two handles of a donkey cart

341

b What type of machine (e.g. lever, pulley, inclined plane) could you use in the following situations?

   i to get a heavy box from the ground to the first floor of a house

   ii to send a flag up a flag pole

   iii to remove a nail that is stuck in a piece of wood.

c How much energy is converted in raising a 300 kg car engine through a height of 0.8 m? (Note: a mass of 1 kg has a weight of approximately 10 N.)

d If a force of 800 N is used to raise the car in **Q4c**, what is the mechanical advantage of the machine used?

e Suggest a suitable simple machine that can be used to lift the engine.

5 We use machines because they can act as force multipliers, allowing us to perform tasks more easily. However, machines are rarely 100% efficient. Explain what is meant by the efficiency of a machine, and describe some factors which can reduce the efficiency of a machine. How can these problems be reduced or overcome? In your answer, refer to levers, pulleys, inclined planes and any other simple machines you consider relevant.

6 Jermain is working at a storage depot. He has to drag heavy sacks up a ramp into the warehouse.
Jermain measures the force needed to pull each sack. It takes 400 N, and the ramp is 6 m long.

a Calculate the work done in moving each sack.

b In lifting each sack, it gains 1800 J of energy. Calculate the efficiency of Jermain's working.
Jermain thinks it would be easier to use a machine to help in moving the sacks. He uses a single pulley to help him. The force he uses is still 400 N, but Jermain finds it easier to move the sacks.

c Explain why this is so.

d Jermain decides to use a small 4-wheel trolley to help move the sacks up the ramp. Explain why this reduces the force he must use.

e Does the use of the trolley increase the efficiency of Jermain's working? Explain your answer.

C4 Conservation of Energy

What is energy?

We use clocks and watches to tell the time. There are different ways of making the hands move round the face of the clock.

▲ **Fig C4.1** Each of these clocks has a different source of energy to make it go

- An old-fashioned wind-up clock has a spring inside. Each day, the spring must be wound up; as it unwinds, the spring makes the hands go round.
- A pendulum clock has a large weight on the end of a chain. Each week, the weight must be pulled up to the top; as it falls, it keeps the pendulum swinging and the hands moving round.
- An electric clock has a battery. Each year the battery must be replaced; electric current from the battery operates a motor which makes the hands go round.

Each type of clock needs a source of energy to make it work. A wound-up spring, a raised weight, a battery – these are all stores of energy that we can use to make a clock work.

The concept of energy

Energy is the ability to make things happen, or to produce a change. In the case of a clock, energy is needed to make the hands move round. When the supply of energy runs out, the hands stop moving.

In Unit C2, you studied the importance of the Sun as a source of energy for most of the things we do on Earth. Without energy from the Sun, the Earth would be a dead place, with no changes happening.

SBA Skills

ORR	D	MM	PD	AI
✓				

To do – Energetic toys

Examine some toys that move. These might be:

- battery-powered toys
- clockwork toys
- toys that run down a slope

1 For each, find out its source of energy.

2 What happens when it runs out of energy?

3 How can it be supplied with more energy?

4 Report your ideas to the class.

 Fig C4.2

Changes produced by energy

Energy can cause many different sorts of changes:

- Energy can cause a *temperature change*. Example: Heating water in a beaker – the temperature of the water rises.
- Energy can make a substance *change state*. Example: Boiling water in a pot – the water changes to steam when it boils.
- Energy can change *chemical composition* by causing a chemical reaction. Example: Cooking an egg – chemical changes in the egg cause it to solidify.
- Energy can produce *motion*. Example: Throwing a ball – the force of the hand gives energy to the ball so that it flies through the air.

Experiment – Energy producing changes

1 Use a candle or a spirit burner to heat a small amount of water in a beaker or can. Use a thermometer to monitor the temperature of the water.

2 Observe what happens until steam leaves the surface of the water. Can you see convection currents in the water?

3 Now discuss how energy originally stored in the candle wax or the spirit is producing changes. Can you identify the four different types of change listed on page 344?

▲ Fig C4.3

Q1 A van is used to transport goods from a shop to the customer. What is the energy supply of the van? What *change* is produced by the use of this energy?

Q2 Give an example of a situation where energy is required to produce a change of state from solid to liquid.

Experiment – Heating naphthalene

Slowly heat some naphthalene in a test tube. Note that the crystals begin to melt at about 80°C, and that the temperature remains constant during the melting stage, although the test tube is still being heated.

The heat supplied during the melting stage was used to change the physical state of the naphthalene; in this case from solid to liquid.

Experiment – The effect of heat on copper(II) carbonate

Strongly heat green copper(II) carbonate in a hard glass test tube. Pass the gas given off into lime water, as shown in the diagram. You should notice:

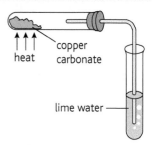

- the lime water turns milky, indicating that carbon dioxide was evolved
- the starting material was green, whereas the residue is black.

These changes show that the heat energy supplied has caused a chemical change to occur. When copper carbonate is heated it is chemically decomposed.

▲ Fig C4.4

SBA Skills

ORR	D	MM	PD	AI
✓				

Experiment – Lime power

Obtain a lime, orange or other citrus fruit and stick two different strips of metal (e.g. a strip of copper and a strip of zinc) into it. Complete the circuit with a voltmeter. Is there a reading on the meter?

SBA Skills

ORR	D	MM	PD	AI
				✓

Experiment – Light energy produces a chemical change

1 Soak two strips of filter paper in a solution of sodium chloride and allow them to dry. Then dip them in a solution of silver nitrate.

2 Cover one of the strips with a piece of black paper. Expose the other to direct sunlight or to light from a lamp and observe it for a period of 15 minutes. At the end of this time compare the two strips.

3 The strip that was exposed to sunlight has become dark. This shows that light can produce a chemical change in some compounds.

4 Explain why one strip must be kept out of the light.

Energy stores

Because energy is useful, we need to be able to identify ways of storing it. We have already seen some of these in Unit C2 – for example, water behind a dam stores energy, and fuels are stores of chemical energy.

potential energy – any form of stored energy (e.g. gravitational, chemical)

Stored energy is often called **potential energy (PE)**. Potential energy is energy that isn't doing anything yet, but that has the *potential* to produce changes in the future. Here are some examples of potential energy:

chemical potential energy – energy stored in chemical substances, released in reactions

Chemical potential energy – Fuels are stores of chemical PE. When they are burned, their energy is released. Burning involves a chemical reaction that releases energy. Batteries are also stores of chemical PE; when the battery is in use, the chemicals inside it react together to make an electric current flow.

gravitational potential energy – energy stored by an object that has been raised above some rest position

Gravitational potential energy – As we saw in Unit C2, water stored behind a dam is a store of gravitational PE. Any object that is raised up has gravitational PE; for example, if you lift a ball, you are giving it gravitational PE.

elastic potential energy – energy stored by an object that has been stretched or squashed

Elastic potential energy – If you stretch an elastic band, you are giving it elastic PE. You could use that energy to catapult a pellet at a target. Similarly, a stretched or compressed spring has elastic PE that can be used to make things move. The shock absorbers in a car are compressed when the car goes over a bump. This absorbs energy for a short while, making for a smoother ride.

nuclear energy – energy stored in the nucleus of an atom

Nuclear energy – This is potential energy stored in the nucleus of atoms. It is released during nuclear reactions. There are two types of nuclear reaction:

- fission reactions in which large unstable atoms are split to yield smaller ones
- fusion reactions in which small atoms are brought together to yield larger ones

Fission reactions are used in nuclear power stations to release nuclear energy stored in uranium atoms. Fusion reactions occur constantly in the Sun to produce light and heat. Nuclear bombs may make use of either fission or fusion reactions.

Chemical PE is a chemical means of storing energy. Gravitational and elastic PE are physical means, because they do not involve chemical reactions.

Q3 Think back to the different types of clock you studied at the beginning of this unit. For each clock, state the type of potential energy it makes use of.

Calculating gravitational potential energy

Fig C4.5 shows a lift moving up and down inside a building. When the lift is on the fourth floor it has more gravitational potential energy than when it is in its rest position in the basement. The higher an object above its rest position (or some other position), the more gravitational potential energy it possesses.

The motor has to work hard to pull the lift to the top of the building. It is transferring energy to the lift. The greater the mass of the lift, the greater the energy required to raise it and so the greater its gravitational potential energy.

This shows that the gravitational potential energy of an object depends on both its mass and its height above some rest position. The actual value of the potential energy stored in a body is given by:

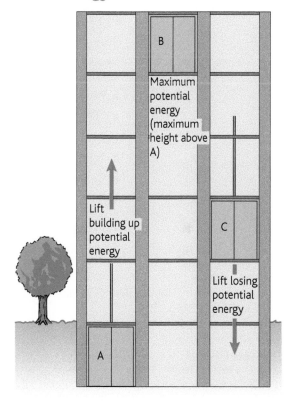

▲ **Fig C4.5** The higher the lift rises above its rest position in the basement (position A), the greater its gravitational potential energy

gravitational potential energy = mass × height × acceleration due to gravity

The acceleration due to gravity has a value of $10 \, \text{m/s}^2$.

Example: A girl of mass 50 kg climbs to the top of a tree that is 8 m tall. Her gravitational potential energy is:

$$GPE = 50 \times 8 \times 10 = 4000 \, \text{J}$$

Note that energy is measured in joules (J).

Q4 Consider the four objects whose masses and heights above ground are shown in the table below.

object	mass / kg	height / m
A	18	3
B	2	5
C	2	4
D	4	2.5

Object A has potential energy of $18 \times 3 \times 10\,\text{J} = 540\,\text{J}$.

a Which object possesses the greatest potential energy?

b Which object possesses the least potential energy?

c What can you say about the potential energies of objects B and D?

Q5 Calculate the gravitational potential energy of the water in a tank in the loft of a house. The water has a mass of 1250 kg and it is 6 m above ground level.

Q6 A boy with a mass of 40 kg is at the top of a slide, 2.5 m above the ground. Calculate his gravitational potential energy:

a at the top of the slide

b halfway down the slide

c at the foot of the slide

Energy inter-conversions

We have seen that energy comes in different forms. When we make use of energy, it changes from one form to another. In this section, we will look at some examples of **energy conversions**.

energy conversion – the conversion or transformation of energy from one form to another

kinetic energy – the energy of a moving object

GPE to KE

As we saw in Unit C2, a moving object has **kinetic energy (KE)**. The amount of KE depends on two factors:

- the object's speed – the greater the speed, the greater the KE
- the object's mass – the greater the mass, the greater the KE

Fig C4.6 shows how the energy of a cricket ball changes when it is hit:

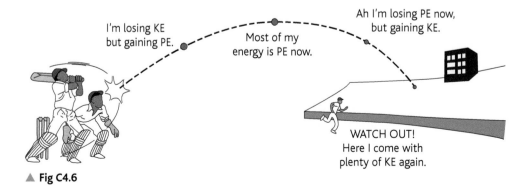

I'm losing KE but gaining PE.

Most of my energy is PE now.

Ah I'm losing PE now, but gaining KE.

WATCH OUT! Here I come with plenty of KE again.

▲ **Fig C4.6**

- The ball has a lot of KE when it leaves the bat.
- As it moves upwards, its GPE increases but its KE decreases. Some of its KE has been converted to GPE.
- As it returns towards the ground, its GPE decreases but its KE increases. Now some of its GPE is being converted back to KE.

Q7 A girl is jumping up and down on a trampoline.

    a Describe how the girl's energy changes as she moves upwards and then downwards.

    b What type of energy is stored by the rubber trampoline when it is stretched by the girl's weight?

More energy conversions

We can represent energy changes by writing an equation that shows how energy changes from one form to another. For example, when a ball is thrown upwards KE is converted to GPE:

$$KE \longrightarrow GPE$$

We will look at some important examples of energy conversion and write equations to represent them.

Photosynthesis is the process by which plants gain energy from sunlight. The green pigment chlorophyll captures the energy of sunlight to provide the energy for chemical reactions in the plant's leaves. These reactions produce the sugar and starch that the plant needs as its food.

$$\text{light energy} \longrightarrow \text{chemical potential energy}$$

A spacecraft may spend several years working in space and so it needs a supply of energy. Although it will take some fuel with it, most of its energy usually comes from sunlight captured by solar cells (photovoltaic cells). These produce electricity that is used to charge batteries on the spacecraft.

$$\text{light energy} \longrightarrow \text{electrical energy} \longrightarrow \text{chemical potential energy}$$

▲ **Fig C4.7** This spacecraft is designed to explore the asteroid belt, far out in space. Its 'wings' are its solar panels that capture the energy of sunlight to generate electricity

An electric motor is supplied with electrical energy. It makes things move.

electrical energy ⟶ kinetic energy

A telephone system is complex. The user speaks into the small microphone; this converts the sound energy into electrical energy. This is transmitted to the listener whose phone has a tiny loudspeaker that converts electrical energy back to sound. In between, it may be converted to a number of different forms of energy, including light and microwaves.

sound energy ⟶ electrical energy ⟶ sound energy

Computers and calculators are similarly complex. They use electrical energy to process information and perform calculations. The user sees the results on a screen or display; sounds may also be produced.

electrical energy ⟶ light energy + sound energy

Q8 Name a device that converts light energy directly to electrical energy.

Q9 Name a device that converts electrical energy directly to light energy.

Q10 Name a device that could be represented by the following energy conversion equation:

electrical energy ⟶ kinetic energy

Q11 Write an energy conversion equation for a battery.

Energy conservation

We have seen that energy can be converted from one form to another. Energy never disappears, it just changes its form. Sometimes, energy gets wasted when it gets changed into a form we do not want, such as the heat energy produced by a light bulb. However, the total amount of energy coming from the light bulb in light and heat is the same as the amount of electrical energy supplied to it.

electrical energy ⟶ light energy + heat energy

At the same time, we cannot create energy out of nothing. It would be useful if we could – we would solve all our energy problems!

principle of conservation of energy – the idea that energy cannot be created or destroyed

These ideas combine together to give the **principle of conservation of energy**:

Energy can be neither created nor destroyed; it can only be converted from one form to another.

This is a very important principle because it allows us to make predictions – for example, how high will a ball go if it is thrown upwards? The ball's KE changes to GPE as it rises, so if we know that it has 100 J of KE when it is thrown, we know that it will have 100 J of GPE when it reaches its highest point.

Q12 An electric motor is used to make a toy car go. If the battery supplies 10 J of energy to the motor, how much energy does the motor produce?

Energy transfers by waves

wave – energy travelling outwards from a source in the form of vibrations

One way of moving energy from place to place is by means of waves. A **wave** on the sea transfers energy. The wave moves across the surface of the water and eventually breaks when it reaches land. If you stand in the sea where waves are breaking, you will feel the energy that they carry.

▲ **Fig C4.8** These children are enjoying the gentle waves on a beach in Grenada

Experiment – Observing waves

Try out some of these ways of observing waves in the classroom or laboratory:

1 Tie one end of a long rope or rubber tube to a firmly fixed point. Hold it taut. Move the free end up and down at a regular rate. You will see waves move along the rope or tube.

2 Lay a slinky spring along a bench so that it is stretched to several times its original length. Hold one end firmly. Move the other end from side to side to see waves travel along the spring.

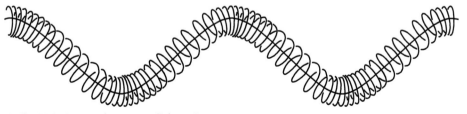

▲ **Fig C4.9** A wave shape on a slinky spring

3 Set up a ripple tank with a shallow depth of water. Drip water into the tank and see the ripples (waves) that spread out. Use a vibrating bar to produce a series of straight, parallel ripples on the surface of the water.

What is a wave?

To make a wave, we need a source that is vibrating. In the case of the stretched rope, it is your hand that is vibrating (moving up and down at a regular rate). A loudspeaker or headphone makes sound waves in air by vibrating. These vibrations spread outwards from the source, and that is what makes a wave.

We can see waves moving up and down on the surface of water, such as waves on the sea. But there are other waves whose vibrations we cannot see:

- sound waves, which we detect with out ears; this is how sound energy is transferred from place to place
- light waves, which we detect with our eyes; this is how light energy travels around
- radio waves, microwaves and other types of electromagnetic radiation, used for radio, TV, mobile phones, etc.

Waves carry energy

Waves are an important way of transferring energy from place to place. For example, light and heat waves from the Sun bring energy to the Earth. Radio waves bring telephone messages to mobile phones and TV programmes to our homes. Microwaves transfer energy into food in a microwave oven.

Q13 How do you know that waves on the ocean carry energy? Where does this energy come from?

Reflecting waves

An important property of waves is that they are reflected when they strike a barrier. You make use of this when you look in a mirror; light waves from your face are reflected back to your eyes by the mirror's shiny surface.

A car's headlamps have shiny reflectors to make sure that as much light as possible is reflected forwards to help the driver see the road ahead. Fig C4.10 shows how a concave mirror like this works. The curved shape of the mirror makes sure that each ray of light from the bulb is reflected forwards.

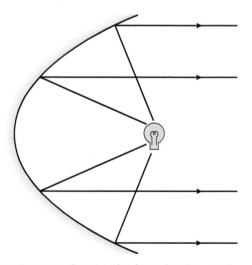

▲ **Fig C4.10** The principle of a car headlamp. Light rays from the bulb are reflected forwards by the concave mirror

Satellite television systems make use of curved 'dish' aerials. The television signals are broadcast down to Earth and are reflected by the dish onto the detector at its focus. From there, an electric current flows into the television receiver.

▲ **Fig C4.11** This café has a dish aerial for receiving television broadcasts. In the background you can see more dish aerials on the mobile phone masts

SBA Skills

ORR	D	MM	PD	AI
✓		✓		

Experiment – Bringing energy to a focus

For this experiment, you will need quite a large concave mirror. The reflector from a car headlamp will do, or you can make a reflector by lining a large bowl with shiny aluminium foil.

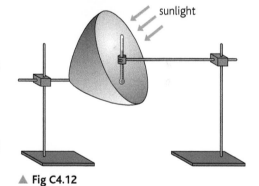

sunlight

1 First, look into your reflector. If it has a smooth reflecting surface, you will see a reflection of yourself; this image is small and upside down.

▲ **Fig C4.12**

2 Next, point your reflector towards the Sun. It will collect the Sun's rays and bring them to a focus – the point where all the rays meet. Anything placed at this point will become hot. Move your hand around in front of the reflector, taking care not to block the light. Can you find a point where your hand gets hot?

3 If you are using a headlamp reflector, place the bulb of a thermometer where the headlamp bulb would have been. Does the temperature rise?

4 Now use your reflector to focus sound waves. Hold the reflector at one side of your head, about 20 cm from your ear. Use a finger to block your other ear. Ask a partner to stand at the opposite side to the reflector, and whisper. Can you use the reflector to focus the sound waves into your ear?

Q14 If you shout into a cave you may hear your voice coming back to you. What do we call this reflected sound wave?

To do – Discovering dishes

1 Look for a dish aerial – for example, one used for a satellite television system. Make a drawing of the dish.

2 Show the detector that is positioned where the waves from the satellite are focused.

3 Show the wire that carries the signal from the detector to the television set.

4 Look at all the dish aerials in your area. Are they all facing in the same direction? Can you say why this might be?

The idea of momentum

An object that is moving can produce a change when it comes into contact with another object that is either stationary or moving. The faster an object is moving, the greater the change it can produce.

momentum – a measure of the motion of an object; equal to mass × speed

A big object that is moving fast is said to have a lot of **momentum**. Momentum is a way of measuring the amount of 'motion' an object has. The greater the mass and the greater the speed, the greater the momentum of a moving object. This means that it will have a greater effect if it is involved in a collision.

Here is how we calculate momentum:

$$\text{momentum} = \text{mass} \times \text{velocity (or speed)}$$

An object with a mass of 5 kg moving at 10 m/s has more momentum than one with a mass of 2 kg moving at 3 m/s.

Q15 Which has more momentum, an object of mass 20 kg moving at 5 m/s or one of mass 10 kg moving at 8 m/s?

The experiment that follows will help you to understand the idea of momentum.

Experiment – Observing collisions

1 Roll a cricket ball along the floor and allow it to collide with a light cardboard box. The cardboard box moves forward.

2 Now increase the speed of the ball. Note that the box moves a greater distance.

3 Place a weight inside the box and repeat the experiment. What effect does this have?

4 Describe an experiment that you could carry out to show that the greater the mass of an object, the greater the change it can produce when moving at a given speed.

Collisions

You may have seen a toy called Newton's Cradle, as shown in Fig C4.13. Five steel balls hang by threads so that they just touch. One ball is pulled to the side and released. It swings down and strikes the line of stationary balls. The ball at the other end of the line receives a jolt and swings up into the air. When it swings back down again, it strikes the line of balls causing the first ball to swing upwards again.

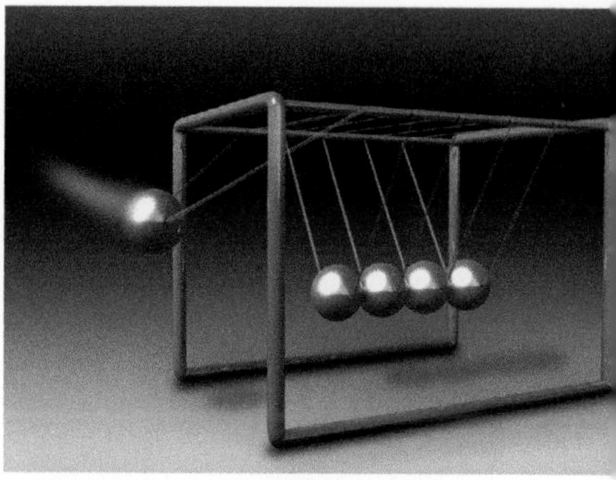

▲ **Fig C4.13** Newton's Cradle illustrates momentum changes during collisions

When two vehicles collide, their speeds change. For example, if one car is stationary and another runs into the back of it, the two cars will both move forward, but their speeds will be less than the speed of the first car.

There are many other situations where objects collide. A footballer may kick a stationary ball in an attempt to score from the penalty spot. A cricketer has to hit a moving ball. Out in space, a meteor may collide with a planet.

These are examples of collisions. We can use the idea of momentum to understand what is happening when objects collide.

SBA Skills

ORR	D	MM	PD	AI
✓		✓		✓

Experiment – Trolley collisions 1 – springy collisions

You will need two identical trolleys (A and B) and some 1 kg masses.

1 First release the springload of trolley A so that when it collides with trolley B the collision is springy (elastic).

2 Push A so that it moves at a steady speed towards B. Observe how the trolleys move after the collision.

3 Now observe a collision in which A and B are moving towards each other at different speeds. What happens?

4 Repeat these experiments but with a 1 kg mass placed first on trolley A and then on trolley B.

5 Make a prediction: Trolley A is stationary and has an extra 2 kg on it; trolley B collides with it. What will happen? Test your idea.

Transferring momentum

In the experiment above, you should have observed the following:

- When B is stationary and A collides with it, B moves off at the same speed as A, leaving A stationary. This shows that all of the momentum of A has been transferred to B.
- When the two trolleys are moving at the time that they collide, they bounce back after the collision. A moves with the original speed of B and B moves with the original speed of A. in other words, the momentum of A has been transferred to B and the momentum of A has been transferred to B.

principle of conservation of momentum – the idea that momentum cannot be created or destroyed

In both cases, the total amount of momentum before and after the collision is the same. We say that *momentum* is *conserved*. This is the **principle of conservation of momentum**:

Momentum cannot be created or destroyed; the total amount of momentum in a closed system is constant.

(A 'closed system' means that there is no external force acting.)

The collisions you have looked at are springy. However, the same rule applies to non-springy collisions where the trolleys stick together.

SBA Skills

ORR	D	MM	PD	AI
✓				

Experiment – Trolley collisions: 2 – sticky collisions

You will again need two identical trolleys (A and B) and some 1 kg masses.

1 First retract the springload of trolley A and stick some Blu-tack to the front of it so that when it collides with trolley B the two trolleys stick together.

2 Push A so that it moves at a steady speed towards B. Observe how the trolleys move after the collision.

3 Now observe a collision in which A and B are moving towards each other at different speeds. What happens?

4 Repeat these experiments but with a 1 kg mass placed on trolley B and then on trolley A.

5 Can you write two sentences to summarise what you have observed?

Understanding collisions

When the moving trolley A collides with the stationary trolley B, the two move off at half the speed of the first trolley. We have twice the mass moving at half the speed, and so the total amount of momentum is unchanged. This shows that the momentum of A has been shared between the trolleys.

If B has twice the mass of A, they will move with one-third of the speed; again, momentum has been conserved.

Q16 A trolley, moving at 4 m/s, collides with an identical stationary trolley. The trolleys stick together. What will be their speed after the collision?

Q17 If the collision between the trolleys in **Q16** had been springy, how would the trolleys have moved after the collision?

The importance of understanding collisions

As we have discussed above, there are many situations where collisions are important. The idea of momentum can help us to understand them. For example, a cricketer should know that, if he uses a heavier bat, he can transfer more momentum to a ball so that it will go farther and faster – provided he has the strength to swing the bat fast enough.

In vehicular collision, momentum is transferred from one vehicle to another. The photograph shows a test centre where newly designed cars are tested to see how they withstand collisions.

▲ **Fig C4.14** A car undergoing collision testing. The figures inside are dummies, wired up with electronic sensors to record their movements

End-of-unit questions

1 Which of the following is *not* a form of stored energy?

  A nuclear energy

  B chemical energy

  C kinetic energy

  D elastic potential energy

2 Which of the following most clearly explains what we mean by energy?

  A the motion of an object

  B food and fuel

  C the ability to produce a change

  D a store of chemical substances

3 In what form is energy stored when the shock absorbers of a car are squashed?

  A heat energy

  B chemical energy

  C kinetic energy

  D elastic potential energy

4 A dish aerial collects radio waves and brings them to a focus.

  a Draw a diagram to show how parallel rays are reflected by a dish aerial and brought to a focus.

  b State and explain one way in which a car headlamp is similar to a dish aerial, and one way in which it differs.

5 In a collision, momentum is conserved.

  a Explain what it means to say that 'momentum is conserved'.

  b Describe an experiment that shows that when two identical trolleys collide momentum is conserved. State what you would expect to observe if the collision is elastic (springy), and if the two trolleys stick together.

6 Mr Andrews lives alone. During the day, he uses energy to help him in different activities. Describe the energy changes which occur during each of the following activities:

  a Heating a pot of soup on a gas stove.

  b Using an electric lamp to help him to see the newspaper.

  c Using a battery-powered shaver.

  d Driving his car to the shops.

C5 Forces

A force is a push or a pull. Forces occur when two objects interact. For example, if you trip over a stone, the stone exerts a force on your foot, and you also exert a force on the stone. If you fall over and land on the ground, you will feel another force acting on you.

In this unit, you will learn about some important forces and how they affect the way things move.

The importance of friction

It is important to know the names of some forces and how they come about.

friction – the force produced when one surface slides over another

Friction is an important force that we use every day. Without friction, you would not be able to walk across the floor. A completely smooth floor has no friction, and that makes it impossible to walk on. We make use of friction when we write with a pencil. Friction with the paper causes small flakes of pencil lead to be left on the paper.

From Unit C3, you will know that friction can be a problem in machines – it reduces their efficiency so that energy is wasted. However, friction can also be a good thing. The tyres of a bicycle or car need friction with the road surface if they are not to slip round and cause a skid. On a smooth or icy road surface there is not enough friction and the wheels will spin round, rather than pushing the car forwards, much like trying to walk on a smooth floor while wearing socks.

drag – the frictional force when an object moves through a fluid

air resistance – the frictional force when an object moves through air

Drag is the name given to friction when an object moves through water or air. In air, it is called **air resistance**. When an object is moving, friction, drag and air resistance always act in the opposite direction, to oppose motion.

Like friction, drag caused by the air or water can slow down a vehicle and make it much less efficient. To reduce the amount of drag acting on a vehicle, such as a car, plane or boat, they may be given a streamlined shape. This means that they cut through the air or water more easily, so less energy is needed to keep them moving and they may be able to achieve a higher top speed.

▲ **Fig C5.1** This Virgin Islands ferry moves fast – it is designed to reduce the effect of the water's drag

Q1 Look at the photograph of the ferry in Fig C5.1. Explain how its design helps to reduce drag.

Q2 Draw outline shapes of two cars, one designed to have a higher top speed than the other. Indicate the design features that make one go slower than the other.

Forces and Newton's laws

Isaac Newton formulated three laws of motion. We will take a brief look at the first two and then concentrate on the third.

1 Newton's first law of motion can be stated as follows:

   If there is no net force acting on a body, then the body either remains at rest or continues to move with the same velocity.

   If you are standing in a bus which then comes to a sudden stop, you will notice that you tend to continue in the forward direction; this is an example of the first law of motion.

2 Newton's second law of motion can be stated as follows:

   The acceleration of a body is directly proportional to the force acting on it; i.e. the greater the force to which a body is subjected, the greater its acceleration will be.

   Acceleration means speeding up or slowing down. A force must be applied to accelerate an object.

3 Newton's third law of motion states:

   For every **action** there is an equal and opposite **reaction**.

   This means that, when one object exerts force on a second object, the second object exerts an equal and opposite force on the first. For example, tap with your knuckles on the table. You are exerting a force on the table (that's the 'action') and you can feel the force it exerts on you (the 'reaction'). If you stand on soft sand, your feet push downwards into the sand and the sand pushes back upwards to support you.

   An important point to note about action and reaction forces: they act on *different* bodies. If two forces are acting on the *same* body, they cannot be action and reaction, even if they are equal and opposite.

action – a force exerted by one object on another

reaction – a force equal and opposite to an action force

SBA Skills

ORR	D	MM	PD	AI
✓		✓		✓

Experiment – Action and reaction

The following experiments will help you to understand the action/reaction principle.

1 Two students stand on skateboards facing each other. Can you predict what will happen when one student pushes the other?

Student A pushes student B (action). At the same time, student A is pushed backwards (reaction). The students move in opposite directions, showing that action and reaction act in opposite directions.

▶ **Fig C5.2**   A   B

2 Collect two spring balances and hook them together, end-to-end. Pull on one balance while your partner pulls on the other. Do the spring balances show the same reading? Explain this observation.

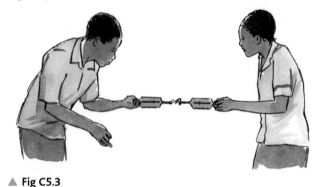

▲ **Fig C5.3**

3 **a** Fill a balloon with air and then release it. As the air rushes out in the forward direction, the balloon moves back (i.e. in the opposite direction).

b Set up a string 'runway' across the room. Tape supporting threads to an inflated balloon and use paper clips to hang it from the runway, as shown in Fig C5.4.

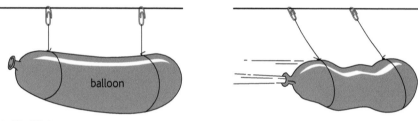

balloon

▲ **Fig C5.4**

What do you think will happen if the air in the balloon is suddenly released? What will be the effect on the balloon's speed of increasing the amount of air in the balloon before it is released?

Walking and jumping

We use the action/reaction principle whenever we walk. To walk, we need a force to push us forwards. So, what do we do? We push *backwards* with our foot on the ground. This makes a backward force (action) on the ground, caused by friction. The ground provides a forward force (reaction), which pushes us forwards.

If you want to jump upwards, what do you do? You bend your knees and then straighten them. The muscles of your legs push down hard on the ground; the ground pushes back upwards on you, and that is the force that pushes you up into the air.

▶ **Fig C5.5** Reaction forces help us to walk and jump

forward push of ground on foot

backward push of foot on ground

upward push of ground on feet

downward push of feet on ground

Rockets and flying

How does a rocket get into space? A rocket burns fuel and this pushes hot gases out of the back, rather like the air escaping from the balloon in the experiment you carried out on page 360. The rocket pushes the gases backwards and the gases push the rocket forwards.

We can understand how an aircraft flies in the same way. It needs **thrust** to push it forwards and **lift** to keep it in the air. Where do these forces (shown in Fig C5.7) come from?

thrust – the forward force on an aircraft caused by the engines

lift – the upward force on an aircraft's wings

▲ **Fig C5.6** Heading for the Moon – you can see the downward moving gases that push the rocket upwards as it is launched from the rocket base at Kourou, Guyane

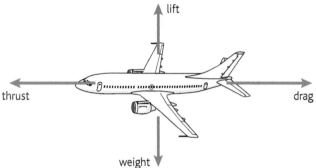

▲ **Fig C5.7** The forces acting on an aircraft in flight

Thrust

Jet engines are like rockets – they burn fuel and push hot gases out of the back. This produces a reaction force, the thrust, which pushes the aircraft forwards. Propeller aircraft push air backwards to have the same effect.

Lift

SBA Skills

ORR	D	MM	PD	AI
✓				

Experiment – Blowing gives lift

You may wonder why an aircraft does not fall towards the ground. It makes use of the force of lift. Try this experiment to discover where lift comes from.

1 Take a narrow strip of paper. Hold it at one end and blow over the top of it. You will find that the paper tends to rise. It experiences an upward force – that's lift.

2 Now use another strip of paper to make a wing shape. This is flat underneath and curved on top, as in Fig C5.8. Put a drinking straw through the middle of it. Put a length of string through the straw and hold it vertically, so that the wing can move up and down.

3 Blow at the front edge of the wing. Can you blow strongly enough to produce lift?

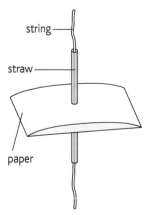

▲ **Fig C5.8** Experimenting with lift

In the previous experiment, the paper rises because of the unequal air speeds above and below it. This is how aeroplanes fly; the shape of the wings causes air to rush faster over the upper surface of the wings than over the lower, pushing air downwards and creating a lift force.

From Fig C5.9, you can see that the wing is shaped and tilted so that, as the aircraft moves forward, the air is pushed downwards. This produces an upward force on the wing. (Sail boats make use of this wind current effect to sail against the wind.)

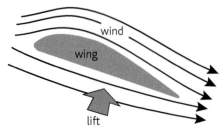

▲ **Fig C5.9** Producing lift on an aircraft wing

If the aircraft stops moving forwards, or moves too slowly, it will not be pushing on the air sufficiently strongly and the force of lift will not be enough to keep it in the sky.

It is easier for an aircraft to take off on a windy day by taking off into the wind. Then the air is already moving over the wings, contributing to the lift. A sideways wind is more difficult – it pushes the aircraft sideways and can even tip it over.

Q3 Look at Fig C5.7 of the aircraft in flight. Which force is pulling it downwards? And which force keeps it up?

Q4 A racing car has a 'wing' at the back which produces a downward force on the car, ensuring that it keeps a good grip on the track. What shape must this 'wing' have if the lift force is to push downwards on the car?

Q5 Why would it be impossible to fly an aircraft on the Moon?

Q6 When a bird flies, its wings do two things: they push it forwards (by pushing backwards) and they give it lift (by pushing downwards). Explain how the bird is making use of the action/reaction principle.

The force of gravity

Every object experiences a force that acts on it and tends to pull it towards the centre of the Earth. This force is known as the force of **gravity**, and it gives every object its **weight**.

Why doesn't gravity pull you downwards, through the floor? Another force pushes upwards. This is the **contact force** of the floor. Whenever two objects press on each other, there is a contact force.

As an object gets farther away from the Earth, the pull of gravity on it decreases so its weight decreases. An object on the

gravity – the force of attraction between any two objects caused by their masses

weight – the downward force of gravity on a body

contact force – the force of one body on another when they touch

▲ **Fig C5.10** The chair provides an upward contact force to cancel out the downward pull of gravity

surface of the Moon experiences a smaller gravitation pull than it would near the surface of the Earth. This happens because the 'pull-back' of the Moon's gravity is one-sixth that of the Earth's. For this reason an object weighs less on the Moon than on Earth, although its mass remains the same.

▲ **Fig C5.11** Astronauts in space may experience weightlessness

SBA Skills

ORR	D	MM	PD	AI
			✓	

Experiment – Experiencing gravity

1 Drop a ball so that it falls towards the floor. It's hard to see, but the ball accelerates (speeds up) as it falls. Fig C5.12 shows this more clearly. What force makes it accelerate?

2 Now throw a ball vertically upwards. Describe and explain its movement.

3 Devise a method to measure the greatest height to which you can throw a ball.

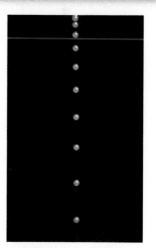

▲ **Fig C5.12** A multiflash photo of a ball falling shows it speeding up

Q7 If you fall downstairs, which force pulls you downwards? Which force pushes on you each time you bump over another step? What force tends to slow you down?

Projectiles

A projectile is any object that is thrown or projected into the air, like the ball you threw into the air in the experiment above. A variety of projectiles are used in sporting activities. The javelin, discus, cricket ball, arrow, football, golf ball, shot-putt, and the human body in a broad jump or high jump are all projectiles.

How do projectiles move through the air? As you will have noticed in the experiment, a projectile follows a curved path through the air. As it moves along horizontally, gravity is pulling it down vertically.

SBA Skills

ORR	D	MM	PD	AI
✓		✓		✓

Experiment – Investigating the range of a projectile

The simple catapult shown in Fig C5.13 can be used to investigate projectile motion.

1 For different values of the angle of projection, find the range (the horizontal distance travelled). To investigate fairly which angle of projection makes the projectile travel farthest, the following things must be done:

- The rubber must be stretched to the same position each time.
- The same projectile must be used each time.
- The projection must be done several times for each angle and the average measurement taken for that angle.

2 Explain why these three precautions are necessary in an investigation like this.

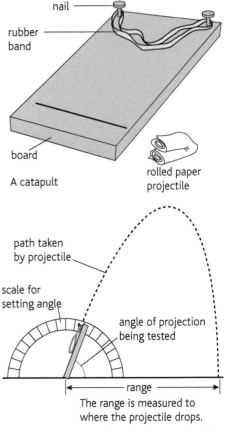

▲ **Fig C5.13** A catapult, and the meaning of 'range'.

Explaining projectile motion

The main reason that projectiles behave as they do is that they are under the influence of gravity as they move through the air. Gravity tends to pull objects vertically downwards towards the ground. The moment a projectile is fired, therefore, it starts to fall – whether it is projected upwards or straight ahead.

In the absence of air resistance, a projectile will travel farthest when it is launched at an angle of 45° to the horizontal. However, air resistance slows it down and therefore reduces its range (the horizontal distance travelled). If a wind is blowing against the direction of travel, the range will be further shortened. However, the

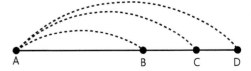

▲ **Fig C5.14** Projectile ranges under different conditions:
AD – no air resistance;
AC – still air;
AB – wind blowing against projectile

range will be lengthened if the projectile is travelling with the wind. Also, if the wind is blowing across the path of the object the direction of travel will be affected.

Escaping gravity

If you throw a ball upwards, gravity pulls on it so that it slows down and then returns to Earth. The faster you throw it, the higher it goes before it returns.

However, if you could throw a ball very, very fast, at over 11 000 m/s, it would escape from the pull of the Earth's gravity and disappear into space. That speed is called the escape velocity.

So, to make an object escape from the Earth's gravity, we must give it enough kinetic energy. Then it will fly off into space, free from the pull of gravity.

Q8 Will the escape velocity for the Moon be greater than that of the Earth, less, or the same? Explain your answer.

Satellites

orbit – the path of one object around another

The Moon orbits the Earth. It is held in its **orbit** by the pull of the Earth's gravity. Without that pull, the Moon would travel off in a straight line and we would never see it again.

satellite – an object travelling around another, held in its orbit by gravity

The Moon is the Earth's natural **satellite**. A satellite is any object that orbits another, held in its orbit by gravity. The Earth and the other planets are all satellites of the Sun. Most planets have moons, which are their natural satellites. A spacecraft travelling around the Earth is an artificial satellite.

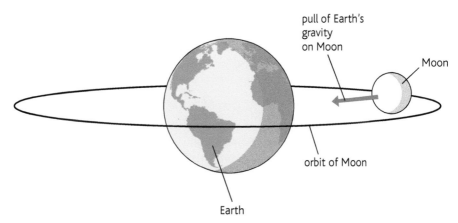

▲ **Fig C5.15** The Earth's gravity pulls on the Moon, keeping it in its orbit

centripetal force – the force, directed towards the centre of an orbit, which holds a satellite in orbit

The Moon's orbit is a rough circle with the Earth at the centre. The Earth's gravitational pull acts towards the centre of the circle. For this reason, it is described as a **centripetal force** ('Centripetal' means 'towards the centre'. The opposite of centripetal is **'centrifugal'**, meaning 'away from the centre'.)

centrifugal force – any force that is directed outwards from the centre of an orbit

Any object moving in a circle must have a centripetal force acting on it, to keep it moving in a circle. For example, if you tie a rubber bung to the end of a string and whirl it round your head, the pull of the string provides the centripetal force needed to keep the bung moving in a circle.

Experiment – The whirling bung

1 Tie a rubber bung with a hole through it securely to the end of a length of string.

2 In an open space, and away from everyone else, whirl the bung around your head in a horizontal circle. You should feel the pull needed to stop the bung from escaping, keeping the bung in orbit.

3 Suddenly release the string so that there is no longer a centripetal force acting on the bung. It will fly off at a tangent to the circle.

▲ **Fig C5.16**

Q9 Draw a diagram to show the Earth in its orbit around the Sun. Add an arrow to show the centripetal force that holds the Earth in its orbit. What causes this force?

Artificial satellites

Isaac Newton suggested that if a projectile is raised above the Earth and given a horizontal velocity of 8000 m/s, it will enter an indefinite orbit around the Earth. Such objects are known as 'artificial satellites'.

The projectile is falling continuously, pulled by gravity, but since it is moving at such a high velocity, it falls only as fast as the curved surface of the Earth falls away beneath it.

Satellites vary in size and weight and they perform many different functions. For example, they are used in weather forecasting, in

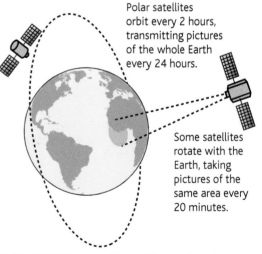

Polar satellites orbit every 2 hours, transmitting pictures of the whole Earth every 24 hours.

Some satellites rotate with the Earth, taking pictures of the same area every 20 minutes.

▲ **Fig C5.17** Some satellites orbit over the poles; others stay above the equator

remote sensing of the Earth's resources, for military and intelligence purposes, in telecommunications and TV broadcasting, and in navigation.

To do – Spacecraft in orbit

1 Select one type of artificial satellite. Find out about:
  - how it is launched
  - what orbits it follows and how fast it moves
  - what instrumentation it uses
  - how it is powered.

2 Prepare a presentation for the rest of the class.

Q10 What force holds an artificial satellite in its orbit around the Earth?

Living in space

Space is a vacuum – there is no air up there. So astronauts in space have to take an air supply with them, along with food and water.

Astronauts also need an energy supply to operate their equipment. Usually this is supplied by photovoltaic cells (solar cells) on the outside of the spacecraft – see Unit C2. They also need a small amount of fuel if they are going to move their craft into a different orbit. For this, they have small rocket motors.

Astronauts also have to learn to live in weightless conditions. Because they are in orbit, they are constantly 'falling' towards the Earth, so they can float around inside their spacecraft. Fig C5.18 shows some astronauts in training for this.

▲ **Fig C5.18** Astronauts practising in weightless conditions

Being weightless means that an astronaut's muscles aren't used much, so they must do lots of exercises to keep fit. The International Space Station has several exercise machines on board.

Centre of gravity

Every object behaves as though its total weight acts from a single point. The point through which the total weight of an object appears to act is called its centre of gravity.

Finding the centre of gravity

There are several ways to find an object's centre of gravity. Most simply, if an object is supported at its centre of gravity, it will balance. So, if we can find where it balances, we can find the centre of gravity of an object.

Fig C5.19 shows a way of locating the centre of gravity of a rule. Balance it on a pencil. When balanced, the centre of gravity of the rule will be directly above the pivot.

▲ **Fig C5.19** Balancing a rule to find its centre of gravity

Experiment – Finding the centre of gravity by balancing

1 Using the method in Fig C5.19, try to find the centre of gravity of the following objects:

   a a pencil   **b** a book

   Are their centres of gravity at their midpoints?

2 Collect an irregularly shaped piece of thick card. Hold your finger vertically and balance the card horizontally on your finger. Its centre of gravity is just above your finger.

3 Here is another method for finding the centre of gravity of a rule. Use a rule, a pencil or prism to act as a pivot, and a weight W_1 that can be moved along the rule to make it balance. When the rule is balanced, the position x of the centre of gravity from the pivot can be calculated if W_1, x_1 and W are known. You will need to use the principle of moments (see Unit C3).

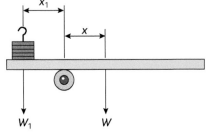

▲ **Fig C5.20** The centre of gravity of the rule is at distance x from the pivot

Symmetrical objects

If an object is symmetrical, it is easy to find its centre of gravity. Simply find its centre of symmetry. For example, for a cylinder, the centre of gravity is in the centre, halfway along its length. Similarly, for a rectangular block, the centre of gravity will be at its exact centre – see Fig C5.21.

For a triangular card, draw a line from each corner to the midpoint of the opposite side. The centre of gravity is at the point where all three lines cross.

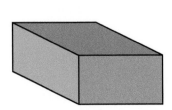
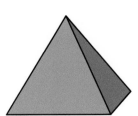

▲ **Fig C5.21** Locating the centre of gravity of a regularly shaped object

Q11 Where will the centre of gravity of a circular disc be?

Q12 Draw a diagram to show how to find the centre of gravity of a triangular card.

SBA Skills

ORR	D	MM	PD	AI
✓				

To do – Balancing symmetrical shapes

1 Cut out the following regular shapes from thick card: a square, a rectangle, a triangle, a circle.

2 Hold your finger vertically. Balance each shape on your finger. Mark the centre of gravity. Is it at the centre of the shape?

Unsymmetrical objects

It is harder to find the centre of gravity of an object that is not symmetrical, e.g. an irregularly shaped cardboard sheet. We can use the fact that the centre of gravity of an object always seeks the lowest possible position when the object is suspended. The following experiment shows how this works.

SBA Skills

ORR	D	MM	PD	AI
		✓		

Experiment – The hanging card

1 Collect a long pin, a clamp and stand, an irregularly shaped piece of card and a plumb line (a string and weight).

2 Clamp the pin so that it is horizontal.

3 Make three holes around the edge of the card.

4 Suspend the card from the pin using one of the holes and allow it to swing freely until it comes to rest. In this position the centre of gravity must be somewhere along a vertical line under the pin.

5 Using the plumb line, mark this line.

6 Repeat the procedure with the pin in each of the other holes.

The centre of gravity is located where all three lines cross since it is somewhere along each line. You can verify this by trying to balance the sheet at the point of intersection on the tip of a pencil.

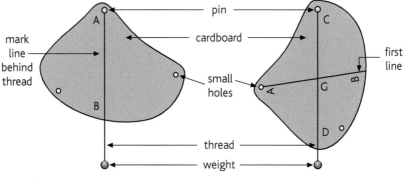

▲ **Fig C5.22**

Equilibrium

equilibrium – the state of being balanced

When an object is balanced, we say that it is in **equilibrium**. The following experiment will help you to understand the three types of equilibrium: stable, unstable and neutral equilibrium.

To do – Keeping your balance

1 Collect a watch glass or other curved dish, and a ball bearing or other small ball.

2 Place the ball bearing in the three positions shown in Fig C5.23, so that it is stationary (in equilibrium).

3 Observe what happens if you displace it slightly from its balanced position.

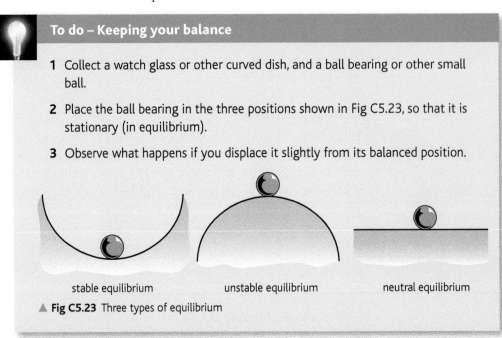

stable equilibrium   unstable equilibrium   neutral equilibrium

▲ **Fig C5.23** Three types of equilibrium

stable equilibrium – when moved slightly, an object returns to its previous position

1 In **stable equilibrium**, the centre of gravity of the ball bearing is at its lowest position. If the ball is moved, the centre of gravity is raised to a higher position. The ball therefore tries to move back to the position of low centre of gravity.

unstable equilibrium – when moved slightly, an object moves away from its previous position

2 In **unstable equilibrium**, the centre of gravity is at the highest point. Any movement of the ball therefore causes the centre of gravity to be lowered. Once the position of its centre of gravity is lowered, an object does not usually return to a position where the centre of gravity is higher – it keeps going down until the centre of gravity is again as low as possible.

neutral equilibrium – when moved slightly, an object remains in its new position

3 In **neutral equilibrium**, the height of the centre of gravity remains the same wherever the object moves to. Move the ball bearing to the side – it stays there.

Q13 The pictures in Fig C5.24 show a conical object on a table. Copy the diagram and label each with the type of equilibrium it shows.

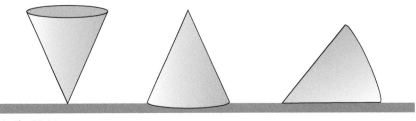

▲ **Fig C5.24**

Centre of gravity and stability

When young children are learning to walk, they often fall over. Why is this? They are used to crawling around, with their centre of gravity close to the ground, which makes them stable. Standing up, their centre of gravity is much higher up. If they tip over slightly, they are in danger of toppling. They are unstable.

▲ **Fig C5.25** This is one way to avoid toppling over

If an object is to remain upright, its centre of gravity must be above its base; that is, above the area that is contact with the ground. The child in the photograph will remain stable because she has spread out her hands and feet to give a wide base. It's harder when you stand up; your base is now the area of your feet on the ground and, if you lean over, your centre of gravity is no longer above your base and you will fall over.

Q14 Explain why a person is more stable when standing with their feet apart.

SBA Skills

ORR	D	MM	PD	AI
				✓

Experiment – The balancing parrot

1 Make a parrot from cardboard. Add a weight such as modelling clay or a coin to its tail.

2 Balance the parrot on a perch (use a horizontal wooden rod or pencil). Will it balance on your finger?

3 If you tip the parrot forwards or back, it returns to its balanced position. Explain why it is stable.

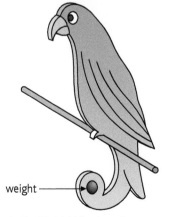

weight

▲ **Fig C5.26** Why does the parrot stay on its perch, even when you push it gently?

SBA Skills

ORR	D	MM	PD	AI
✓				

Experiment – Investigating stability

1 Collect a plank or board (to use as an inclined plane), a block of wood and a protractor.

2 Place the block on the plane, as shown in Fig C5.27. The base of the block is the area in contact with the plane.

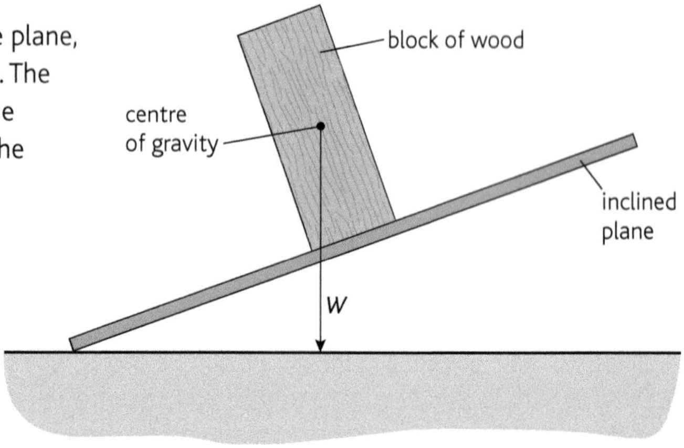

block of wood

centre of gravity

inclined plane

W

▲ **Fig C5.27**

3 Slowly lift one end of the plane. Eventually the block will topple over. Measure the angle of the plane at which this happens.

4 Replace the block with a different face on the plane. Repeat the experiment. In which position is the block most stable? In which is it least stable?

Avoiding instability

Tall, thin objects tend to be unstable. So do objects whose weight is mostly high up, so that their centre of gravity is high above the base.

Instability can lead to problems. Vehicles tend to topple over more easily if they are loaded so that their centre of gravity is too high. If a lot of people are standing in a bus or a truck the vehicle can easily topple, especially when swinging around corners. This is why there are traffic rules such as 'Not more than 10 standing'. If too many people or too much of the load is on one side of the vehicle it will also topple over very easily. This is especially noticeable in boats.

For safety, large vehicles such as buses and trucks are usually labelled to show the maximum loading capacity (how much load they can carry) and the tare (their weight before any load is added).
The total weight of the vehicle = tare + load.

Q15 Look at the picture of the men in the boat in Fig C5.28. Where is the centre of gravity of the boat likely to be? What is likely to happen, and why?

▲ **Fig C5.28**

End-of-unit questions

1 When an aircraft flies, there is an upward force on its wings. This force is called:

   A thrust

   B drag

   C lift

   D air pressure

2 The object shown in Fig C5.29 is in unstable equilibrium. Which dot marks its centre of gravity?

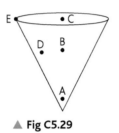

▲ **Fig C5.29**

3 Choose the correct statement:

A racing car should have wheels that are far apart…

   A to increase the friction between the tyres and the road.

   B to give it a streamlined shape, reducing air resistance.

   C to provide a bigger centripetal force when cornering.

   D to increase its stability when cornering.

4 What type of equilibrium is illustrated by each of the following? Explain your answer by reference to the centre of gravity of the object concerned.

   a a simple pendulum, suspended vertically

   b an irregularly shaped sheet of cardboard balanced on the point of a pencil

   c a balanced compass needle

5 There are many artificial satellites in orbit around the Earth. Explain how a satellite can remain in orbit despite the pull of the Earth's gravity, and outline some uses of satellites.

6 A rocket takes off from a launch pad in a Caribbean country. It is carrying a spacecraft which will be put into orbit around the Earth.

   a Explain why the rocket must burn a lot of fuel to get the spacecraft into its orbit.

   b The spacecraft (without the rocket) travels around the Earth in a circular orbit. What force holds it in its orbit?

   c What name is given to a force which holds an object in a circular path?

   d If there was no force acting on the spacecraft, what path would it follow? What can you say about its speed?

   e The spacecraft is to be used for broadcasting satellite television programmes to receivers on the Earth. Give two other uses for satellite in orbit around the Earth.

The School-Based Assessment (SBA)

Below are some examples of sample SBA experiments with suggested mark schemes. An extensive planning and design experiment and some further experiments with mark schemes can be found on our website:

www.pearsoncaribbean.com/heinemannintscicsec

To investigate osmosis in living tissue

Procedure

1 Cut five fresh potato chips using a cork borer. Make sure each potato chip is the same length.

2 Place each chip into a different concentration of sugar solution ranging from 0.0 Molar (pure water) to 1 Molar sugar solution.

3 Leave the chips for one hour then carefully re-measure the length of each chip. Calculate the change in length.

4 Plot your results as a graph showing change in length of chip on the Y axis and concentration of solution on the X axis.

Manipulation and Measurement (MM)

a	All five strips are exactly same length.	(2 marks)
b	The edges of strips are straight to ensure accurate measurement.	(2 marks)
c	Strips immersed completely in various solutions.	(2 marks)
d	Strips placed in various solutions at the same time.	(1 mark)
e	Similar volumes of solution used.	(2 marks)
f	Accurate measurement of time.	(1 mark)

To determine whether chlorophyll is needed for photosynthesis

Procedure

For this test you will need a variegated leaf that has been left attached to a plant in bright sunlight for at least three hours. A variegated leaf is a leaf that is different colours because not all of the leaf contains chlorophyll. A leaf that is both green and white works well with this test.

1 Draw a diagram of the pattern on the leaf and then carry out the starch test.

2 When you have completed the test make a second drawing of the leaf.

Observation, Recording and Reporting (ORR)

Organisation and Conciseness

a	The sequencing of the report is logical.	(1 mark)
b	Each section is named.	(1 mark)
c	The style of reporting is concise.	(1 mark)
d	The form of reporting is appropriate.	(1 mark)

Accuracy of Recording Observation

a	Appropriate type of data is recorded.	(2 marks)
b	Detailed recording of data, e.g. solution turned green, leaf became white when boiled, leaf became stiff after boiling in ethanol, the green part of leaf became blue black.	(4 marks)

Analysis and Interpretation (AI)

Summarised data

    a Accurately identify the relationships between the green and non green parts of leaf and starch (photosynthesis).   (2 marks)

    b Explain why the leaf was boiled and why it was placed in ethanol.   (2 marks)

Evaluation Data

    a State at least two precautions necessary in conducting the experiment.   (2 marks)

    b Explain why alcohol removed the chlorophyll.   (1 mark)

    c Explain why it was necessary to remove the chlorophyll.   (2 marks)

    d Identify linkage between chlorophyll and photosynthesis.   (1 mark)

To draw and label diagrams of food storage organs

Procedure

Examine and draw an example of a runner, bulb, corm, tuber and rhizome.
Label each diagram and explain how it reproduces by vegetative propagation.

Drawing (D)

Clarity of Drawing

    a Drawing is large.   (1 mark)

    b Drawing is two dimensional.   (1 mark)

    c Clear and accurate lines are present.   (1 mark)

    d The title is accurately stated.   (1 mark)

    e The drawing is labelled accurately.   (1 mark)

Accuracy of Drawing

    a Drawing is in proportion.   (1 mark)

    b Magnification is correctly stated.   (1 mark)

    c View is correctly stated.   (1 mark)

Labelling

    a Straight lines are neatly drawn.   (1 mark)

    b Labelling lines do not cross.   (1 mark)

Page numbers in **bold** indicate Key Word definitions.